Multivariable Model-Building

Multivariable Model-Building

A pragmatic approach to regression analysis
based on fractional polynomials for
modelling continuous variables

PATRICK ROYSTON
*Cancer and Statistical Methodology Groups,
MRC Clinical Trials Unit, London, UK*

WILLI SAUERBREI
*Institute of Medical Biometry and Medical Informatics,
University Medical Center, Freiburg, Germany*

John Wiley & Sons, Ltd

Other Wiley Editorial Offices

John Wiley & Sons Inc., 111 River Street, Hoboken, NJ 07030, USA

Jossey-Bass, 989 Market Street, San Francisco, CA 94103-1741, USA

Wiley-VCH Verlag GmbH, Boschstr. 12, D-69469 Weinheim, Germany

John Wiley & Sons Australia Ltd, 42 McDougall Street, Milton, Queensland 4064, Australia

John Wiley & Sons (Asia) Pte Ltd, 2 Clementi Loop #02-01, Jin Xing Distripark, Singapore 129809

John Wiley & Sons Canada Ltd, 6045 Freemont Blvd, Mississauga, ONT, L5R 4J3

Wiley also publishes its books in a variety of electronic formats. Some content that appears in print may not be available in electronic books.

Library of Congress Cataloging in Publication Data

Royston, Patrick.
 Multivariable model-building : a pragmatic approach to regression analysis based on
 fractional polynomials for continuous variables / Patrick Royston, Willi Sauerbrei.
  p. cm.
 Includes bibliographical references and index.
 ISBN 978-0-470-02842-1 (cloth : acid-free paper)
 1. Regression analysis. 2. Polynomials. 3. Variables (Mathematics)
 I. Sauerbrei, Willi. II. Title.
 QA278.2.R696 2008
 519.5′36—dc22   2008003757

British Library Cataloguing in Publication Data

A catalogue record for this book is available from the British Library

ISBN 978-0-470-02842-1 (H/B)

Typeset#in#10/12pt#Times#by#Integra#Software#Services#Pvt.#Ltd,#Pondicherry,#India#

Contents

Preface

Multivariable Model-Building: a pragmatic approach to regression analysis based on fractional polynomials for modelling continuous variables is principally written for scientists (including statisticians, researchers and graduate students) working with regression models in all branches of application. Our general objective is to provide a readable text giving the rationale of, and practical advice on, a unified approach to multivariable modelling which aims to make such models simpler and more effective. Specifically, we focus on the selection of important variables and the determination of functional form for continuous predictors. Since our own background is in biostatistics and clinical research, inevitably there is a focus on applications in medicine and the health sciences, but the methodology is much more widely useful. The topic of multivariable model-building is very broad; we doubt if it is possible to cover all the relevant topics in a single book. Therefore, we have concentrated on what we see as a few key issues. No multivariable model-building strategy has rigorous theoretical underpinnings. Even those approaches most used in practice have not had their properties studied adequately by simulation. In particular, handling continuous variables in a multivariable context has largely been ignored. Since there is no consensus among researchers on the 'best' strategy, a pragmatic approach is required. Our book reflects our views derived from wide experience. The text assumes a basic understanding of multiple regression modelling, but it can be read without detailed mathematical knowledge.

Multivariable regression models are widely used in all areas of science in which empirical data are analysed. We concentrate on normal-errors models for continuous outcomes, logistic regression for binary outcomes and Cox regression for censored time-to-event data. Our methodology is easily transferred to more general regression models. As expressed in a very readable paper by Chatfield (2002), we aim to 'encourage and guide practitioners, and also to counterbalance a literature that can be overly concerned with theoretical matters far removed from the day-to-day concerns of many working statisticians'. The main focus is the modelling of continuous covariates by using fractional polynomials. The methods are illustrated by the analysis of many datasets, mainly from clinical epidemiology, ranging from prognostic factors in breast cancer and treatment in kidney cancer to risk factors in heart disease.

WHAT IS IN OUR BOOK

Our main concern is how to build a multivariable regression model from several candidate predictors, some of which are continuous. We are more interested in explanatory models (that is, in assessing the effects of individual variables in a multivariable context) than in deriving a 'good' predictor without regard to its components. The basic techniques that are dealt with in many textbooks on regression analysis are not repeated.

Chapters 2 and 3 deal mainly with the selection of variables and coping with different types of variables in a modelling context. Relationships with continuous covariates are assumed linear. The importance of the coding chosen for categorical covariates is discussed. Chapters 4 and 5 provide a reasonably comprehensive account of univariate fractional polynomial (FP) models, our preferred method of working with continuous predictors. We introduce the function selection procedure (FSP). Chapter 6, the heart of the book, introduces multivariable FP (MFP) modelling, combining backward elimination with the FSP. In Chapter 7, FP modelling is extended to include interactions between predictors, both categorical-by-continuous and continuous-by-continuous. Chapter 8 looks at techniques for assessing the stability of multivariable models. Bootstrap resampling is the key tool here. Chapter 9 briefly outlines spline models. We introduce two multivariable modelling procedures in which the FSP is adapted for splines, and we compare the results with FP models in several examples. Chapter 10 is a fairly self-contained guide to working with MFPs, taking a problem-oriented approach using an artificial but realistic dataset. A practitioner with some experience in regression modelling should be able to take in the principles and practice of MFP modelling from this chapter. As throughout the book, frequent use is made of model criticism, particularly of plots of fitted functions and of smoothed residuals, and of techniques for assessing the effects of influential observations on the selected model. Chapter 11 is a brief tour of further applications of FP methodology. Chapter 12 gives our recommendations for practice, briefly discusses some topics not dealt with in our book, and points to further research.

We lay stress on deriving parsimonious models that make sense from a subject-matter viewpoint. We are more concerned with getting the 'big picture' right than in refining the minor details of a model fit.

HOW TO READ OUR BOOK

The chapters have been organized such that the ideas unfold in a logical sequence, with Chapter 1 providing motivation and a flavour of what is to follow. However, to grasp the core ideas of our book more rapidly, we suggest as a bare minimum reading the following segments:

- Section 1.7 defines our approach to modelling in general terms.
- Section 2.6 discusses stepwise and other procedures for selecting variables.
- Sections 4.2–4.10 introduce FP functions and show how they are used in modelling a single continuous predictor. Section 4.14 contains a worked example. For an experienced modeller, the example may be a sufficient guide to the main principles.
- Sections 6.1, 6.2, 6.3 and 6.5 describe the key parts of the MFP method of multivariable model-building.
- Chapter 10 is particularly recommended to the practitioner who wants an appreciation of how to use MFP. We include material on some of the pitfalls that may be avoided using simple diagnostic techniques. Sections 10.5.6 and 10.8.3 on interactions may be omitted at a first reading.
- Chapter 12 summarizes some recommendations for practice.

SOFTWARE AND DATA

For practical use it is important that the necessary software is generally available. Software for the basic MFP method has been implemented in Stata, SAS and R. Special-purpose programs for Stata are available on our book's website http://www.imbi.uni-freiburg.de/biom/Royston-Sauerbrei-book for all the extensions we describe. In some of the examples, we show that use of the software is simple if basic principles of the methodology are understood. To assist the reader in developing their own experience in multivariable model-building, some of the datasets used in the book are available on the website.

EDUCATIONAL RESOURCES

Supplementary materials, including datasets, software, exercises and relevant Web links, are available on the website. Many of the issues in multivariable model-building with continuous covariates discussed in our book are explored in Chapter 10, where an artificial dataset (the 'ART study') is described and analysed. The dataset and details of the design and Stata programs used to create it are available on the website, allowing the data to be modified for different purposes. Exercises based on the ART study are suggested. We would encourage others to extend and develop the exercises as part of the material used to teach MFP methodology. Slide presentations are available as a starting point for preparing talks and teaching the material.

ACKNOWLEDGEMENTS

We are indebted to our colleagues at the MRC Clinical Trials Unit, London, and at the Institute of Medical Biometry and Medical Informatics, University Medical Center Freiburg for discussion and encouragement. Our research and the book have benefited from constructive comments from many people, including particularly the following: Doug Altman, Harald Binder, Gareth Ambler, Carol Coupland, Christel Faes, David Hosmer, Tony Johnson, Paul Lambert, Rumana Omar, Michael Schemper, Martin Schumacher, Simon Thompson, and Hans van Houwelingen. We thank the following for kind permission to use their valuable datasets in our book: Lyn Chitty (fetal growth), Tim Cole (triceps), John Foekens and Maxime Look (Rotterdam breast cancer), Amy Luke (research body fat), John Matthews and Maeve O'Sullivan (nerve conduction), Alastair Ritchie and Mahesh Parmar (kidney cancer), Philip Rosenberg (oral cancer), Martin Shipley (Whitehall I). We are grateful to Lena Barth, Karina Gitina, Georg Koch and Edith Motschall for technical assistance. Finally, we owe very special thanks to the director and staff of the Mathematisches Forschungsinstitut Oberwolfach, Germany. The excellent atmosphere and working conditions during visits there over several years were conducive to the development of many research ideas and papers which led up to our book.

London and Freiburg
November 2007

CHAPTER 1

Introduction

1.1 REAL-LIFE PROBLEMS AS MOTIVATION FOR MODEL BUILDING

Data are collected in all areas of life. In research, data on several variables may be collected to investigate the interrelationships among them, or to determine factors which affect an outcome of interest. An example from medicine is the relationship between the survival time of a patient following treatment for cancer and potentially influential variables (known in this context as prognostic factors), such as age, size of the tumour, its aggressiveness (grade), and so on. Often, effects of more than 10 potentially influential variables must be considered simultaneously in a single model. Our main emphasis is on examples from the health sciences, particularly clinical epidemiology. We discuss statistical methods to develop a model which best tries to answer specific questions in the framework of regression models. Although nearly all of the examples discussed have a background in the health sciences, the methods we describe are also highly relevant in other areas where multivariable regression models with continuous variables are developed.

1.1.1 Many Candidate Models

With today's fast computers and sophisticated statistical software, it is nearly trivial to fit almost any given model to data. It is important to remember that all models are based on several more or less explicit assumptions, which may correspond more or less well to unknown biological mechanisms. Therefore, finding a good model is challenging. By 'good' we mean a model that is satisfactory and interpretable from the subject-matter point of view, robust with respect to minor variations of the current data, predictive in new data, and parsimonious. Therefore, a good model should be useful beyond the dataset on which it was created.

In fitting regression models, data analysts are frequently faced with many explanatory variables, any or all of which may to some extent affect an outcome variable. If the number of variables is large, then a smaller model seems preferable. An aim of the analysis, therefore, is the selection of a subset of 'important' variables that impact on the outcome. For this task, techniques for stepwise selection of variables are available in many statistical packages and are often used in practical applications (Miller, 2002). Despite the importance of methods of variable selection and the enormous attention paid to the topic, their properties are not well understood. All are criticized in the statistical literature.

Multivariable Model-Building   Patrick Royston, Willi Sauerbrei
© 2008 John Wiley & Sons, Ltd

1.1.2   Functional Form for Continuous Predictors

A second obstacle to model building is how to deal with nonlinearity in the relation between the outcome variable and a continuous or ordered predictor. Traditionally, such predictors are entered into stepwise selection procedures as linear terms or as dummy variables obtained after grouping. The assumption of linearity may be incorrect. Categorization introduces problems of defining cutpoint(s) (Altman et al., 1994), overparametrization and loss of efficiency (Morgan and Elashoff, 1986; Lagakos, 1988). In any case, a cutpoint model is an unrealistic way to describe a smooth relationship between a predictor and an outcome variable.

An alternative approach is to keep the variable continuous and allow some form of nonlinearity. Hitherto, quadratic or cubic polynomials have been used, but the range of curve shapes afforded by conventional low-order polynomials is limited. Box and Tidwell (1962) propose a method of determining a power transform of a predictor. A more general family of parametric models, proposed by Royston and Altman (1994), is based on fractional polynomial (FP) functions and can be traced back to Box and Tidwell's (1962) approach. Here, one, two or more terms of the form x^p are fitted, the exponents p being chosen from a small, preselected set of integer and noninteger values. FP functions encompass conventional polynomials as a special case.

To illustrate the main issues considered in our book, we start with two characteristic examples. In the first example, we consider a simple regression model for a continuous outcome variable with a single, continuous covariate. In the second example we illustrate several approaches to modelling the simultaneous effect of seven potential prognostic factors on a survival-time outcome with censoring. As with most real-life research questions, the prognostic factors are measured on different scales and are correlated. Finding satisfactory multivariable models which include continuous predictors is still a great challenge and is the main emphasis of our book.

1.1.3   Example 1: Continuous Response

We start with a simple illustration of how researchers try to cope with the problems of modelling a continuous covariate in real-life studies. Luke et al. (1997) described the relationship between percentage body fat content (pbfm) and body-mass index (bmi) in samples of black people from three countries (Nigeria, Jamaica and the USA). See Appendix A.2.1 for further details. The authors aimed to find out how well bmi predicted pbfm.

'Standard' Analysis (Polynomial Regression)
Percentage of body fat and body-mass index were highly correlated, with Spearman correlation coefficient $r_S(\text{bmi, pbfm}) = 0.92$. The authors stated that the relationship between pbfm and bmi was 'quadratic in all groups except Nigerian men, in whom it was linear'. No indication was given as to how the quadratic (or linear) model was arrived at. The left panel of Figure 1.1 shows the raw data and the authors' fitted quadratic curve for the subsample of 326 females from the USA. Although the fit looks reasonable, we see some minor lack of fit at the lower and upper extremes. More important, the quadratic curve turns downwards for $\text{bmi} > 50\,\text{kg m}^{-2}$, which makes little scientific sense. We would not expect the fattest women to have a lower body fat percentage than those slightly less obese. Quadratic functions always have a turning point, but it may or may not occur within the range of the observed data. We return to the critical issue of the scientific plausibility of an estimated function in Section 6.5.4.

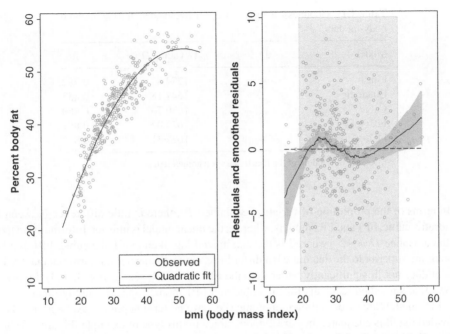

Figure 1.1 Research body-fat data. Left panel: raw values of pbfm and the fitted quadratic curve. Right panel: raw and smoothed residuals with 95% confidence interval (CI; see section 1.4 for further explanation). Lack of fit of the quadratic is seen in the right panel.

Further evidence of lack of fit is seen in the right panel of Figure 1.1, which shows smoothed residuals from the quadratic curve together with a 95% pointwise CI. We used a locally linear (running-line) smoother here (Sasieni and Royston, 1998). A pattern somewhat characteristic of a cubic polynomial is apparent in the smoothed mean residual; indeed, adding a cubic term in bmi to the model improves the fit significantly ($P = 0.0001$). A quartic term is not significant at $P < 0.05$, so conventionally one would stop elaborating the model there.

Fractional Polynomial Functions
As an alternative, we also selected a curve from the family of degree-one FPs, also known as FP1 functions. FP1 functions are an extension of an *ad hoc* approach often taken by applied statisticians in the past and examined in some detail by John Tukey (Tukey, 1957; Mosteller and Tukey, 1977). A power p or logarithmic transformation is applied to a predictor x, giving a model whose systematic part is $\beta_0 + \beta_1 x^p$ or $\beta_0 + \beta_1 \ln x$. The power p for FP1 functions is restricted to the predefined set $S = \{-2, -1, -0.5, 0, 0.5, 1, 2, 3\}$ proposed by Royston and Altman (1994) for practical use. FP1 functions are easily extended to higher order FPs (FP2, FP3, ...) in which combinations of power transformations of x are used, resulting in a family of flexible functions. The full definition is given in Section 4.3. Selection of the best-fitting model is discussed in Sections 4.8 and 4.10.

The best-fitting FP1 function for the research body-fat data, among the eight transformations in set S, turns out to have power $p = -1$, for which the formula is $\beta_0 + \beta_1 x^{-1}$. The best-fitting FP2 function has powers $(-2, -1)$, for which the formula is $\beta_0 + \beta_1 x^{-2} + \beta_2 x^{-1}$. For comparison with traditional approaches, the fit of five models (linear, quadratic, cubic, FP1, FP2) is summarized in Table 1.1.

Table 1.1   Goodness-of-fit statistics for five models for the research body-fat data.

Model	d.f.[a]	Deviance D	R^2
Linear	1	1774.87	0.782
Quadratic	2	1647.18	0.853
Cubic	3	1630.70	0.860
FP1(-1)	2	1629.02	0.861
FP2(-2, -1)	4	1627.31	0.862

[a] Degrees of freedom of the (fractional) polynomial terms.

In terms of the proportion of variation explained R^2, there is little difference between the quadratic, cubic, FP1 and FP2 models, whereas the linear model is inferior. Informally, in terms of the deviance $D = -2l$ (where l is the maximized log likelihood), the cubic, FP1 and FP2 models are superior to the quadratic but differ little among themselves. It turns out that the FP2 model does not fit significantly better than the FP1 ($P = 0.4$; see Section 4.10.3). Figure 1.2 shows that smoothed residuals for the cubic and FP1 models appear roughly random. Note that FP1 models are by definition monotonic (i.e. have no turning point – see Section 4.4); so, provided the fit is adequate, they are a good choice in this type of example. Figure 1.3 shows the fitted curves from the models in Table 1.1.

The main lessons from this example are the benefits of a systematic approach to model selection (here, selection of the function) and of the need to assess the results critically, both

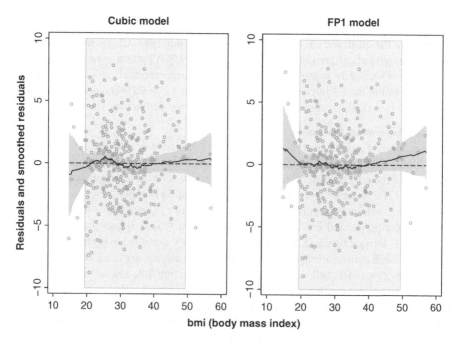

Figure 1.2   Research body-fat data. Left and right panels show residuals and smoothed residuals with 95% CIs from the cubic and FP1 models respectively.

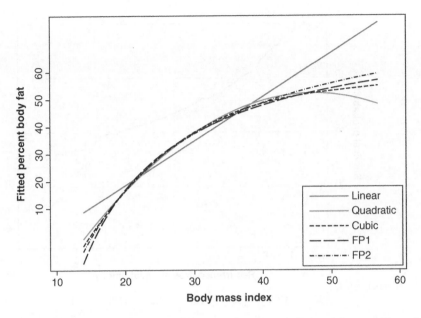

Figure 1.3 Research body-fat data. Fitted lines for five models. For raw data, see Figure 1.1.

from statistical and subject-matter perspectives. A plot such as Figure 1.1 clearly shows that a quadratic fits reasonably well, and equally clearly that it is not an ideal model – the curve is implausible. In purely statistical terms, an examination of smoothed residuals shows the deficiencies of the quadratic model and gives a hint on how to remedy them. The cubic, FP1 and FP2 models are better and fit the data about equally well, but the parsimony principle leads us to prefer the FP1 model. In our experience, simple models are generally more robust and generalize better to new data than complex ones.

1.1.4 Example 2: Multivariable Model for Survival Data

In a study of primary node positive breast cancer patients, seven standard prognostic factors (age, menopausal status, tumour size, tumour grade, number of positive lymph nodes, progesterone receptor status, and oestrogen receptor status) were considered in the development of a prognostic model for recurrence-free survival (RFS) time. For further details, see Appendix A.2.2. Figure 1.4 shows a Kaplan–Meier survival plot. Median RFS time was 4.9 years (95% CI, 4.2 to 5.5 years). Cox proportional hazards modelling (see Section 1.3.3) was used to investigate the simultaneous effect of the factors on the relative hazard of recurrence.

For continuous covariates, we used two standard methods. First, we assumed a linear relationship between the factor and the log relative hazard of an event (the 'linear approach'). Second, we categorized the factor into two or three groups according to predefined cutpoints (the 'step approach'). FPs were used as an additional method (the 'FP approach').

Univariate Models for age
To illustrate differences between the methods, we first consider the prognostic effect of age (age) in univariate analysis. Whether age is really a prognostic factor was controversial at the time of the original analysis of the GBSG study in the mid 1990s (see discussions in

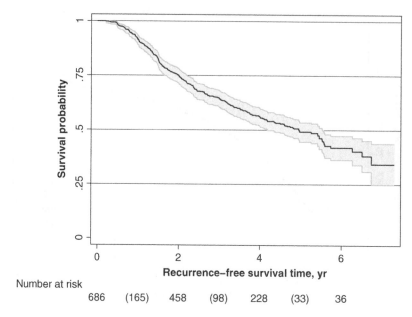

Figure 1.4 German Breast Cancer Study Group (GBSG) breast cancer data. Kaplan–Meier plot of RFS probabilities with 95% pointwise confidence band. Numbers at risk and (in parentheses) failing in each two-yearly interval are tabulated beneath the plot.

Sauerbrei et al. (1997; 1999)). In Table 1.2 we give deviance differences from the null model and P-values for the univariate effect of age, according to several types of model.

According to the linear approach, age has no apparent effect on RFS ($P > 0.4$) and the slope $\hat{\beta}$ on age is small (see Figure 1.5). As is usual with the Cox model, the effect of age is expressed as the log relative hazard with respect to an unspecified baseline hazard function. With a quadratic model, the P-value is 0.01. With the two predefined cutpoints used in the original analysis, the P-value is 0.15. Following discussions during the last decade and various data-driven searches for 'optimal' cutpoints, some researchers have argued in favor of 40 years as a good cutpoint for age. The analysis based on this cutpoint suggests an effect of age with

Table 1.2 GBSG breast cancer data. Deviance differences from the null model and P-values from univariate models assuming different functional forms.

Model	Deviance difference	d.f.	P-value
Linear	0.6	1	0.45
Quadratic	9.0	2	0.011
FP1(-1)	6.4	2	0.041
FP2($-2, -1$)	17.6	4	0.002
Categorized (1)[a]	3.8	2	0.15
Categorized (2)[b]	5.3	1	0.021

[a] Predefined cutpoints 45 and 60 years.
[b] Selected cutpoint 40 years.

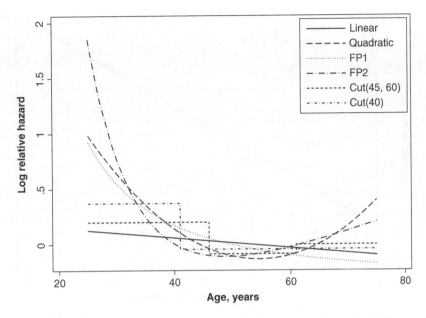

Figure 1.5   GBSG breast cancer data. Fitted functions from six Cox models for age.

a P-value of 0.02. Interpretation is obviously difficult. The model implies that patients aged 30 or 39 years have the same risk, whereas the risk decreases substantially (estimated relative hazard 0.66) for a patient aged 41 years. A patient of age 65 years has the same risk as a 41-year old, according to this model.

Use of the transformation age^{-2} provides evidence that age really is a prognostic factor. This transformation gave the best fit within the class of eight FP1 transformations proposed by Royston and Altman (1994). Within the more flexible class of FP2 transformations, a further improvement in fit is obtained with the FP2 function $\hat{\beta}_1 age^{-2} + \hat{\beta}_2 age^{-0.5}$. The overall P-value for age in this model is 0.002. The functions from the six models shown in Figure 1.5 display major differences.

Multivariable Model-Building

So far, we have illustrated several problems when investigating the effect of one continuous variable. However, in the real world the norm is that several covariates must be considered simultaneously. Although univariate analyses of explanatory variables are a good starting point, ultimately a multivariable analysis is required.

Sauerbrei and Royston (1999) developed three multivariable models for the GBSG data by using backward elimination (BE) with a nominal P-value of 0.05. For continuous variables, they considered linear functions, step functions with predefined cutpoints, and functions from the FP class. We defined two dummy variables, gradd1 and gradd2, from the ordered categorical factor grade. gradd1, nodes and pgr (progesterone receptor) were selected in all three models. The multivariable FP (MFP) procedure identified age as an additional prognostic factor. Its fitted curve is similar to the FP2 function from the univariate analysis (see Figure 1.5). The deviances for the linear, step and FP models are 3478.5, 3441.6 and 3427.9 respectively, showing that the FP2 model fits best. The lower the deviance is, the better the model fit is.

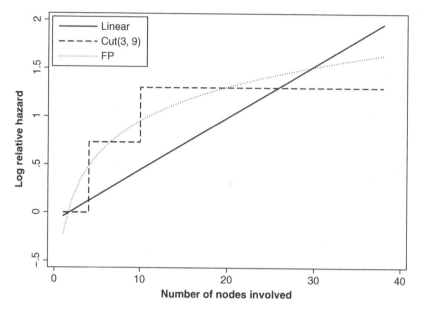

Figure 1.6   GBSG breast cancer data. Fitted functions from three models for `nodes`, estimated within multivariable Cox models. (Adapted from Sauerbrei et al. (1999) with permission from Macmillan Publishers Ltd. British Journal of Cancer, copyright 1999.)

Besides `age`, the models differ in the functional form for `nodes` and `pgr`, whereas `graddl` had a similar effect. The number of positive nodes has been identified as the most important prognostic factor in many studies of early breast cancer. The hazard of an event increases with the number of nodes. Such medical knowledge may be incorporated in the modelling process by restricting the candidate functions to be monotonic. Here, we allowed only FP1 functions (guaranteed monotonic) for `nodes`, the resulting function being log(`nodes`) – see model II* of Sauerbrei and Royston (1999).

The three functions are shown in Figure 1.6. The step function is a rough approximation to the FP (log) function; with more cutpoints, the approximation would be closer, but 'noisier'. The linear function seriously underestimates the hazard for a small number of nodes and overestimates it for a large number.

The functions derived for the continuous variable `pgr` also differ (see Sauerbrei et al. (1999)). In this example, the three ways of handling continuous variables in a multivariable analysis give different results and raise some general issues that are an important topic of our book.

1.2   ISSUES IN MODELLING CONTINUOUS PREDICTORS

1.2.1   Effects of Assumptions

An assumption of linearity may prevent one from recognizing a strong effect of a variable (e.g. `age`), or to lead one to mismodel the effect (e.g. `nodes`). The popular cutpoint approach introduces several well-known difficulties. Data-dependent determination of the cutpoint, at the most extreme by using the 'optimal' cutpoint approach, results in P-values that are too small and in an overestimation of the effect (Royston and Altman, 1994; Royston et al., 2006).

In other words, severely biased results can be obtained. Predefined cutpoints may not reflect the true, but unknown, functional relationship, and the number of cutpoints required is unknown. Furthermore, cutpoint approaches do not use the full information from the data, and step functions may conflict with biological principles, which demand smooth(er) functions. For further discussion, see Section 3.4.

With the FP approach, much more information from the data is utilized. The question of whether a nonlinear transformation improves the fit of a model is assessed systematically, with further advantages of flexibility in handling continuous variables, predicated on established statistical principles, transparency and ease of use.

1.2.2 Global versus Local Influence Models

Several approaches to modelling continuous covariates other than linear and FP functions are available. In general, it is useful to distinguish between regression functions for a continuous variable with the property of either global or local influence. For a function of x with the global-influence property, the fit at a given value x_0 of x may be relatively unaffected by local perturbations of the response at x_0, but the fit at points distant to x_0 may be affected, perhaps considerably. This property may be regarded by proponents of local-influence models as a fatal flaw; see the discussion and the example given by Hastie and Tibshirani in Royston and Altman (1994). A rigorous definition of the global-influence property has not to our knowledge been framed, but such models are usually 'parametric' in nature. Examples include polynomials, nonlinear models such as exponential and logistic functions, and FPs. By contrast, functions with the local-influence property, including regression splines (de Boer 2001), smoothing splines (Green and Silverman 1994), and kernel-based scatter-plot smoothers such as lowess (Cleveland and Devlin, 1988), are typically 'nonparametric' in character. Perturbation of the response at x_0 usually greatly affects the fit at x_0 but hardly affects it at points distant to x_0. One key argument favouring functions with global influence is their potential for use in future applications and datasets. Without such an aim, functions with local influence might appear the more attractive (Hand and Vinciotti, 2003).

Although FP functions retain the global-influence property, they are much more flexible than polynomials. Indeed, low-dimensional FP curves may provide a satisfactory fit where high-order polynomials fail (Royston and Altman, 1994). FPs are intermediate between polynomials and nonlinear curves. They may be seen as a good compromise between ultra-flexible but potentially unstable local-influence models and the relatively inflexible conventional polynomials. The title of our book makes it clear that our main emphasis is on FPs, but we also compare and discuss results from global- and local-influence models in several examples.

1.2.3 Disadvantages of Fractional Polynomial Modelling

Modelling with FP functions and our MFP approach also has difficulties. Perhaps the most important aspects are insufficient power to detect a nonlinear function and the possible sensitivity to extreme values at either end of the distribution of a covariate. An example of the latter is a 'hook' in the best FP2 function for nodes (see model II of Sauerbrei and Royston (1999)). In the analysis in Section 1.1.4, we used only FP1 functions for nodes, whereas in the original paper we discussed advantages of working with the preliminary transformation $enodes = \exp(-0.12 \times nodes)$, guaranteeing a monotonic function with an asymptote. This transformation was used in all our subsequent analyses of the breast cancer data and is also often used in our book.

The power issue has several aspects. Owing to insufficient sample size (in survival data, too few events), variables with a modest or weak effect may not be selected; or, by default, linear effects may be chosen instead of more realistic nonlinear functions. See Section 4.16 for some simulation results on the topic. Our approach to multivariable model-building may, therefore, add to the problem of low power inherent in all types of statistical modelling in small samples (see also Section 6.9.2). It is questionable whether variable selection makes sense at all in small samples. In medium-sized studies, loss of power is a justifiable cost, balanced by the benefits of combining variable selection with the selection of the functional form for continuous variables.

1.2.4 Controlling Model Complexity

We prefer a simple model unless the data indicate the need for greater complexity. A simple model transfers better to other settings and is more suited to practical use. The simplest dose–response relationship is the linear model, and that is our default option in most modelling situations. We would use an FP model if prior knowledge dictated such a model. In the absence of prior knowledge, we would use an FP model if there was sufficient evidence of nonlinearity within the data. The FP approach contrasts with local regression modelling (e.g. splines, kernel smoothers, etc.), which often starts and ends with a complex model.

1.3 TYPES OF REGRESSION MODEL CONSIDERED

For the types of model that we use, textbooks describing all aspects in great detail are available. Therefore, we introduce the models only very briefly, referring as necessary to other sources. A modern text addressing many detailed issues in model building is Harrell (2001); however, it appears to have been written for people already expert in regression modelling. Other recommendable textbooks on regression analysis, referenced in our book, are Cohen et al. (2003), DeMaris (2004), Vittinghof et al. (2005) and Weisberg (2005). The first two are based in behavioural and social sciences.

Although we work with multiple linear regression models, in our own research and in examples we are more concerned with important generalizations, including logistic and Cox models, and generalized linear models (GLMs), of which logistic regression is a special case. Such types of model are familiar tools nowadays, allowing the analyst to deal flexibly with many different types of response variable.

Methods of variable selection and related aspects have usually been developed and investigated for the multiple linear regression model. However, the methods, or at least their basic ideas, are commonly transferred to GLMs and to models for survival data. Additional difficulties may then arise; for example, obtaining satisfactory definitions of residuals or of equivalents of the proportion of explained variation R^2.

1.3.1 Normal-Errors Regression

For an individual with response y, the multiple linear regression model with normal errors $\varepsilon \sim N(0, \sigma^2)$ and covariate vector $\mathbf{x} = (x_1, \ldots, x_k)$ with k variables, may be written

$$y = E(y) + \varepsilon = \beta_0 + \beta_1 x_1 + \cdots + \beta_k x_k + \varepsilon = \beta_0 + \mathbf{x}\beta + \varepsilon \qquad (1.1)$$

The linear predictor or 'index', $\eta = \beta_0 + \mathbf{x}\beta$, is an important quantity in multivariable modelling (see also Section 1.3.5). Throughout our book, (1.1) is called the 'normal-errors model'. Although, as presented here, the model is linear in the covariates, models that are nonlinear in $\mathbf{x}$ come under the same heading.

Suppose we have a set of n observations

$$(y_1, \mathbf{x}_1), \dots, (y_n, \mathbf{x}_n) \equiv (y_1, x_{11}, x_{12}, \dots, x_{1k}), \dots, (y_n, x_{n1}, x_{n2}, \dots, x_{nk})$$

conforming to Equation (1.1). Here (but not in general), for simplicity in presenting the equation for $\widehat{\beta}$, each covariate is assumed to have been centered around its observed mean. Thus, for the jth covariate ($j = 1, \dots, k$) we have $\sum_{i=1}^{n} x_{ij} = 0$. The principle of ordinary least squares (OLS) estimation leads to the estimated regression coefficients

$$\widehat{\beta} = \left(\widehat{\beta}_1, \dots, \widehat{\beta}_k\right)^{\mathrm{T}} = \left(\mathbf{X}^{\mathrm{T}}\mathbf{X}\right)^{-1}\mathbf{X}^{\mathrm{T}}\mathbf{y}$$

where the (i, j)th element of the matrix $\mathbf{X}$ is x_{ij} and $\mathbf{y} = (y_1, \dots, y_n)^{\mathrm{T}}$. The fitted or 'predicted' values are

$$\widehat{y}_i = \widehat{\beta}_0 + \mathbf{x}_i\widehat{\beta}$$

where

$$\widehat{\beta}_0 = \overline{y} = \frac{1}{n}\sum_{i=1}^{n} y_i$$

Details of the theory of multiple linear regression may be found in a standard textbook such as Draper and Smith (1998) or Weisberg (2005).

Transformation of the Response Variable

A major topic in our book is the use of transformed predictors to accommodate nonlinear regression relationships. In contrast, we do not consider transformation of the response to improve fit, e.g. using the well-known method of Box and Cox (1964). In all the examples involving normal-errors models (see Tables A.1 and A.2), we assume that an appropriate transformation of the response has already been derived if required. Thus, the symbol y denotes the response variable in a given model, possibly after suitable transformation.

Residuals

OLS residuals are defined as

$$r_i = y_i - \widehat{y}_i$$

Raw residuals are typically too 'noisy' to be helpful; therefore, in our book we are generally concerned with smoothed residuals (see also Section 1.4). Mismodelling in Equation (1.1) may appear as a systematic pattern in the local mean residual as a function of some covariate x or the index η, and may be revealed by a smoothed scatter plot of the residuals on x (e.g. Cleveland and Devlin, 1988, Sasieni and Royston, 1998). The mean residual at a given value of x is interpretable as an estimate of the bias for Equation (1.1) in the mean value of y at x, a quantity of direct interest. Usually, 95% pointwise CIs are shown on the plot, to help one judge whether observed patterns are likely to be 'real' or due to chance.

1.3.2 Logistic Regression

Logistic regression, a special case of a GLM (see Section 1.3.4), is concerned with modelling binary responses. The multiple linear logistic regression model with covariates $x_1, \ldots, x_k$ asserts that the probability π of occurrence of a binary event y of interest, e.g. death or 'caseness' in a case-control study, may be represented by

$$\text{logit } \pi = \log \frac{\pi}{1-\pi} = \beta_0 + \sum_{j=1}^{k} \beta_j x_j \qquad (1.2)$$

$\pi/(1-\pi)$ is known as the odds of an event. Suppose y takes the values 1 for an event and 0 for a nonevent. If (1.2) is correct, then y has a Bernoulli distribution with probability parameter (and expected value) π.

In a model with just a single binary covariate x, taking the values 0 and 1, then logit $\pi = \beta_0$ when $x = 0$ and $\beta_0 + \beta_1$ when $x = 1$. Let $\pi_{(1)} = \text{logit}(\pi | x = 1)$ and $\pi_{(0)} = \text{logit}(\pi | x = 0)$. It follows that

$$\text{logit } \pi_{(1)} - \text{logit } \pi_{(0)} = \log \left(\frac{\pi_{(1)}}{1 - \pi_{(1)}} \right) - \log \left(\frac{\pi_{(0)}}{1 - \pi_{(0)}} \right) = (\beta_0 + \beta_1) - \beta_0 = \beta_1$$

$$= \log \left[\left(\frac{\pi_{(1)}}{1 - \pi_{(1)}} \right) \Big/ \left(\frac{\pi_{(0)}}{1 - \pi_{(0)}} \right) \right]$$

This shows the well-known result that the log odds ratio of an event when $x = 1$ compared with that when $x = 0$ equals the regression slope β_1 (or that the odds ratio equals $\exp \beta_1$).

Having now a sample of n observations $(y_1, \mathbf{x}_1), \ldots, (y_n, \mathbf{x}_n)$, estimates of β_0 and β are found by maximum likelihood. Let $\widehat{\eta}_i = \widehat{\beta}_0 + \mathbf{x} \widehat{\beta}_i$ be the index (linear predictor) from Equation (1.2). The probability that the ith observation is an event, i.e. $\Pr(y_i = 1 | \mathbf{x}_i)$, is estimated by

$$\widehat{\pi}_i = \frac{\exp(\widehat{\eta}_i)}{1 + \exp(\widehat{\eta}_i)}$$

Further details of the theory and practice of logistic regression may be found in Hosmer and Lemeshow (2000) or Collett (2003a).

Residuals

Several types of residual are available in logistic regression (e.g. Pearson and deviance). We work with the simplest and most accessible, the raw residuals $r_i = y_i - \widehat{\pi}_i$. As with normal-errors regression, mismodeling in (1.2) may be revealed by a scatter plot smooth of the residuals on x with 95% pointwise CIs, conceptually similar to Copas's (1983a) suggestion of 'plotting p against x'. In logistic regression, the smoothed residual at x is an estimate of the bias for (1.2) in the probability of an event given x.

1.3.3 Cox Regression

The 'Cox model' (Cox, 1972), also known as the proportional hazards model, is designed for modelling censored survival data. In its simplest form, the Cox model with covariates $x_1, \ldots, x_k$ describes the hazard of an event of interest at a time $t > 0$ after a starting point or time origin $t = 0$:

$$\lambda\left(t;\mathbf{x}\right) = \lambda_0\left(t\right)\exp\left(\sum_1^k \beta_j x_j\right)$$

For a single binary covariate x with regression coefficient β_1, the Cox model implies that

$$\frac{\lambda\left(t;1\right)}{\lambda\left(t;0\right)} = \frac{\lambda_0\left(t\right)\exp\left(\beta_1 \times 1\right)}{\lambda_0\left(t\right)\exp\left(\beta_1 \times 0\right)} = \exp\beta_1$$

The quantity $\lambda(t;1)/\lambda(t;0)$ is known as the *hazard ratio* (HR) for $x=1$ compared with $x=0$. More generally, the HR is the hazard at $\mathbf{x}$ divided by that at $\mathbf{x} = 0$. The HR plays a central role in survival analysis, since it is a convenient summary of the relationship between two entire survival curves. The crucial assumption of proportional hazards (PHs) is equivalent to saying that the HR is independent of t. If a non-PH is detected, then the Cox model may be extended in various ways to accommodate it (Therneau and Grambsch, 2000). A strategy for assessing potential non-PH and modelling it is described in Section 11.1.

A sample of n observations for survival analysis by a multivariable Cox model takes the form $(t_1, \mathbf{x}_1, \delta_1), \ldots, (t_n, \mathbf{x}_n, \delta_n)$, where δ_i is the 'censoring indicator'. δ_i takes the value 1 when t_i is an observed failure time and 0 when t_i is right-censored (i.e. when the precise time-to-event is unobserved, but is known to be $\geq t_i$). All times t_i must be positive; values of zero make no contribution to the estimation process. The parameter vector β is estimated by maximum partial likelihood. Many theoretical and practical details of survival modelling may be found in Hosmer and Lemeshow (1999) and Collett (2003b).

Residuals

We use martingale residuals and scaled Schoenfeld residuals. For details, see Hosmer and Lemeshow (1999, pp. 163, 198) or Therneau and Grambsch (2000, pp. 80, 85). Unscaled martingale residuals give a local estimate of the difference between the observed and predicted number of events. The pattern of the (smoothed) martingale residuals provides information on the functional form of a continuous covariate x in a model (Therneau et al., 1990). For a proposed function of x, systematic patterns seen in a plot of the smoothed martingale residuals against x indicate lack of fit, and may suggest how the chosen function of x may be improved. Note that, for comparability with the function estimated from a Cox model, the martingale residuals should be scaled by dividing by the ratio of the number of events to the number of individuals (Therneau et al., 1990). Such scaling does not affect the *pattern* of martingale residuals, only their *magnitude*, and is not applied in the relevant examples in our book.

Scaled Schoenfeld residuals are based on score residuals and are useful in a visual assessment of the PH assumption. Under PH, the mean of these residuals is zero, and is independent of time. A systematic pattern in the smoothed residuals when plotted against time suggests a time-varying effect of the covariate. The Grambsch–Therneau test (Grambsch and Therneau 1994) may be applied to test the PH assumption formally, for specific covariates or globally over all the covariates in the model.

1.3.4 Generalized Linear Models

A GLM (McCullagh and Nelder, 1989) comprises a random component, a systematic component and a link function which connects the two components. The response y is assumed to have a probability density function from the exponential family, namely

$$\exp\left[\frac{y\theta - b(\theta)}{a(\phi)} + c(y, \phi)\right]$$

where θ is known as the natural parameter, ϕ is a dispersion (scale) parameter, and $a(.)$, $b(.)$ and $c(.,.)$ are known functions. This density function defines the random component. The model also prescribes that the expectation μ of y is related to covariates $x_1, \ldots, x_k$ by $g(\mu) = \eta$ where $\eta = \beta_0 + \sum_1^k \beta_j x_j$. The index or linear predictor η is the systematic component and $g(.)$ is known as the link function.

The mean μ is related to θ by $\mu = db/d\theta$. A convenient link for a given member of the exponential family is the canonical link, in which $g(\mu)$ is chosen so that $\eta = \theta$. For the Bernoulli distribution, which underlies logistic regression with a binary outcome, we have $E(y) = \Pr(y = 1)$. The usual choice in data analysis is the canonical link, which for logistic regression is the logit (see Equation (1.2)); sometimes the probit or the complementary log–log link is used. Standard practice is to define the model in terms of μ and η, so that θ plays no further part.

Given choices for the random and systematic components and the link function, and a sample y of n observations, estimation of the model parameters is done iteratively by maximum likelihood.

Residuals

Again we use the simplest and most interpretable residuals for GLMs, the raw residuals $r_i = y_i - \widehat{\mu}_i$. The r_i are conceptually identical to the raw residuals $y_i - \widehat{\pi}_i$ in logistic regression. They are in fact scaled Pearson residuals (see McCullagh and Nelder (1989, equation (2.11), p. 37)). Variations in $\text{var}(r_i)$ as a function of a covariate or the index η_i are not of concern, since they are accommodated by the varying width of pointwise CIs given by a scatter-plot smoother applied to the raw residuals.

1.3.5 Linear and Additive Predictors

A different type of generalization of Equation (1.1) is from models with a linear predictor or index $\beta_0 + \sum_1^k \beta_j x_j$ to those with an *additive* predictor (also called an index) of the form $\beta_0 + \sum_1^k f_j(x_j)$, where $f_j(x_j)$ is a more complicated function of x_j than $\beta_j x_j$. These models apply only when x_j is a continuous covariate, such as age or blood pressure. Examples of types of $f_j(x_j)$ include polynomials, FPs, regression splines, smoothing splines, wavelets, Fourier series, and so on. Hastie and Tibshirani (1990) devote a whole book to an approach to regression modelling based on various types of additive function. Convenient classes of functions to use are cubic smoothing splines (Green and Silverman, 1994) and regression splines (de Boer 2001).

Note that $f_j(x_j)$ can be broken down into simpler parts, sometimes leading again to a linear predictor. For example, the quadratic regression model for a single predictor x may be written in additive format as $\beta_0 + f(x)$, where $f(x) = \beta_1 x + \beta_2 x^2$, or in linear format as $\beta_0 + \beta_1 x + \beta_2 x^2$. Be clear that the 'linearity' in the linear format relates to the *two* variables x and x^2. The quadratic model is additive in x, nonlinear in x, and linear in (x, x^2).

1.4 ROLE OF RESIDUALS

1.4.1 Uses of Residuals

Residuals have many roles in statistics. Most of them are connected with some form of model criticism (e.g. see Belsley et al. (1980)). In our book, we use residuals almost exclusively as a graphical tool to study the (lack of) fit of a function of a continuous predictor.

Preferred types of residual for the normal-errors, logistic and Cox models were discussed in Section 1.3. There are arguments favouring different residuals for different purposes. For example, residuals which are identically distributed under the assumption that the model is correct are particularly suitable for detecting outliers in the response. Residuals which are easily interpretable when smoothed are advantageous for detecting meaningful lack of fit of a function. For further discussion of residuals, please refer to Belsley et al. (1980).

Examples of smoothed residuals from models with a single predictor have already been presented (e.g. Figures 1.1 and 1.2). Such plots are useful for picking up anomalies in the fit. We use a univariate running-line smoother (Fan and Gijbels, 1996), implemented for Stata in the command `running` (Sasieni et al., 2005). We use the default amount of smoothing provided by `running`. Running-line smoothers of residuals provide a detailed picture of the relationship. As a result, they can give quite 'noisy' results, but the message from the data, of a lack of fit or otherwise, is usually sufficiently clear.

1.4.2 Graphical Analysis of Residuals

Generically, our favoured graphical analysis of residuals for a continuous predictor x, exemplified in Figures 1.1 and 1.2, has the following elements combined into a single plot:

1. A smooth of the residuals as a function of x, plotted as a solid line.
2. A pointwise 95% CI for the smooth, plotted as a shaded area.
3. A lightly shaded box within the plot region, bounded vertically by the 2.5th and 97.5th centiles of the observed distribution of x and horizontally by a convenient range of values. The box shows where most (95%) of the observations of x lie. The aim is to down-weight the visual impact of extreme values of x on the estimated function of x. The data are usually sparse near extreme values of x, and 'end effects' (unstable estimates of the function) are most likely to occur there.
4. A horizontal line representing $y = 0$, the expected value of the residuals if the model is correct.
5. (Optionally) a scatter plot of the raw residuals against x. This component may be omitted if the variation among the residuals visually overwhelms the smooth and its CI. To enhance legibility, martingale residuals from time-to-event models that are less than -1 may be truncated at -1.

Note that the pointwise 95% CIs should be interpreted cautiously, since they do not represent a global confidence region. For example, the value 0 may be excluded for some small range of x values, even when there is no serious lack of fit.

In a multivariable model, a plot of smoothed residuals may be drawn for every relevant predictor, perhaps including those not selected in the model. When feasible, a composite plot showing all such smooths in a single graphic is helpful. An example is Figure 6.5 (see Section 6.5.2).

1.5   ROLE OF SUBJECT-MATTER KNOWLEDGE IN MODEL DEVELOPMENT

The consensus is that subject-matter knowledge should generally guide model building. A study should be carefully planned, guided by the research questions and taking into account the methods envisaged for analysing the data. In randomized trials, the main research question is precisely specified – for example, is the new treatment better than the current standard with a hazard ratio of < 0.75? A detailed analysis plan is written before the analysis starts.

The situation is more complex and difficult with observational studies. At first glance, the research question may still be simple, e.g. whether there is an association between an exposure and the probability of developing a disease, adjusting for known confounders. However, this simple question may pose at least two serious challenges to the analyst. First, the relationship between a continuous exposure and the disease probability may have many possible functional forms. Does subject-matter knowledge support a particular functional form? That is unlikely when a 'new' exposure for a disease is under investigation. Assuming linearity is certainly a good starting point, but it must be checked and should be abandoned if the data 'say' otherwise. Second, what are the 'known' confounders in a given setting, and in what form should they be included in the model? If they are indeed 'known', why does it often happen that different sets of confounders are used in different studies? A small number of confounders may be prespecified, but why are many more variables typically collected? The potential problems of the statistical analysis increase if it is necessary to determine which of a large number of variables are associated with the outcome in a multivariable context (a typical question in studies of prognosis). Because subject-matter knowledge is typically limited or at best fragile, data-dependent model building is necessary (Harrell, 2001).

Unless stated otherwise, we assume in our book that subject-matter knowledge is so limited that it does not affect model building. However, when such knowledge does exist, analyses using fractional polynomials can easily be adapted to include it. For example, in Sauerbrei and Royston (1999) we noted that an FP function seemed biologically implausible and suggested a way to ensure that the estimated function was monotonic with an asymptote. Variables should sometimes be included in a model without being statistically significant, or should be excluded despite statistical significance. The former situation is more common with 'known' confounders, whereas the latter can happen if an 'expensive-to-measure' variable adds little to the model fit or to the explained variation. These two situations can easily be handled by 'forcing' variable(s) into or out of the model.

Sometimes, subject-matter knowledge may require a restricted class of nonlinear functions for certain variables (e.g. monotonic functions). When this is the case, the power to detect variables with a weak influence is increased by choosing the FP1 class as the most complex allowed functional form. See some simulation results in Section 4.16 and a fuller discussion in Section 6.9.2.

Most analyses of observational studies rely on a blend of subject-matter knowledge and data-dependent decisions. Initial decisions include grouping of categorical variables or, for continuous variables, considering how to handle extreme values or outliers (see Section 2.3). The concerns are specific to a dataset, but where possible the decisions should be based on subject-matter knowledge. The aims in general are to obtain an 'optimal' fit to the data, interpretable covariate effects, consistency with subject-matter knowledge where available, general usability by others, and transportability to other settings (external validation). In summary, multivariable model-building has elements of art in the attempt to provide satisfactory answers to more or less vague questions through the analysis of the data at hand under much uncertainty.

1.6   SCOPE OF MODEL BUILDING IN OUR BOOK

The techniques discussed in our book are intended and appear to work best for model building under certain conditions, as summarized in Table 1.3.

Table 1.3   Issues in building regression models, when the aim is to identify influential variables and to determine the functional form for continuous variables.

Issue	Assumption in our book (unless stated otherwise)	Reason for the assumption
Subject matter knowledge	No knowledge	Subject-matter knowledge should always be incorporated in the model-building process or should even guide an analysis. However, often it is limited or nonexistent, and data-dependent model building is required
Number of variables	About 5 to 30	With a smaller number of variables, selection may not be required. With many more variables (e.g. high-dimensional data), the approaches may no longer be feasible or will require (substantial) modification
Correlation structure	Correlations are not 'very' strong (e.g. correlation coefficient below 0.7)	Stronger correlations often appear in fields such as econometrics, less commonly in medicine. For large correlations, nonstatistical criteria may be used to select a variable. Alternatively, a 'representative', e.g. a linear combination of the correlated variables, may be chosen
Sample size	At least 10 observations per variable	With a (much) smaller sample size, selection bias and model instability become major issues. An otherwise satisfactory approach to variable and/or function selection may fail, or may require extension (e.g. shrinkage to correct for selection bias)
Completeness of data	No missing data	Particularly with multivariable data, missing covariate data introduces many additional problems. Not considered here
Variable selection procedure	Only sequential and all-subsets selection strategies are considered	Stepwise and all-subsets procedures are the main types used in practice. BE and an appropriate choice of significance level gives results similar to all-subsets selection
Functional form of continuous covariates	Full information from the covariate is used	Categorizing continuous variables should be avoided. A linear function is often justifiable, but sometimes may not fit the data. Check with FPs or splines whether non-linear functions markedly improve the fit
Interaction between covariates	No interactions	Investigation of interactions complicates multivariable model-building. Investigation of interactions should take subject-matter knowledge into account

(Adapted from Sauerbrei et al. (2007a, Table 1) and Sauerbrei et al. (1999) with permission from John Wiley & Sons Ltd.)

The restriction to no interactions is applied rigorously in Chapters 1–6, but is lifted in Chapter 7 and subsequently. Relaxation of the assumptions is possible in some cases, e.g. the number of variables may be about 40, or 9 observations per variable are acceptable. The limitations make it easier to answer the main research questions, but do not seriously reduce the scope of what may be done and the recommendations for practice (see Section 12.2). Also, in some of our examples we violate some of the assumptions.

1.7   MODELLING PREFERENCES

1.7.1   General Issues

The application of complex statistical methods for the development of regression models has greatly increased in recent years. Advances in statistical methodology now allow us to create and estimate more realistic models than ever before, and the necessary computer programs are often available. By contrast, however, the properties of many model-building procedures are still unknown, and the few comparisons that do exist tend to be based on (small) simulation studies. This unfortunate situation is a key reason why tastes in model building vary so much between statisticians.

The aims of an investigation play an important role in every statistical analysis. At least in the health sciences, there seems to be a consensus that subject-matter knowledge must be incorporated in such analyses. With some minor modifications, this can usually be done with all the procedures discussed in our book. However, practical experience shows that, in most analyses of observational studies, data-driven model building still plays an important role. Some variables are inevitably chosen mainly by statistical principles – essentially, P-values for including or excluding variables, or information criteria. The definition of a 'best' strategy to produce a model that has good predictive properties in new data is difficult.

1.7.2   Criteria for a Good Model

It is important to distinguish between two main aims when creating a model. The first is prediction, with little consideration of the model structure; the second is explanation, where we try to identify influential predictors and gain insight into the relationship between the predictors and the outcome. Much published research focuses on prediction, in which model fit and mean-square prediction error are the main criteria for model adequacy. With our background in clinical epidemiology, the second aim is more appropriate. Studies are done to investigate whether particular variables are prognostically or diagnostically important, or are associated with an increased risk of some outcome. For continuous predictors, the shape of the function is often of interest, e.g. whether there is an increasing trend or a plateau at high values of x. Because disease causation is invariably multifactorial, such assessments must be done in a multivariable context. In reality, many variables may be considered as potential predictors, but only a few have a relevant effect. The task is to identify them. Often, generalizability and practical usefulness must also be kept in mind when developing a model. Consider, for example, a prognostic model comprising many variables. All constituent variables would have to be measured in an identical or at least in a similar way, even when their effects are very small. Such a model is impractical, therefore 'not clinically useful' and likely to be 'quickly forgotten' (Wyatt and Altman, 1995). In reality, a model satisfying the second aim, although not providing an optimal predictor in the sense of minimizing mean-square error or similar criteria, typically has only slightly inferior performance. A model that fits the current dataset well may be too data driven to reflect the underlying relationships adequately.

The distinction between prediction and interest in the effects of individual variables was stressed by Copas (1983b). He noted that the loss functions are different, and stated that a good predictor 'may include variables which are not significant, exclude others which are, and may involve coefficients which are systematically biased'. Such a predictor would clearly fail to satisfy the explanatory aim of many studies. Apart from these general considerations, no clear guidance on how to develop a multivariable model fulfilling such an aim appears to be available.

1.7.3   Personal Preferences

Our general philosophy is based on experience in real applications and simulation studies, and on investigations of model stability by bootstrap resampling (Efron, 1979; Sauerbrei and Schumacher, 1992; Sauerbrei, 1999; Royston and Sauerbrei, 2003; Ambler and Royston, 2001). It is influenced by the potential use of our models in future applications, an important aim of most of the models we have developed so far. These considerations have led us to prefer simple models unless the data indicate the need for greater complexity. In the context of time-series forecasting, Chatfield (2002) states that the cost of achieving an excellent fit to the current data may be a poor fit to future data, and 'this emphasizes the importance of checking any model with new data ... and explains my preference for simple models'. Hand (2006) expresses similar views regarding classification methods.

A distinctive feature of FP modelling is the availability of a rigorous procedure for selecting variables and functions. The principles for selecting an FP function are easily explained. Combination with BE results in a procedure applicable without detailed expert knowledge.

Arguments favouring BE over other stepwise methods have been given by Mantel (1970). Sauerbrei (1999) argued that BE(0.157) (i.e. BE using a nominal significance level of 0.157) may be used as a substitute for all-subsets procedures with C_p or Akaike's information criterion (AIC). His conclusion was based on asymptotic and simulation results on the significance level for all-subsets procedures, on simulation results for the stepwise methods, and on empirical comparisons in particular datasets (Teräsvirta and Mellin, 1986; Sauerbrei, 1992; Sauerbrei, 1993). Models selected with BE(0.157) and AIC usually have at most minor differences (Blettner and Sauerbrei, 1993; Sauerbrei, 1993). For further discussion, see Section 2.6. We consider BE to be a good candidate for a sensible variable selection strategy. We also believe that the class of FP functions is a good candidate for finding nonlinear relationships with continuous covariates and at the same time for generating interpretable and transferable (generally useful) models (Royston and Sauerbrei, 2005; Sauerbrei et al., 2007a).

The MFP procedure combines these two components (selection of variables and functions). It is computationally not too demanding, statistically comprehensible, and may be applied to most types of regression model. Furthermore, it addresses the two main tasks in multivariable model-building: elimination of 'unimportant' variables and selection of a 'reasonable' dose–response function for continuous variables. We are well aware that every model can only be a crude approximation to the complex relationships existing in reality. We do not aim to fit the data in any sense 'optimally'. A model that includes at least the strong predictors and whose unknown functional form seems to be 'roughly' modelled in a plausible way is, from our point of view, acceptable. Since the MFP modelling process is comprehensible without detailed expert knowledge, the resulting models are interpretable and transferable. We consider MFP modelling to be an important pragmatic approach to determine multivariable models for continuous variables. In a similar vein, Hand (2006) observes that 'more complicated models often require tuning...and, in general, experts are more able to obtain good results than are inexperienced users. On the other hand, simple models can often be applied successfully by inexperienced users'.

In the MFP approach, the nominal significance levels are the tuning parameters, which largely determine the nature of the resulting model with respect to both the number of variables chosen and the complexity of any selected functions. Depending on the aim of a study, significance levels, which may be different for selection of variables and of complexity of functions, may be chosen. For example, when determining adjustment factors in an epidemiological study, a nominal P-value of 0.2 may be sensible, whereas in a study developing a multivariable diagnostic index a P-value of 0.01 may be more appropriate.

MFP has been progressively extended to perform wider tasks, e.g. modelling interactions between a categorical and continuous covariate (Royston and Sauerbrei 2004a), and determining time-varying functions of regression coefficients in the Cox model (Sauerbrei et al., 2007c). We are well aware that data-dependent modelling ignores the uncertainties of the model-building process and leads to potentially biased estimates of parameters and underestimation of their standard errors.

1.8   GENERAL NOTATION

Here, we provide a concise explanation of general notation used in our book. We have kept mathematical exposition and notation to an absolute minimum throughout.

In general, x denotes a predictor (covariate, independent variable, explanatory variable, risk factor, etc.) and y an outcome variable (response, dependent variable, etc.). We use lowercase letters (e.g. x, β) to denote scalar quantities. Uppercase letters are used sparingly, sometimes denoting models (e.g. M_1). Lowercase bold letters (e.g. $\mathbf{x}$, $\boldsymbol{\beta}$, $\mathbf{p}$) are used for (row) vectors. The expression $\eta = \beta_0 + \mathbf{x}\boldsymbol{\beta}$ denotes the 'index' of a model (see Section 1.3.5), where $\mathbf{x}$ and $\boldsymbol{\beta}$, each vectors with k elements, are explanatory variables and regression parameters of the model respectively, and β_0 is the intercept. Strictly speaking, $\mathbf{x}\boldsymbol{\beta}$ should be written as $\mathbf{x}\boldsymbol{\beta}^{\mathrm{T}}$, where the superscript T denotes vector or matrix transpose, but no ambiguity results from omitting the transpose.

Expectation is denoted by $E()$, variance by var(), standard deviation by SD or SD() and standard error by SE or SE(), and ninety-five percent confidence interval by 95% CI. The distribution of y is sometimes indexed by $\mu = E(y)$, or by an equivalent parameter $g(\mu)$, where g is a monotonic function known as the link function (see Section 1.3.4).

The quantity D denotes the 'deviance' or minus twice the (maximized) log likelihood of a model. A Gaussian or normal distribution with mean μ and variance σ^2 is denoted by $N(\mu, \sigma^2)$. The chi-squared distribution with d degrees of freedom (d.f.) is denoted by χ_d^2, and $F_{m,n}$ refers to the F-distribution with m and n degrees of freedom. A chi-square test statistic obtained from a likelihood ratio test is denoted by χ^2.

When describing examples and case studies, variable names such as age and pgr are written in typewriter font to distinguish them from the rest of the text. Sometimes, for brevity, variable names are given in algebraic form (e.g. x_1, $x_5 - x_8$).

Notation relating to model selection algorithms with nominal significance levels α or (α_1, α_2), e.g. BE(α), FSP(α), MFP(α_1, α_2), is introduced when required in Chapters 2, 4 and 6. Notation specific to FPs is introduced in Section 4.3.

When no confusion can arise, the same notation may be used for different quantities. For example, x^* denotes scaled x in Section 4.11 and a negative exponential transformation in Section 5.6.2.

CHAPTER 2

Selection of Variables

Summary

1. Because subject-matter knowledge in observational studies is usually limited, data-driven model selection has an important role.
2. Multivariable modelling has different possible goals; the main distinction is between predictive and explanatory models. The former aims for a good predictor, whereas the latter aims to identify important variables.
3. Despite claimed theoretical advantages, the full model is not a practical proposition in most studies.
4. Interpretability and practical usefulness are essential attributes of explanatory models. Simple models are more likely to have such properties.
5. Several model selection algorithms have been suggested. The most used are stepwise methods, the best of which is BE. Methods based on information criteria (AIC and the Bayesian information criterion (BIC)) are the main competitors.
6. With stepwise methods, the key tuning parameter is the nominal P-value for selecting or eliminating a variable. Larger P-values produce larger models.
7. Replication instability of selected models should be assessed by bootstrap.
8. Parameter estimates of selected models, irrespective of the strategy used, are prone to different types of bias. With small samples the bias can be large.
9. Techniques combining selection of variables with shrinkage of their parameter estimates reduce bias, but their properties require further exploration.

2.1 INTRODUCTION

In this chapter, the key issue of how to select variables for a 'final' multivariable model is considered. To simplify the discussion, handling possible nonlinearity in the functional forms for continuous predictors is deferred until Chapter 6. For the time being, therefore, all effects of continuous variables are assumed to be linear. Categorical variables are assumed to be represented by one or more dummy variables.

Multivariable Model-Building   Patrick Royston, Willi Sauerbrei
© 2008 John Wiley & Sons, Ltd

2.2 BACKGROUND

In our book, we address several issues in building multivariable regression models. In fact, any model derived on a given dataset is predicated on several (implicit or explicit) assumptions that the analyst must make before starting any formal or informal model-building process. These assumptions include data interpretation, for example, definition and coding of variables, and how to deal with missing values. In addition, a model class must be chosen. For example, with censored survival data there may be a decision between the Cox model, a fully parametric model, or other approaches; with data suited to generalized linear modelling, the distribution of the response and the link function must be chosen. Many other preliminary decisions are needed before deriving a model. Any of them may influence the resulting model considerably. Some background on these premodelling decision processes in a particular case-control study is given by Blettner and Sauerbrei (1993). For a discussion of further aspects, see Chatfield (2002).

Although subject-matter knowledge should guide model building (see Section 1.5), it is often limited and data-dependent model building is necessary (Harrell, 2001). This is the situation that is discussed here. Generally, subject-matter knowledge can be incorporated in a straightforward way.

We do not consider how to check the required assumptions (e.g. constant variance and normality in the normal-errors model) – it is assumed here that all is well. Detection of outliers and influential observations are considered specifically in the context of FP modelling (see Section 5.2). Otherwise we assume that outliers are not an issue in our analyses.

Our book is mainly about how to identify important predictors for inclusion in a multivariable model, and how to determine a suitable functional form for continuous predictors. In this chapter, however, we address only the process of selecting influential variables. Data-driven model building then reduces to the question of including or excluding a variable.

First, we describe in a bit more detail some of the preliminary decisions required before modelling activities start. We next consider possible aims of multivariable modelling, and this leads to two key questions. First, is the analyst mainly interested in deriving a good predictor (e.g. risk score), with little consideration of the components of the model, or in the effects of individual variables? Second, is a parsimonious ('small') model required, or is a complex ('large') model with many variables allowed or even preferred?

These questions bring us to the fundamental point of how complex a model should be. We discuss the choice between a 'full' model, including all available variables, and a reduced model, in which variables have been eliminated according to statistical criteria. For the selection of variables, we introduce stepwise methods based on repeated significance tests, and information-based procedures utilizing the AIC or the BIC. For stepwise approaches, we emphasize the role of the nominal significance level.

Because variable selection introduces several biases, we discuss the combination of variable selection with shrinkage of parameter estimates. The methodology has evolved in the normal-errors model and is not fully developed for generalized regression models or models for censored survival data. We include a short section (Section 2.8) on this issue because we believe that its importance and relevance will increase in future. Currently, it is an area of active research.

Finally, before the discussion, we stress the necessity of assessing the stability of a selected model under repeated resampling of the data (bootstrapping). With present-day computer power there is no excuse for ignoring this aspect of model checking. It is also a first step towards incorporating model uncertainty into predictive modelling, another relevant topic of increasing interest, but one which is beyond the scope of our book.

2.3   PRELIMINARIES FOR A MULTIVARIABLE ANALYSIS

We consider different methods of choosing a suitable model reflecting the relationship between potential predictors $x_1, \ldots, x_k$ and an outcome y of interest. We assume that for each of n observations all variables are measured, that a 'reasonable' model class is available, and that appropriate statistical methods are used to derive suitable model(s) within the predefined class. Unless stated otherwise, a 'sufficient' sample size is assumed. More precise specification of sample size depends on the study aims (see Harrell et al., 1996; Schmoor et al., 2000; Vittinghoff and McCulloch, 2007). The consensus seems to be that stepwise and other variable-selection procedures have limited value as tools for model building in small studies (Steyerberg et al., 2000) (see also Section 2.9.1).

Unfortunately, owing to the mathematical intractability of the problem, theory can answer only rather limited questions about the characteristics of heuristic model selection procedures. To get any answers at all, unrealistic assumptions are required (e.g. independence between covariates). In general, therefore, comparisons between techniques are limited either to simulation studies or to empirical work with real datasets. The latter are often taken from the literature (e.g. textbooks) or, more recently, from the Web. This not only saves much time, it also hugely simplifies the problems of data analysis. For example, the data are assumed to be clean, the 'relevant' potential predictors are supposedly given, and the number of predictors is manageable. We do not consider microarray data and other special situations with very large numbers of variables.

It is not questioned whether the data were sampled from a defined population, nor even whether relevant data are available. The issues of representativeness and relevance of the sample are major constituents of what Mallows (1998) calls the 'zeroth problem' of data analysis. One must be aware that published data have usually undergone rigorous preselection of variables. It is doubtful that estimates from a model including all available variables have the nice statistical properties of a prespecified 'full' model, because the estimates do not allow for the uncertainties associated with preliminary selection of variables and other data manipulations that are invariably performed by the original researchers before the data are made publicly available.

To someone responsible for the initial multivariable analysis of a new study, it soon becomes apparent that many things must be done before model building can begin in earnest. These preliminary activities greatly affect the model-building strategy adopted. Clearly, considerable common sense is required. Mallows (1998) states that

> *Statisticians must think about the real problem, and must make judgements as to the relevance of the data in hand, and other data that might be collected, to the problem of interest . . . one reason that statistical analyses are often not accepted or understood is that they are based on unsupported models. It is part of the statistician's responsibility to explain the basis for his assumption.*

Typically, the preliminaries for a multivariable analysis are not described in publications. In the analysis of a case-control study on potential risk factors for adult brain tumour, Blettner and Sauerbrei (1993) illustrated several preliminary steps needed before model building can begin. For example, they assumed that data cleaning and associated decisions had been successfully completed beforehand, that the logistic regression model was an appropriate way to assess the influence of risk factors in their case-control study, and that the matching variables of age and sex should always be included in the model. They then discussed five problems remaining for the analyst in this particular example:

- Problem 1. *Excluding variables prior to model building.* Two hundred items were coded and further variables were derived from them. They decided to restrict the analysis to 29 variables, representing different classes of risk factor.
- Problem 2. *Variable definition and coding.* The distribution of variables and their correlation may require careful collapsing of categories, creation of new variables and a judicious choice of coding for ordinal variables (see Section 3.3).
- Problem 3. *Dealing with missing data.* A generally recognized difficulty in multivariable analysis.
- Problem 4. *Combined or separate models.* They discuss whether subgroups should be ana- lysed separately. Knowing that the statistical power to detect interactions is generally low, it may be tempting to argue on subject-matter grounds for separate models in subgroups.
- Problem 5. *Choice of nominal significance level and selection procedure.* A key issue in our book.

Blettner and Sauerbrei (1993) showed through sensitivity analyses that different preliminary decisions would have changed the results. As is often the case, the scientific background knowledge was weak and could support different attitudes towards Problems 1–5. Subjective judgement is certainly required.

One essential component of our book, modelling the functional form of the relationship between a continuous covariate and the outcome, played a negligible part in the analysis and in Blettner and Sauerbrei (1993). It is one aspect of Problem 2. Further relevant preliminaries could have been raised; for example, see Chatfield's (2002) section 'Tackling real life statistical problems' and the importance of an *initial data analysis* as a vital preliminary to model building and inference, or the discussion in Mallows (1998), who states:

> *In considering the application of statistical methods to a real problem, many such judge- ments must be made; some may be based on actual data, others may require us to guess what data would be like if it were possible to make the observations. Thinking in these terms helps to clarify what data needs to be collected, and how it relates to the question of interest.*
>
> (Reproduced by permission of American Statistical Association)

In summary, the real-life problem must determine the aims of a study and, therefore, the aims of a multivariable modelling exercise. Together, they must determine the data analysis strategy.

2.4 AIMS OF MULTIVARIABLE MODELS

It is important to distinguish between two main goals when creating a model (Copas, 1983b; Sauerbrei et al., 2007a). The first is prediction, with little consideration of the model structure;

the second is explanation, where we try to identify influential predictors and gain insight into the relationship between the predictors and the outcome. Much published research focuses on prediction, in which model fit and mean-square prediction error are the main criteria for model adequacy. With our background in clinical epidemiology, the second aim is more appropriate. Studies are done to investigate whether particular variables are prognostically or diagnostically important, or whether they are associated with an increased risk of some outcome. Because disease causation is invariably multifactorial, such assessments must be done in a multivariable context. In reality, many variables may be considered as potential predictors, but only a few have a relevant effect. The task is to identify them. Often, generalizability and practical usefulness must also be kept in mind when developing a model. In reality, a model satisfying the second goal, although not providing an optimal predictor in the sense of minimizing mean-square error (MSE) or similar criteria, typically has only slightly inferior performance.

Multivariable regression models perform many different tasks in all areas of science in which empirical data are analysed. Tasks include

- prediction of an outcome of interest;
- identification of 'important' predictors;
- understanding the effects of predictors ('explanatory' aim);
- adjustment for predictors uncontrollable by experimental design;
- stratification by risk.

Different tasks imply different aims for a model under development. Whether or not a model is judged adequate depends on the aim(s). In a broad sense, most of the modelling processes we have experience of fall into one of the following six classes:

1. The model is predefined. All that remains is to estimate the parameters and check the main assumptions.
2. The aim is to develop a good predictor. The number of variables should be small.
3. The aim is to develop a good predictor. Limiting the model complexity is not important.
4. The aim is to assess the effect of one or several (new) factors of interest, adjusting for some established factors in a multivariable model.
5. The aim is to assess the effect of one or several (new) factors of interest, adjusting for confounding factors determined in a data-dependent way by multivariable modelling.
6. Hypothesis generation of possible effects of factors in studies with many covariates.

Estimate of the treatment effect in a randomized trial, with or without adjustment for pre-specified covariates, is a standard example of class 1. Models in the other classes are all data driven, sometimes with a subject-matter component. Although the advantages are always stressed, it is rather an exception that sufficient subject-matter knowledge exists to determine a model in biostatistics, epidemiology and many other fields (Harrell, 2001). For one example where subject-matter knowledge modified a component of a data-driven model, see Sauerbrei and Royston (1999) or Section 6.5.4.

Combinations of the two main components – (1) estimating the effect of a predictor or of a single factor or (2) requiring a parsimonious model with only a few predictors, or a complex model with many predictors allowed – give rise to classes 2–5. There is some overlap between the classes, and some extensions may be required by the problem to hand. For example, the possible prognostic effect of the urokinase marker UPA and its subcomponents

PAI-1 and PAI-2 is a topic of interest in early breast cancer (Foekens et al., 2000). The additional prognostic value of these three correlated variables is of concern, but assessment of the value calls for a model that includes standard prognostic factors established in early breast cancer for some considerable time. Each of the three variables could be considered in turn, adjusting for established factors. Each such analysis belongs to class 4. An extension of class 4 would be a detailed analysis of the joint effect of the three factors, adjusting for the other factors. Trying to derive a good predictor from all of these variables is an example of class 2. The statistical analysis must be adapted according to the various different aims for this dataset.

A more transparent example of class 4 is a new diagnostic marker whose ability to discriminate between patients with and without a particular disorder is assessed with adjustment for age and one or two other variables. An example of class 2 may be a prognostic index to assess the risk of disease recurrences for a cancer patient. Such risk assessment often plays a role in the choice of treatment. For this purpose, a requirement for hospital use is for a limited number (say, between three and six) of easily measured variables.

Sometimes, it is important to have a good predictor, irrespective of the number of variables included and whether or not the effects of these variables are expressed as a complicated function. An example is weather forecasting, which may be based on a complex computer simulation, the details of which are of no relevance to the consumer. The SE or MSE of the predictor are important criteria from a statistical point of view, whereas other aspects, such as the 'meaning' of individual components, the interpretability of the model, or the cost of measuring all the required variables on a new subject, are less important. This example belongs to class 3.

To try to overcome the problems (e.g. unrealistically narrow CIs) caused by inference conditional on a single selected model, the idea of incorporating model selection uncertainty has attracted attention in recent years (Buckland et al., 1997; Hoeting et al., 1999; Burnham and Anderson, 2002; Augustin et al., 2005). Interpretation of the effect of individual variables is essentially ignored when model averaging is used to improve a predictor and to assess its SE more realistically. This is another example of class 3, but model averaging is not considered in general in our book (see also Section 1.7.3). In Section 11.3.2 we discuss a recent approach to averaging functions.

To understand the aetiology of a disease better and to improve the prognostic assessment of patients in many diseases, an enormous amount of work has been done to identify new factors and to assess their influence on the hazard, survival time or other outcome criteria. It is recognized that such an assessment must be done in a multivariable context including other factors considered important for the question under investigation. Usually, there is a complex correlation structure between the factors and some of them have no influence on the outcome in a multivariable context. Assessing the role of a new prognostic factor in a multivariable model including well-established factors belongs to class 4 or 5.

To assess a risk factor for a disease, epidemiologists usually rely on observational studies, such as cohort studies or case-control studies. Control of confounding is a key issue here. Because the effects of confounders are of little interest in themselves, the usual practice seems to be to include a long list of confounders in the model. Overfitting the data by including too many potential confounders is not considered to be a serious issue; the estimate of the risk factor and its variance are mainly of interest. Such a model belongs to class 5.

Class 6 represents another important aim of regression modelling. A prominent recent example is the analysis of genetic marker data (Simon et al., 2003). Other examples are the search for risk factors in epidemiological studies (Thomas et al., 1985; Greenland, 1989; Harrell et al., 1996), and searching for interactions (see Chapter 7).

Here, we restrict ourselves to classes 2–5. Model building (the main topic of our book) is less relevant in regression models from class 1. Trying to extract pertinent information from studies in which thousands of markers are measured in tens or hundreds of patients is the most challenging problem in class 6, requiring methodology not considered here. However, it may be that parts of our MFP approach can be incorporated in the proposed new strategies, or replace parts of them. For hypothesis generation with, say, tens of variables in hundreds of observational units, the strategies discussed here may be used with minor modifications.

Validation of proposed models, prognostic indices, etc. is another important aspect of multivariable regression modelling. Internal validation can be done by resampling methods (Harrell, 2001). External validation requires independent data; the topic is beyond the scope of our book. Such issues are not considered here, as we concentrate on the development of regression models.

2.5   PREDICTION: SUMMARY STATISTICS AND COMPARISONS

As regards prediction, the main 'output' from a multivariable model is its index $\widehat{\eta}$. The predictive ability of a normal-errors model, for example, may be summarized through its explained variation (also known as the index of determination) R^2. Among many equivalent formulae for R^2 is the squared correlation between y and $\widehat{\eta}$. See Hardin and Hilbe (2007, pp. 58–63) for some of these formulae and for a discussion of extensions of R^2 to generalized linear models.

Two models may differ with respect to their covariates and/or the types of function of continuous variables that are used. Given such differences, it is natural to ask how closely they agree with respect to prediction. Their R^2 values can be compared directly, and further information may be obtained by a comparison of their indexes. This amounts to a method-comparison study. Bland and Altman's (1986) widely accepted limits-of-agreement approach to method comparison involves a plot of the individual differences against the individual means of the two methods, and a calculation of the mean and SD of the individual differences across the sample. The limits of agreement are the mean difference $\pm 2SD$.

Limits of agreement are scale dependent and, therefore, are not comparable across types of model and modelling scenarios. As a summary statistic for agreement between the indexes from two models, we prefer the intraclass correlation coefficient (ICC), as discussed for example by Dunn (2004). The ICC r_I for indexes $\widehat{\eta}_1$ and $\widehat{\eta}_2$ may be calculated by counting each point in the sample twice: once with the first member of each pair $(\widehat{\eta}_1, \widehat{\eta}_2)$ as X and once with the first member as Y. r_I is the Pearson correlation between X and Y in the 'doubled' sample. Fleiss and Cohen (1973) have shown that r_I is approximately equivalent to a well-known measure of agreement, namely Cohen's κ (with quadratic disagreement weights).

2.6   PROCEDURES FOR SELECTING VARIABLES

In this section, a linear effect is assumed for all continuous variables, perhaps following a preliminary transformation such as $\log x$. Basic modelling assumptions, such as proportional hazards in the Cox model, are taken for granted. Subject-matter knowledge is unavailable and cannot play a role in determining the model. Model selection then reduces to a decision to include or exclude each variable. For k candidate variables there are 2^k possible models.

Many procedures for selecting variables have been proposed. Often they do not lead to the same solution when applied to the same problem. There seems to be no consensus among modellers as to the advantages and disadvantages of the various procedures. All procedures are criticized, but for different reasons (Harrell, 2001; Burnham and Anderson, 2002; Sauerbrei et al., 2007a). In practical terms, however, either the full model must be fitted or one of the strategies for selecting variables must be adopted.

There are two main types of strategy for variable selection. Sequential strategies, such as forward, stepwise or BE procedures, are based on a sequence of tests of whether a given variable should be added to the current model or removed from it, or selection should stop. A nominal significance level for each of these tests is chosen in advance and largely determines how many variables end up in the model. In the second type, the all-subsets strategies, all 2^k possible models are fitted and the best model is chosen by optimizing an information criterion derived from the likelihood function. AIC (Akaike, 1973) and BIC (Schwarz, 1978) are often used. They compare models based on goodness of fit penalized by the complexity of the model. In the normal-errors model, the error sum of squares is taken as a measure of goodness of fit and the penalty term is the number of variables in the model multiplied by a constant. The penalty constant for BIC is larger than for AIC, resulting in smaller models. Several modifications of AIC and BIC exist but are not considered here. For further information, see Teräsvirta and Mellin (1986), Burnham and Anderson (2002), Burnham and Anderson (2004), and Kuha (2004).

Methods for variable selection are usually developed in the normal-errors model and transferred by analogy to more general models, such as GLMs or models for censored survival data. In practical terms, computational difficulties may limit the use of a preferred approach. For example, at the beginning of the 1990s even the software package SAS had no routine for all-subsets regression with the Cox model. In this regard, test-based sequential procedures have an advantage, which probably accounts for their popularity. See also Section 2.6.4.

By applying the procedures to different datasets, we illustrate the point that variation in the nominal significance level is the main reason for selecting different models. However, if the same nominal significance level is used, then under the assumptions of Table 1.3 the resulting models are similar. In view of the similarities of the results, the differences between the procedures would appear to have been overemphasized in the literature (Sauerbrei et al., 2007a). This may be a type of publication bias, reflecting apparently alarming differences between methods in special cases or in studies with small sample sizes. For an extreme example, see Kuk (1984) and our discussion in Section 2.7.1.

2.6.1 Strength of Predictors

An important concept in the present chapter and beyond is the 'strength' of a predictor x in a model. For present purposes, we define a 'strong' predictor to be one that is significant at $P < 0.001$ in a given model, and a 'weak' predictor to be one that is not significant the 5% level. A 'medium(-strength)' predictor lies in between. Assuming a normally distributed test statistic and two-tailed testing, the equivalent cut-offs on $Z = |\widehat{\beta}|/\mathrm{SE}(\widehat{\beta})$ are $Z \geq 3.29$ and $Z < 1.96$ respectively. If the expected value of Z in repeated sampling from the same population was 3.29 (or 1.96), then the probability of x being statistically significant at the 0.1% or at the 5% level would be approximately 91% or 50% respectively.

Note that since, other things being equal, $SE(\widehat{\beta})$ decreases with the sample size n, the strength of x is a function of $\widehat{\beta}$, of its estimated variance and (through the latter) of n. In a multivariable context, the strength of x varies according to which other variables are in the model, so it is not a fixed quantity. Strength plays an important role in variable selection and its associated biases (see Section 2.8.2) and in model stability (see Chapter 8).

The importance and, therefore, the classification of a given variable as a strong, medium or weak predictor changes with the aims of the analysis and the consequent nominal significance level, α.

2.6.2 Stepwise Procedures

For clarity, we describe BE and forward selection (FS) algorithms, together with stepwise variants in which omitted variables can be reincluded and included variables can be removed. Unfortunately, different names are found in the literature for the same procedure, and the same name may be used for different variants. Unless specified otherwise, the procedures without reinclusion or re-exclusion (BE-only, FS-only; see Table 2.1) are not considered. First, nominal significance levels α_1 for including variables (FS) and α_2 for removing them (BE) are chosen. In software packages, α_1 is sometimes called the 'P-to-enter' and α_2 the 'P-to-remove'. In practice they are often taken as equal ($\alpha_1 = \alpha_2 = \alpha$). In the following description (and the rest of our book) we do not distinguish between them.

Consider covariates $x_1, \ldots, x_k$. Let us review in detail the logic of the first step of the BE procedures (BE, BE-only). The logic is

1. Estimate the model on $x_1, \ldots, x_k$ (i.e. the full model).
2. For $j = 1, \ldots, k$, consider dropping x_j.
3. Find the variable that is least significant. Remove it if its P-value is $\geq \alpha$.

The logic of the first step of FS and FS-only is

1. Estimate the null model (i.e. $\eta = \beta_0$).
2. For $j = 1, \ldots, k$, consider adding x_j.
3. Find the variable that is most significant. Include it if its P-value is $< \alpha$.

Stepwise procedures (BE, FS) can also alternate between adding and removing terms. The full logic is given in Table 2.1. As an alternative to specifying a nominal significance level, one may use the AIC or BIC as a stopping rule (see also Section 2.6.3).

Unfortunately, no analytical results on the operating characteristics of such procedures are available. Sauerbrei (1992; 1993) concluded from simulation studies, at least in the normal-errors model and for a sufficient sample size, that for an uncorrelated uninfluential variable the true type I error probabilities of stepwise methods are only slightly higher than the nominal level; they increase with the correlation of that variable with influential variables. When the ratio of the sample size to the number of variables is small, the methods can select the wrong predictors (Freedman, 1983). These problems increase with the amount of random variation in the response variable, an increasing number of predictors and higher correlations between them. As already mentioned, we assume at least a moderate sample size and common sense to handle the issue of highly correlated predictors. In such a situation, stepwise procedures may not be seriously affected by these difficulties (Marubini and Valsecchi, 1995, p. 309).

Table 2.1   Logic for four variants of stepwise model selection procedures.[a]

Procedure	Logic
BE-only	Estimate full model on $x_1, \ldots, x_k$. Repeat: while the least significant term has $P \geq \alpha_2$, remove it and re-estimate the model
BE	Estimate full model on $x_1, \ldots, x_k$. If the least significant term has $P \geq \alpha_2$, remove it and re-estimate; otherwise stop.
	Again: if the least significant term has $P \geq \alpha_2$, remove it and re-estimate; otherwise stop
	Repeat
	• if most significant excluded term has $P < \alpha_1$, add it and re-estimate;
	• if least significant included term has $P \geq \alpha_2$, remove it and re-estimate;
	until neither action is possible.
FS-only	Estimate null model. Repeat: while the most significant excluded term has $P < \alpha_1$, add it and re-estimate.
FS	Estimate null model. If the most significant excluded term has $P < \alpha_1$, add it and re-estimate; otherwise stop.
	Again: if the most significant excluded term has $P < \alpha_1$, add it and re-estimate; otherwise stop.
	Repeat
	• if least significant included term has $P \geq \alpha_2$, remove it and re-estimate;
	• if most significant included term has $P < \alpha_1$, add it and re-estimate; until neither action is possible.

[a]Adapted from description of `stepwise` routine in the Stata reference manual (StataCorp, 2007).

In contrast to FS-only, FS allows a variable which was included in an earlier step to be eliminated. Because of this advantage, FS-only is now rarely used. From our experience of many analyses using BE, excluded variables are rarely re-entered. In contrast, re-exclusion is more common with FS.

Mantel (1970) argues strongly in favour of BE over FS, especially when collinearity is present. FS starts by considering k univariate models, all of which underfit the data and have a large residual variance. In the first step the best of k potentially bad models is selected. In contrast, BE starts with the full model. It is likely that some of the variables in the full model have no effect, but it is at least a reasonable starting point. If collinearity between predictors is low, then FS and BE frequently select the same model. With larger amounts of collinearity, FS has a higher probability of stopping without considering better models including members of a cluster of correlated covariates.

FS is defensible when possible models for the data become complex, e.g. in high-dimensional data or when many interactions may be considered. We use FS to build models with two-way interactions (section 7.11.2) or with interactions with time (Section 11.1.2). Criticisms of sequential procedures are discussed in Section 2.7.

2.6.3   All-Subsets Model Selection Using Information Criteria

According to this approach, to select a model an information criterion must be minimized over the 2^k available models with up to k variables. For a given model M, the information criterion IC is of the form

$$IC = -2l + a \dim (M) \tag{2.1}$$

where l is the maximized log likelihood of M, $\dim(M)$ is the number of estimated parameters and a is a penalty constant. The penalty constant $a = 2$ gives the AIC, independent of sample size. BIC has a penalty constant $a = \log n$ and, therefore, depends on the sample size. For censored survival data, the sample size n may be replaced by the number of events (effective sample size, see Volinsky and Raftery (2000)). See also Section 2.6.4.

Despite the apparent simplicity of Equation (2.1), such information criteria are not *ad hoc* statistics but are derived from explicit theoretical considerations. Model selection based on such a criterion can thus be regarded, at least approximately, as a choice of the best model according to some underlying definition of optimality (Kuha, 2004).

As discussed by Burnham and Anderson (2004, p. 299), AIC and BIC have different types of target model; therefore, a direct comparison may not be appropriate. The target model of AIC is one that is specific for the sample size at hand. It is the fitted model that minimizes the expected estimated Kullback–Leibler information loss (Kullback and Leibler, 1951) when the fitted model is used to approximate full reality. This target model becomes more complex with increasing sample size.

The classical derivation of BIC assumes that a single true model M_0 generated the data independently of n. M_0 is taken to belong to the set of 2^k models under consideration, and to be the target model for selection by BIC. However, selection of M_0 with unit probability only occurs in the limit as n gets very large, and that as n increases the set of models is kept fixed. The original derivation of BIC has been relaxed. Now, such convergence is seen as justifying only an inference of a *quasi-true model* (the most parsimonious model closest in Kullback–Leibler information to the truth).

In practical terms, all that matters is the value chosen for the penalty constant a, which determines the complexity of a model. The aim of a study determines whether a more or less complex model is preferable (see Section 2.4).

For the normal-errors model, Teräsvirta and Mellin (1986) present a helpful review of penalty constants and corresponding asymptotic significance levels for the inclusion of one additional variable. The asymptotic significance level for AIC is 0.157, and equals the upper tail area of the χ^2 distribution on one d.f. corresponding to a deviate value of $a = 2$. Similar calculations may be made for BIC for different sample sizes; for example, for $n = 100$ (400) the BIC significance level is 0.032 (0.014).

2.6.4 Further Considerations

Here, some practical questions concerning the use of stepwise and all-subsets selection methods are discussed briefly.

Three types of hypothesis test are employed in sequential strategies: likelihood ratio (LRT), score and Wald tests. All may be computed with any type of regression model. Minor variations are encountered across statistical packages, e.g. use of the t-distribution instead of the normal distribution. Some years ago, computation time may have precluded the implementation of certain more complex test statistics, but that is no longer relevant today – at least for the analysis of a single dataset, computation time is not usually significant. Lee et al. (1983) compared the three statistics in the Cox model and concluded that they perform similarly. The LRT has the best statistical properties overall, followed by the score and Wald tests. For a more detailed discussion, see Harrell (2001, p. 189). We prefer the LRT, but we do not regard it as serious if a software package does not provide it. In real applications with a sufficient sample size, the choice of test statistic is not critical. However, selection of

FP functions by using our closed test procedure (see Section 4.10) uses a sequence of LRTs. MFP combines BE with the selection of FP functions. Implementation of MFP modelling (see Chapter 6), therefore, requires the software to evaluate likelihoods of all relevant models.

Several modifications, e.g. the use of conditional P-values (Pinsker et al., 1987), have been suggested with stepwise procedures, but if the sample size is sufficiently large then they make little difference. The idea of reporting for each step both the variable selected and 'close alternatives' (Hauck and Miike, 1991) has remained undeveloped in the literature. However, to incorporate subject-matter knowledge in the selection process, such an approach may be sensible and may lead to modification of the model selected automatically. Subject-matter knowledge is a fundamental part of a proposal by Allen and Cady (1982) to rank the candidate variables by importance (e.g. by forming groups). Some variables may be forced to be in the model irrespective of their P-value, and others are selected according to their *a priori* ranking. In theory, the approach has better statistical properties; unfortunately, subject-matter knowledge is usually insufficient to rank the variables meaningfully. Forcing certain variables (e.g. confounder variables in epidemiological studies) into a model is obligatory if required by the design of the study. Most computer programs have an option to force variables in.

The development of efficient algorithms has made it possible to determine the 'best' model in an enormous class. For example, with 20 variables as many as $2^{20} = 1\,048\,576$ models must be screened in the all-subsets approach. For the normal-errors model, the leaps-and-bounds algorithm offered such a possibility more than three decades ago (Furnival and Wilson, 1974). For GLMs and models for survival data, the conceptual basis is the same but additional computational barriers must be overcome and certain restrictions may apply. Lawless and Singhal (1978) proposed an algorithm for nonnormal models, such as logistic regression, with linear covariate effects. Kuk's (1984) approach is appropriate for the proportional hazards model. Numerical instability may be an issue in extreme data situations, but in general does not restrict the use of the all-subsets approach in many types of regression model. Stepwise procedures, particularly BE, may be regarded as a simple computational approximation to the all-subsets algorithm. There is some probability that all-subsets methods will find a better-fitting model not considered by BE; but if equivalent significance levels are used and correlations are modest, then the models are often the same.

Because it is simple to program, BE is sometimes used when the real aim is to select the best all-subsets model optimizing a particular criterion. For example, Stata does not provide all-subsets selection, but BE and FS are fully implemented for a wide range of regression commands. Similarly, SAS's new procedure GLMSELECT offers only stepwise algorithms and does not search for the best-fitting among all models. AIC or BIC can be chosen as the stopping criterion. Models selected in this way are termed BE(AIC) or BE(BIC). Likewise, the R package `step` does not provide an all-subsets procedure.

Common sense and an awareness of several simple issues are essential when using any of the selection strategies. First, consider the scale of measurement. For example, an ordinal or categorical variable coded 1 to 5 can be handled as a continuous variable; and unless the analyst specifies otherwise, a linear effect is assumed. With stepwise procedures, most packages have default values of α_1 and α_2 with which to select a model. Sometimes α_1 and α_2 may be unequal, default values may vary, and defaults for the normal-errors model may differ from those for the Cox model. As we have emphasized, the nominal significance level is the key tuning parameter affecting the development of a model fulfilling a given aim. We suggest that the user should *always* specify α_1 and α_2 explicitly. Other parameters, e.g. convergence criteria for models fitted iteratively, may be varied, but the default values supplied are almost always satisfactory.

2.7   COMPARISON OF SELECTION STRATEGIES IN EXAMPLES

As we have noted, the literature is replete with criticism of selection strategies. In reality, however, the analyst not favouring the full model must use one of them. When model-building procedures are compared in simulation studies or according to their mathematical properties, the full model is often a strong competitor and has good properties according to some criteria. However, as discussed in Section 2.3, we doubt that there is a meaningful full model in most real situations. Data manipulation and preselection of variables have either been performed already by someone else (as with publicly available data), or are required anyway as part of the initial data analysis. We compare results from the full model – despite the problems with this term, we use it as a convenient shorthand – with those from models derived by FS and BE with different significance levels, and by using the two information criteria, AIC and BIC. We show a range of examples illustrating different issues, but in which the type of regression model is essentially irrelevant. Examples come from normal-errors models, logistic regression and the Cox model.

2.7.1   Myeloma Study

Sixteen potential prognostic factors were considered in a small study of 65 patients with multiple myeloma, 48 of whom died. Details of the 16 covariates, which for brevity are simply called $x_1, \ldots, x_{16}$, may be found in Krall et al. (1975, Table 1). The survival times of the patients have been analysed several times in the literature by using a Cox model. Here, we consider variables included by different selection strategies. In Table 2.2 we summarize the

Table 2.2   Myeloma study (65 patients, 48 events). Results of applying variable selection strategies.[a]

Variable	Nominal significance level, α						All-subsets
	0.05			0.157			
	Full	BE	FS	Full	BE	FS	AIC
x_1	*	✓	✓	*	✓	✓	✓
x_2			✓			✓	
x_3	*	✓		*	✓		✓
x_4		✓		*	✓		✓
x_5							
x_6		✓		*	✓		✓
x_7	*	✓		*	✓		✓
x_8				*	✓		✓
x_9							
x_{10}							
x_{11}							
x_{12}	*	✓		*	✓		✓
x_{13}	*	✓		*	✓		✓
x_{14}				*			
x_{15}							
x_{16}							

[a] BE(0.01) and FS(0.01) selected only one variable, x_1; * denotes that a variable is significant at the relevant α-level in the full model.

results from the full model using a Wald statistic to determine significance of each variable, all-subsets using the AIC criterion, and BE and FS using $\alpha = 0.05$ and $\alpha = 0.157$.

This example was used by Kuk (1984) to illustrate the advantages of all-subsets regression over stepwise selection in the analysis of survival data. With the all-subsets approach, a model with eight variables was selected, whereas FS(0.05) stopped after the inclusion of x_1 and x_2. In contrast, BE(0.05), which was not considered in the paper by Kuk, selected a model with seven of the 8 variables. Using 0.157 as significance level, BE included the same eight variables as the AIC model. Even with a significance level of 0.157, FS selected only x_1 and x_2. This is the most extreme example of differences between BE and FS models that we have come across in about 20 years of working with stepwise procedures. Likely reasons for the differences are the very small events per variable ratio of 3 (16 variables and 48 events), and some fairly high correlations between the predictors. Even in this unusual example, AIC and BE(0.157) selected the same model.

2.7.2   Educational Body-Fat Data

Brief details of the educational body-fat data are given in Appendix A.2.3. Here, the data are analysed omitting one highly influential observation (case 39, see Section 5.3.1). The aim is to predict pcfat, the percentage of body fat in 251 men, from 13 predictors, most of which are dimensions of body components. Table 2.3 gives details of the full model and that selected by BE(0.05). For other procedures and significance levels, Table 2.4 indicates only which variables were selected.

The standardized estimated coefficients from the full model indicate that x_6 and x_{13} are strong predictors, with x_6 dominating. The effect of x_1 (age) is significant at the 5% level. The two variables with a large effect, x_6 (abdomen) and x_{13} (wrist), are included in all models, whereas selected variables differ across the other models. With a larger significance level, more variables are included and there are more differences between BE and FS, probably caused

Table 2.3   Educational body-fat data. Full model and that selected by BE(0.05). Final three columns give details of the full model excluding x_6.

Variable	Full model			BE(0.05)			Full model excl. x_6		
	$\widehat{\beta}$	SE	$\widehat{\beta}$/SE	$\widehat{\beta}$	SE	$\widehat{\beta}$/SE	$\widehat{\beta}$	SE	$\widehat{\beta}$/SE
x_1	0.074	0.032	2.31	0.056	0.024	2.35	0.211	0.034	6.20
x_2	−0.019	0.067	−0.28				0.227	0.074	3.08
x_3	−0.249	0.191	−1.30	−0.322	0.121	−2.65	−0.915	0.212	−4.32
x_4	−0.394	0.234	−1.68				−0.378	0.278	−1.36
x_5	−0.119	0.108	−1.10				0.150	0.124	1.21
x_6	0.901	0.091	9.90	0.774	0.033	23.26	–	–	–
x_7	−0.146	0.144	−1.02				0.163	0.166	0.98
x_8	0.178	0.146	1.22				0.231	0.173	1.33
x_9	−0.041	0.245	−0.17				−0.095	0.291	−0.33
x_{10}	0.185	0.220	0.85				−0.053	0.259	−0.21
x_{11}	0.178	0.170	1.04				−0.066	0.200	−0.33
x_{12}	0.277	0.207	1.34				0.058	0.244	0.24
x_{13}	−1.830	0.529	−3.46	−1.943	0.406	−4.78	−2.692	0.620	−4.34

Table 2.4 Educational body-fat data. Variables selected by several procedures.[a]

| Variable | Full model | Nominal significance level, α | | | | | | | | All-subsets | |
| | | 0.01 | | 0.05 | | 0.10 | | 0.157 | | | |
		BE	FS	BE	FS	BE	FS	BE	FS	AIC	BIC
x_1	*			✓		✓		✓	✓	✓	
x_2			✓		✓		✓		✓		✓
x_3		✓		✓		✓		✓		✓	
x_4								✓		✓	
x_5								✓		✓	
x_6	*	✓	✓	✓	✓	✓	✓	✓	✓	✓	✓
x_7											
x_8									✓		
x_9											
x_{10}											
x_{11}							✓		✓	✓	
x_{12}								✓			
x_{13}	*	✓	✓	✓	✓	✓	✓	✓	✓	✓	✓

[a] * denotes a variable that is significant at the 0.05 level in the full model.

by the inclusion of weaker effects and by stronger correlations between the measurements $x_2 - x_{13}$. x_1 (age) is nearly uncorrelated with other variables, x_3 (height) is moderately correlated with others, but all other variables have a high correlation with pairwise correlation coefficients ranging up to 0.9. At the 0.01 significance level only one additional variable is included: BE(0.01) includes x_3 (height), whereas FS(0.01) includes x_2 (weight). The two variables are fairly highly correlated (0.51) and the model fit is nearly identical. If a nominal significance level of 0.05 is used, then the FS model does not change, whereas BE includes x_1 in addition.

Parameter estimates with standard errors for the BE(0.05) model including x_1, x_3, x_6, x_{13} are given in Table 2.3. Comparing the selected model with the full model, the estimate for x_{13} is slightly larger in absolute value, that for x_3 increases by about 30%, whereas the other two estimates decrease: x_1 by about 25% and the dominating variable x_6 by about 14%. More striking is the decrease in the standard errors: for x_6 by about 63% and for the other three variables, by between 23% and 36%.

The fit of the full model is similar to that of the BE(0.05) model, the respective values of R^2 being 0.75 and 0.74. According to a Bland–Altman plot (Bland and Altman, 1986), the fitted values from the two models agree well (see Figure 2.1). The intraclass correlation coefficient (see Section 2.5) is 0.992, confirming the close agreement. The right panel of Figure 2.1 reveals a systematic difference between the predictors, but note that the scale is much magnified compared with the scatter plot on the left. Because of the close agreement, it is not surprising that the fit is also similar for the other models selected with the larger significance levels or with all-subsets strategies using AIC. BE(0.157) selected the second-best AIC model, which included x_{11} instead of x_{12} and had a slightly smaller AIC (732.567 versus 732.575). BIC selected a much smaller model than AIC; in fact, the same three-variable model as FS(0.01).

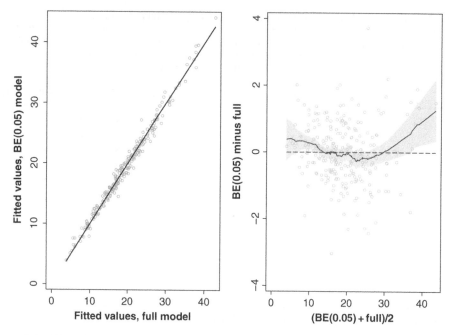

Figure 2.1 Educational body-fat data. Predictors ($\hat{\eta}$) from the full and BE(0.05) models. Left panel: scatter plot with line of identity. Right panel: Bland–Altman plot for difference against mean, with smoothing and 95% pointwise CI.

In Section 2.9 we discuss the issue of variable selection with a focus on comparing a selected model with the full model. Suppose for some reason that the dominating variable x_6 (abdomen) were not present, e.g. the investigators had not measured it or had opted not to make it publicly available. The full model now comprises x_1–x_5 and x_7–x_{13}. Estimates for this model are given in Table 2.3.

The explained variation decreases substantially (R^2 from 0.75 to 0.65) but is still large. More striking are the major changes in the parameter estimates for most of the variables. x_2 and x_3 were previously nonsignificant with a negative sign, but both now have strong effect, x_2 with a positive sign and x_3 with a negative sign. Whereas x_2 is highly correlated with x_6, x_3 is nearly uncorrelated with it. The effect of x_1 increases approximately threefold, making it now highly significant. The estimate of the other important variable, x_{13}, increases by about 50%. None of the variables x_4, x_5, x_7–x_{12} becomes important, but the sign of the parameter estimate changes for some of them. The new analysis indicates that x_1, x_2, x_3 and x_{13} are important predictors if x_6 is not available, but other variables have at most a weak effect. This model with four variables is selected by BE(0.05). The parameter estimates (not shown) are similar to those from the full model excluding x_6, but differ considerably from the model selected by BE(0.05) on all variables including x_6.

2.7.3 Glioma Study

Brief details of the glioma (brain cancer) study are given in Appendix A.2.4. The outcome is overall survival. The data are analysed by Cox regression. Since essentially nothing new

Table 2.5   Glioma data. Variables selected by several procedures.[a]

Variable	Full model	Nominal significance level, α								All-subsets
		0.01		0.05		0.10		0.157		
		BE	FS	BE	FS	BE	FS	BE	FS	AIC
x_1							✓			
x_2										
x_3	*	✓	✓	✓	✓	✓	✓	✓	✓	✓
x_4						✓		✓	✓	✓
x_5	*	✓	✓	✓	✓	✓	✓	✓	✓	✓
x_6	*	✓	✓	✓	✓	✓	✓	✓	✓	✓
x_7										
x_8	*	✓	✓	✓	✓	✓	✓	✓	✓	✓
x_9								✓	✓	✓
x_{10}										
x_{11}						✓		✓	✓	✓
x_{12}				✓	✓	✓	✓	✓	✓	✓
x_{13}										✓
x_{14}						✓		✓	✓	
x_{15}										

[a] * denotes a variable that is significant at the 0.05 level in the full model.

arises in our analyses of this dataset, we only summarize the results; further details are given by Sauerbrei (1999). The results of model selection are given in Table 2.5.

In the full model, four variables (x_3, x_5, x_6 and x_8) have standardized regression coefficients ($|\widehat{\beta}|/\mathrm{SE}$) that exceed 2, i.e. are significant at the 0.05 level. These four, together with x_{12}, are selected by BE(0.05). Regression coefficients for selected variables are similar in the full and selected models, whereas standard errors are slightly lower in the selected model. BE and FS select the same model as each other when the nominal significance level is 1% (x_3, x_5, x_6 and x_8), 5% (these four plus x_{12}) and 15.7% (these four plus x_4, x_9, x_{11}, x_{12} and x_{14}). Slightly different models are selected at the 10% level. The all-subsets model selected by AIC differs from the BE and FS models obtained with significance level 15.7% with respect to just one of the nine selected variables. The four strongest predictors (x_3, x_5, x_6 and x_8), three of which have standardized coefficients exceeding 3, are selected in all models. Whether or not weaker predictors enter the model depends on the nominal significance level. The results are consistent with our experience in many such analyses.

At the 10% significance level, the models selected by BE and FS are slightly different. They have in common x_3, x_5, x_6, x_8 and x_{12}, but FS selects in addition x_1 (the only time this variable is included in any model), whereas BE selects in addition x_4, x_{11} and x_{14}. Thus, with the exception of x_9, BE(0.10) finds the same model as BE(0.157) and FS(0.157) do. Model selection using BE gives consistent results in the sense of including more and more variables as the nominal significance level becomes less stringent. In contrast, FS(0.10) selects a rather peculiar model, since x_1 does not enter any other model and its standardized coefficient in the full model is only -1.36. This result suggests that FS is more likely to select poorer models, especially with respect to variables with a weak effect.

2.8   SELECTION AND SHRINKAGE

2.8.1   Selection Bias

It is well-known that variable selection introduces bias into parameter estimates if the same data are used for both selection and estimation. In an excellent overview on problems introduced by variable selection, Miller (1984) concluded that 'the most important unresolved problem is that of estimation'. Estimating regression parameters by least squares or maximum likelihood after choosing a model introduces a 'selection bias' (Miller, 2002).

An intuitive explanation was given by Copas and Long (1991): 'The choice of the variables to be included depends on estimated regression coefficients rather than their true values, and so x_j is more likely to be included if its regression coefficient is overestimated than if its regression coefficient is underestimated'. Miller subdivides the selection bias into 'competition bias' and 'stopping-rule bias'. The former is the bias introduced by finding the best subset for a fixed number of parameters, whereas the latter arises from the criterion used to determine the number of variables. For example, simpler models are obtained with BIC or a low significance level such as 0.01, and more complex models with AIC or a high significance level such as 0.20. As is clear from Copas and Long's (1991) explanation, selection bias is mainly a problem when variables with weak effects are present. Variables with a strong effect are (almost) always included in the model, irrespective of whether the regression coefficient is underestimated or overestimated in a particular sample. The bias can be pronounced for variables with a weak effect. Examples are given by Schumacher et al. (1997) for a cutpoint model and by Sauerbrei (1999) for models selected by BE.

2.8.2   Simulation Study

We illustrate the effect of selection bias in a small simulation experiment of the case with one variable of interest, x. To be a little more realistic, the simulation also includes four noise variables $x_1, \ldots, x_4$, uncorrelated with x. The underlying model is

$$y = \beta_0 + \beta x + \beta_1 x_1 + \beta_2 x_2 + \beta_3 x_3 + \beta_4 x_4 + \varepsilon$$

where $\beta_j = 0$ and $x_j \sim N(0, 1)$ for all $j = 0, \ldots, 4$, and x and ε are both distributed as $N(0, 1)$. We chose $\beta = 0, 0.2, 0.4$ and 0.8 to represent zero, weak, medium and strong effects of x respectively, in samples of size 50. The interpretation of a 'weak' or 'strong' effect relates to the absolute standardized parameter estimate, $|\hat{\beta}|/\text{SE}(\hat{\beta})$ (see Section 2.6.1).

Increasing the sample size from 50 to 200 reduces $\text{SE}(\hat{\beta})$. A true value of $\beta = 0.2$, a weak effect for the smallest sample size, becomes a medium effect for $n = 200$. In terms of standardized coefficients, the medium effect $\beta = 0.4$ at $n = 50$ doubles to become a strong effect at $n = 200$. With the smaller sample size, selection of x is more likely if $|\hat{\beta}| > |\beta|$ by chance in a particular replication (overestimation), and becomes unlikely if by chance $|\hat{\beta}| < |\beta|$. With the larger sample size, a variable for which $\beta = 0.4$ is nearly always selected. According to theory, $\hat{\beta}$ from the full model has a normal distribution around β with a standard error that depends on the sample size.

Figure 2.2 shows the distribution of estimated regression coefficients $\hat{\beta}$ in samples of size $n = 50$ from 1000 replications of the full model and that selected by BE(0.05). When $\beta = 0$, x is selected in about 5% of replications, and only when $\hat{\beta}$ is extreme ($|\hat{\beta}| > 0.2$ approximately). The remaining estimates are shown as zero in Figure 2.2. When x is selected, the value of $\hat{\beta}$ is close to that in the full model. For $\beta = 0.2$ and 0.4, x is selected more often, and a similar pattern

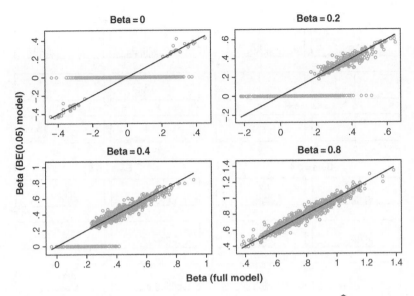

Figure 2.2   Simulation experiment on variable selection. Each panel compares $\widehat{\beta}$ for the full model with $\widehat{\beta}$ in the model selected by BE(0.05) in 1000 replications of a multivariable model with five predictors, four of which are 'noise'. Effect strength is governed by β. Sample size is $n = 50$.

is seen in the estimates. In samples in which the regression coefficient is underestimated, x tends not to be selected. When $\beta = 0.8$, then x has a strong effect and is always selected; the values of $\widehat{\beta}$ agree with those from the full model, apart from random variation induced by chance inclusion of one or more of the noise variables $x_1, \ldots, x_4$. In the normal-errors model, these principles also apply to more complex variable selection situations if x is orthogonal to the other variables. Complications arise in more general models, but the basic considerations are still valid. Simulation results are given by Sauerbrei (1992) for the normal-errors model and the Cox model.

The distribution of estimates for several scenarios is given in Figure 2.3. As expected, in the full model the distribution of $\widehat{\beta}$ is an unbiased normal. With selection of variables, the left-hand portion of the distribution closer to zero is missing, since underestimating β results in nonselection of x. Of course, selection bias is more severe for BE (0.05) than for BE(0.157), because the latter selects variables at a more liberal significance level. With BE(0.05), x with a weak effect is selected in 29% of replications, whereas with a medium effect it is selected in 78% (see Table 2.6).

Corresponding figures for AIC are 48% and 91%. As discussed in Section 2.6.3, selecting a variable with AIC is equivalent to selecting it at the 15.7% significance level. The smaller the inclusion frequency, the larger the value of $|\widehat{\beta}|$ if x is included. When $\beta = 0$, BE(0.05) selects x in about 5% of the replications, as expected. However, in replications in which x is selected $|\widehat{\beta}|$ becomes large, particularly with $n = 50$. When $\beta = 0$, $\widehat{\beta}$ is positive in about half of the replications and negative in the rest.

With correlated variables things are more difficult because exclusion of a variable for which $|\widehat{\beta}| > 0$ introduces an 'omission' bias for correlated included variables, whether or not such variables themselves have an effect. Selection bias and omission bias cannot occur in the full model. Some further simulation results are given by Sauerbrei (1993); for details, see Sauerbrei (1992). For further reading, see Miller (2002).

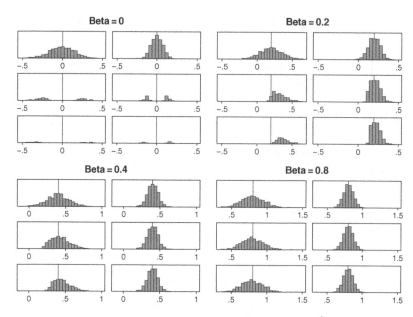

Figure 2.3 Simulation experiment on variable selection. Distribution of $\widehat{\beta}$ for 24 different scenarios. Each group of six plots shows results for a different value of β. In each group, the first row represents the full model, the second that selected by BE(0.157), and the third BE(0.05). The first column of each group is for $n = 50$ and the second for $n = 200$. The true value of β is shown by a vertical line.

Table 2.6 Simulation experiment on variable selection. Inclusion fraction and mean estimated regression coefficients for different effect strengths (governed by β), model selection methods and sample sizes. For $\beta = 0$, the mean absolute regression coefficients are also given. $\overline{\widehat{\beta}}$ and $\overline{|\widehat{\beta}|}$ are calculated conditional on x being included. For $\beta = 0.2$, negative values of $\widehat{\beta}$ were observed in a small number of replications.

Model	n	$\beta=0$			$\beta=0.2$		$\beta=0.4$		$\beta=0.8$	
		Incl.	$\overline{\widehat{\beta}}$	$\overline{|\widehat{\beta}|}$	Incl.	$\overline{\widehat{\beta}}$	Incl.	$\overline{\widehat{\beta}}$	Incl.	$\overline{\widehat{\beta}}$
Full	50	100	−0.01	0.12	100	0.20	100	0.40	100	0.80
	200	100	0.00	0.05	100	0.20	100	0.40	100	0.80
BE(0.157)	50	15.1	−0.04	0.27	47.5	0.32	90.5	0.43	100	0.80
	200	14.3	0.01	0.13	91.3	0.21	100	0.40	100	0.80
BE(0.05)	50	4.2	−0.04	0.33	28.5	0.37	78.0	0.45	100	0.80
	200	4.0	0.03	0.16	80.4	0.22	100	0.40	100	0.80

2.8.3   Shrinkage to Correct for Selection Bias

Shrinkage factors offer a possible route towards improving prediction accuracy and correcting for selection bias. One unsatisfactory way of handling a continuous variable x in a regression model is to categorize it into two groups by searching for the so-called 'optimal' cutpoint on x. This is the cutpoint which minimizes the P-value between the two groups (see Section 3.4.1 for further information). The $\widehat{\beta}$ associated with an optimal cutpoint is an overestimate (heavily biased away from the true value). As a simple model-building problem with one variable, Schumacher et al. (1997) investigated whether several shrinkage factors for $\widehat{\beta}$ could correct for selection bias. Methods included cross-validation, bootstrap resampling, and an heuristic estimate which was a function of the parameter estimates and their variances. They showed that all approaches considered could to some extent correct for overestimation. In a subsequent paper, Holländer et al. (2004) showed that the application of a shrinkage procedure to correct for bias, together with bootstrap resampling for estimating the variance, yielded CIs for the effect of x with the desired coverage. While these results are promising, they relate only to a very limited scenario.

For more relevant situations with a larger number of variables, ridge regression (Hoerl and Kennard, 1970) is a relatively old technique intended to improve prediction in the normal-errors model by adjusting the least-squares regression coefficients. However, ridge regression has not proved popular or useful in clinical studies. Once a model has been obtained, van Houwelingen and le Cessie (1990) proposed a post-estimation shrinkage factor c based on calibrating the linear predictor η by using cross-validation (see Section 2.8.4). As with ridge regression, all parameter estimates are shrunk uniformly by multiplying them by c. More recently, Breiman (1995) introduced the garotte, shortly followed by Tibshirani (1996) with the lasso (Least Absolute Shrinkage and Selection Operator), two procedures which not only select variables but also simultaneously shrink the regression coefficients of variables remaining in the model. Therefore, they combine variable selection with shrinkage. For further details, please see the original papers or Vach et al. (2001).

Selection bias mainly concerns variables with a weak effect. Regression coefficients for variables with a strong effect do not need shrinkage. Recognizing this, Sauerbrei (1999) extended the cross-validation calibration approach of van Houwelingen and le Cessie (1990) to compute parameterwise shrinkage factors for a given model (see Section 2.8.5). As with the garotte and the lasso, some estimates are shrunk considerably by a factor much smaller than one, whereas others remain nearly unshrunk. For the normal-errors model Vach et al. (2001) summarize similarities and differences between the approaches.

Figure 2.4 illustrates the basic principles schematically. To understand the differences between garotte and lasso on the one hand, and of pure shrinkage or selection methods on the other hand, it is useful to consider the orthonormal case in which each x is normally distributed with mean zero and unit variance, and the x are uncorrelated. In this case, the estimates from the procedures considered can be expressed as functions of the OLS estimates (Tibshirani, 1996). Figure 2.4 shows how an arbitrary regression coefficient β behaves after applying the different procedures. The new estimates from each procedure are represented as functions of the OLS estimates, the dotted line indicating the OLS estimates as a reference. With variable selection procedures, an estimate coincides with the OLS estimate if it is large, and is set to zero otherwise. Shrinkage procedures like ridge regression or post-estimation calibration shrink the estimate uniformly with a constant factor regardless of its value. Lasso and garotte combine variable selection and shrinkage, but differ with respect to the amount of shrinkage. The lasso decreases all estimates by a fixed amount, whereas the garotte shrinks larger estimates only slightly.

Although used in some examples, it seems that none of the approaches has gained popularity in applications. However, with the recent challenge of extracting relevant information from high-dimensional data (e.g. gene expression data), the concept of combining selection

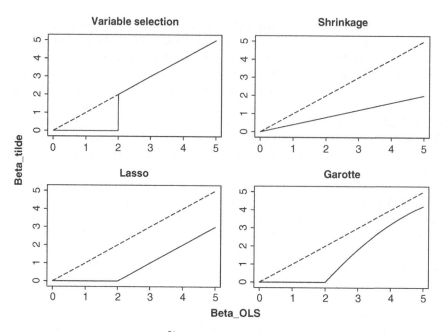

Figure 2.4   The relationship between $\tilde{\beta}$ (beta_tilde, regression coefficient from each of four special model estimation procedures) and $\hat{\beta}$ (beta_OLS, regression coefficient for the full model). The dashed lines represent $\hat{\beta}$ and are included as a reference. (Adapted from Tibshirani (1996) with permission from John Wiley & Sons Ltd.)

with shrinkage appears one of the most promising approaches for tackling the complex problem of multivariable model-building in this setting. Several methods have been suggested, many of them based on the lasso (Efron et al., 2004; Zou and Hastie, 2005; Meinshausen and Bühlmann, 2006). An aim is to correct for the overshrinkage of large effects by the lasso. Because the garotte shrinks large β values much less than the lasso, it seems the more satisfactory approach. However, the garotte starts with the OLS estimate, making it infeasible in high-dimensional data.

Methods are usually developed for the normal-errors model. Transfer to GLMs or models for censored survival data often introduces additional difficulties. Nevertheless, some methods are available, e.g. lasso for survival data (Tibshirani, 1997). This is an area of intensive research.

So far, no method has been proposed which combines variable selection and shrinkage with a systematic search for possible nonlinear effects of continuous variables. As illustrated in Sauerbrei and Royston (1999), the method of parameterwise shrinkage factors is easy to use in conjunction with MFP modelling. Because of its simplicity, it should be usable for all types of regression model. Further discussion of selection and shrinkage, therefore, is restricted to this two-stage proposal and its origins in van Houwelingen and le Cessie (1990).

2.8.4   Post-estimation Shrinkage

An approach to estimating a shrinkage factor, based on leave-one-out cross-validation of the likelihood function, was proposed for generalized linear models and survival time models by van Houwelingen and le Cessie (1990), and extended by Verweij and van Houwelingen (1993).

It was motivated by the predicted residual sum of squares (PRESS) approach, where $\beta_{(-i)}$ means the parameter vector β in a subsample with the ith observation eliminated. Denoting the log likelihood by $l(\beta)$, van Houwelingen and le Cessie (1990) took $l_{(-i)}(\beta)$ as the value with the ith observation eliminated. The contribution of observation i to the log likelihood is defined as $l_i(\beta) = l(\beta) - l_{(-i)}(\beta)$. Maximizing $l_{(-i)}(\beta)$ leads to the estimate $\widehat{\beta}_{(-i)}$. The cross-validated log-likelihood, $cvl = \sum_{i=1}^{n} l_i(\widehat{\beta}_{(-i)})$, was proposed as a measure of the predictive value of the model.

Van Houwelingen and le Cessie (1990) and Verweij and van Houwelingen (1993) also proposed a shrinkage factor based on cross-validation to calibrate a predictor. For a given model, the index estimated for the ith patient is $\widehat{\eta}_i = \mathbf{x}_i \widehat{\beta}$, with $\mathbf{x}_i$ the vector of included variables for patient i and $\widehat{\beta}$ the vector of estimated regression coefficients. The index can be used as the only covariate in a further Cox model. Then $c = 1$ maximizes the log likelihood $l(c)$, where c denotes the regression parameter for the index. Maximizing instead $l^*(c)$, where l^* is the log likelihood using the cross-validated index $\widehat{\eta}_{(-i)} = \mathbf{x}_i \widehat{\beta}_{(-i)}$ as the only covariate, leads to a regression estimate, c^*, interpreted by Verweij and van Houwelingen (1993) as a shrinkage factor. Usually, $c^* < 1$. A value of c^* close to one may indicate that hardly any overfitting has occurred, whereas a small value should indicate overfitting. Additionally, they proposed $l^*(1)$ as a measure of the fit to new data and, therefore, as another measure of the predictive ability of a model. The cross-validated log likelihood $l^*(1)$ may be used as a model selection criterion.

Cross-validation of only the selected model may be criticized for ignoring uncertainty due to variable selection, thereby affecting the predictive ability of the fitted model. Allowing for variable selection would require incorporation of cross-validation or another resampling technique in the model selection procedure.

2.8.5   Reducing Selection Bias

Sauerbrei (1999) suggested adapting van Houwelingen and le Cessie's (1990) shrinkage method to compensate for selection bias. A global factor c^* shrinks each regression coefficient $\widehat{\beta}_j$ of an index $\widehat{\eta} = \widehat{\beta}_1 x_1 + \ldots + \widehat{\beta}_k x_k$ by the same constant. Because selection bias is negligible for a variable with a large effect (Sauerbrei, 1992; Schumacher et al., 1997) (see also Section 2.8.2), multiplying its regression coefficient by c^* may bias its $\widehat{\beta}$ towards zero. To attempt to overcome the selection bias in a given model, Sauerbrei (1999) estimated a parameterwise shrinkage factor (PWSF) for each predictor individually. PWSFs are estimated by maximizing the likelihood for the model

$$\eta_i(c_1, \ldots, c_k) = c_1(\widehat{\beta}_{1(-i)} x_{i1}) + \ldots + c_k(\widehat{\beta}_{k(-i)} x_{ik})$$

by regression on the cross-validated partial predictors $x_{ij} \widehat{\beta}_{j(-i)}$. Variables whose regression parameter estimates are not biased due to model selection should have a PWSF of about one. To estimate PWSFs, the predictors must first be standardized to zero mean and unit standard deviation.

For models derived by variable selection, PWSFs may estimate the amount of selection bias in each variable. Since no selection is involved, PWSFs are not appropriate for the full model. PWSFs are a heuristic extension of the approach of van Houwelingen and le Cessie (1990) and suffer from the same problem of ignoring the uncertainty from the variable selection process mentioned above.

2.8.6 Example

We use the educational body-fat data to compare likelihoods and PWSFs of the full model and models selected with BE. Using significance levels of 0.01, 0.05 and 0.157, BE selects models with three, four and seven variables respectively (see Table 2.7).

See also Tables 2.3 and 2.4. The full model with 13 variables has the largest likelihood, whereas for the others the likelihood is slightly lower. If the cross-validated log-likelihood $l^*(1)$ is used as a measure of the predictive value, then the full model is the worst of all. The full model has the smallest global shrinkage factor, but all the shrinkage factors are very close to unity. Larger differences were found in other studies (Sauerbrei, 1999), but here the small global shrinkage may be a result of the dominating effect of x_6 (see Table 2.3). Removing x_6 from the list of candidate variables decreases the shrinkage factors slightly.

PWSFs illustrate the issue discussed above for models selected with BE. Strong factors selected in models with significance levels 0.01 or 0.05 need hardly any shrinkage, and the PWSFs are close to unity. PWSFs for variables additionally included when the significance level is 0.157 are all much smaller than unity, suggesting substantial selection bias. PWSFs for the strong factors are stable over all selected models.

The results from this example are consistent with experiences from other studies for binary and survival time outcomes (Sauerbrei, 1999; Sauerbrei and Royston, 1999). Global shrinkage factors were much smaller in the two studies discussed by Sauerbrei (1999). Estimates for selected models lay between 0.90 and 0.95, whereas the estimates from the full model were slightly above 0.80.

Table 2.7 Educational body-fat data. Parameterwise and global shrinkage factors for the full model and those selected by BE with different nominal significance levels. $l(1)$ denotes the log likelihood of each model and $l^*(1)$ is the cross-validated log likelihood.

Variable	All variables				Excluding x_6		
	Full[a]	0.157	0.05	0.01	Full[a]	0.157	0.05[b]
x_1		0.85	0.89			0.95	0.98
x_2						0.99	0.99
x_3		0.91	0.91	0.93		0.99	0.99
x_4		0.69				0.63	
x_5		0.71					
x_6		0.96	1.00	0.99		–	–
x_7							
x_8						0.71	
x_9							
x_{10}							
x_{11}							
x_{12}		0.59					
x_{13}		0.98	0.98	0.97		1.00	0.96
$l(1)$	−712.1	−714.4	−718.1	−720.9	−755.5	−756.9	−759.3
$l^*(1)$	−726.0	−722.0	−723.2	−725.1	−768.5	−763.9	−764.4
Global	0.981	0.989	0.992	0.993	0.968	0.984	0.988

[a] PWSFs are not relevant for the full model.
[b] Results for BE(0.01) identical to results for BE(0.05).

2.9   DISCUSSION

2.9.1   Model Building in Small Datasets

In search of an effective model-building strategy in small datasets, Steyerberg et al. (2001) conclude that one should 'apply shrinkage methods in full models that include well-coded predictors that are selected based on external information'. Their conclusion implicitly acknowledges that data-dependent selection in small studies will not produce a good model.

Hjorth's (1989) perspective on data-driven model selection was that 'with a limited set of data and a large set of models, asymptotic results for model selection typically no longer apply'. In practical terms, one cannot expect sensible or reliable results from a limited set of observations if a large class of candidate models is searched. Model selection strategies applied to small samples produce unreliable and unstable models that are worse than useless, because they serve only to increase the uncertainty about the impact of candidate variables. Properties of model selection procedures have too often been studied in small samples, where it is easy to demonstrate difficulties (Steyerberg et al., 2000). For a more detailed discussion, see Sauerbrei et al. (2007b). Such studies do not provide much-needed guidance on model building. We showed in a simple simulation study (Section 2.8.2) that selection procedures often fail to identify variables with a weak effect if the sample size is small. We also demonstrated that the parameter estimates for weakly influential variables that happen to enter the model are subject to selection bias. Different procedures suffer from these problems to a varying extent.

Sample-size recommendations based on the event-per-variable (EPV) relationship have appeared in the literature, derived from practical experience and simulation studies (Sauerbrei, 1992; Concato et al., 1995; Peduzzi et al., 1995; Peduzzi et al., 1996; Harrell et al., 1996). More precisely, it is the number of events per model parameter that matters, but this is often overlooked. The recommendations range from 10 to 25 events per model parameter (Schumacher et al., 2006). In the following, we assume a sufficient sample size. For example, 10 observations (events) per variable (or model parameter), Harrell's rule of thumb, may be a lower limit for obtaining adequate estimates for the important variables. Smaller sample sizes may be sufficient for prediction (goal 1). Vittinghoff and McCulloch (2007) stated that an EPV of 10 may be too conservative and argued for relaxing the rule. However, they focused on just one primary predictor, other variables being confounders. Their results are not so relevant to multivariable model-building with an explanatory aim.

2.9.2   Full, Pre-specified or Selected Model?

We described in Section 2.3 some of the preliminaries that are necessary before starting the analysis of a new dataset. Decisions taken in this early phase can affect the results of multivariable analyses, and raise the question of whether the full model is even a meaningful concept. As a theoretical construct, the full model avoids several complications and biases introduced by model building, and allows some mathematical properties of a model to be derived. In practice, the full model would require pre-specification of all aspects of the model. In principle, even without a preliminary inspection of the data to look for the effect of any potential regressor on the outcome, several decisions are required. These include deciding which candidate variables to choose, how many categories to have for categorical and ordinal variables (see Section 3.5), and what the functional form for a continuous variable should be (see Section 3.6). Such decisions are feasible and pre-specification of the model seems to have become standard for the analysis of randomized trials. However, the main aim of randomized trials is an unbiased estimate of a treatment effect, other variables playing only a minor role. Typically, a small number of adjustment factors are pre-specified.

In an observational study, however, pre-specifying all aspects of a full model is unrealistic. The gain of removing bias by avoiding data-dependent model building is offset by the high cost of ignoring models which may fit the data much better than the pre-specified one. Also, what is to be done if assumptions of the pre-specified model are violated? In addition, interpretation of individual effects is difficult, and applying the predictor in other settings may be nearly impossible. How many researchers would be willing to measure a variable at a high cost when the original analysis indicated that it had only a small effect on the outcome? Breiman (1992) states

> When a regression problem contains many predictor variables, it is rarely wise to try to fit the data by means of a least squares regression on all of the predictor variables. Usually, a regression equation based on a few variables will be more accurate and certainly simpler.

Variable selection methods aim to eliminate all uninfluential variables. Data-dependent model building certainly introduces several types of bias, but the more relevant question is how much these biases matter. Issues such as replication stability, selection bias and overestimation of the predictive ability of a model are raised. These are critically important in particular for weak factors, but 'weak' must be seen in the context of the sample size. In Section 2.8.2 with $n = 50$ we saw that $\beta = 0.2$ or 0.4 represented a factor with a weak or medium effect respectively. Selection bias was large for the smaller effect size and still an issue for the larger one. Increasing the sample size to $n = 200$ converted $\beta = 0.2$ to a medium effect and $\beta = 0.4$ to a strong effect (in the context of model building). With $\beta = 0.4$ the x was always selected and selection bias disappeared. From other simulation studies with more candidate variables and correlations between them, it is known that strong effects do not suffer from selection bias (Sauerbrei, 1992; Schumacher et al., 1997; Holländer et al., 2004). Consequently, strong effects are always detected with a reasonable variable selection procedure, and pronounced replication instability is unlikely. Eliminating variables with a weaker effect (meaning, for large n, that β is close to zero) may not matter much. Omission bias, another criticism of selection procedures, has been mentioned only briefly (see Section 2.8.2), but similar considerations apply there.

We have shown in several examples that predictors from simple models including only strong factors are closely correlated with the predictor derived from the full model. Furthermore, using the leave-one-out cross-validated likelihood as a criterion of predictive ability, the full model was worse than any selected model in two examples. However, in such an assessment, the additional uncertainty due to selecting a model in each subset of the data is not allowed for, perhaps resulting in optimism.

Clearly, simple models are much more interpretable and transferable than the full model, which is a very important advantage. In Chapter 6 we show that the use of simple models makes it relatively easy to incorporate nonlinear effects of continuous predictors. Adopting the full model with all effects linear may result in marked nonlinearity being overlooked. In our view, this represents more serious mismodelling than the elimination of a few predictors with a weak effect.

2.9.3   Comparison of Selection Procedures

With small sample sizes, all selection procedures are unstable, in that many subsets of variables may give about the same fit to the data and that a small perturbation of the data may change the selected model. We assume that a study was planned sensibly and has a sufficient sample size. Most work on variable selection concentrates on predictive models. In approaches based on all-subsets variable selection, a compromise is sought between a good fit and a simple model, using a penalty for each additional variable as part of the selection criterion. Because Mallows' C_p and AIC have an asymptotic significance level of about 16% they usually select relatively complex models. This contrasts with the BIC, whose penalty per variable increases with sample size and whose significance level is almost always much smaller than that of AIC or C_p (Teräsvirta and Mellin, 1986). Modifications of the basic procedures have been suggested (Burnham and Anderson, 2002), but they make virtually no difference when the sample size is reasonable. The nominal significance level, which controls the complexity of a selected model, is the main issue. It is known that many models with different variables have similar apparent predictive ability. Searching for the 'best' model among those with a given number of variables adds only a little extra bias compared with the bias introduced by the choice of the number of variables; see also Hjorth (1989).

In contrast to all-subsets procedures, stepwise procedures are criticized for not being optimal with respect to any specific statistical criterion. Nevertheless, they have several practical advantages. A significance level reflecting the aim of the study can be chosen. As stated earlier, BE(0.157) may be used as a proxy for the all-subsets approach with AIC. In general, BE has advantages over FS procedures. It starts with the full model and eliminates only variables with little influence on the model fit. Strong predictors are selected with high probability, selection bias is not a serious problem for them, and the resulting fit is often comparable to that of the full model. As described in Section 6.3, BE can easily be combined with a systematic investigation of nonlinear effects for continuous variables in the MFP procedure. Computational intensiveness would increase substantially if an all-subsets approach were to replace BE in MFP. Furthermore, BE can be applied without modification to GLMs, models for survival data and other types of single-index model.

2.9.4   Complexity, Stability and Interpretability

We have insisted that complexity is a central issue of model building. In normal-errors models, underfitting increases the residual variance. Complex models with many variables may overfit the data. Apart from difficulties in interpreting the contribution of each predictor, overfitting increases the variance of parameter estimates and of the index η. The increase depends on how much multicollinearity is present. To control the complexity of a final model, a penalty per additional variable can be used as a tuning parameter in all-subsets selection procedures, in much the same way as the nominal significance level is used in stepwise procedures. Using a low significance level or a high penalty per variable results in simple models including only strong factors. These models are more stable, easier to interpret and more generally useful.

Regarding how many parameters may reasonably be estimated, in the context of time-series forecasting Box and Jenkins (1970) invoked the principle of parsimony to suggest that a model should have '... the smallest possible number of parameters for adequate representation of the data'. Burnham and Anderson (2002) discuss over- and under-fitting of a model in relation to a 'best approximating model'. In their view, the aim of modelling is to '... approximate the structural information in the data that is replicable over [similar] samples'. In the context of the present chapter, omitting a variable with a moderate or strong effect would be to miss an 'important replicable structure' (underfitting). In the context of Chapter 6, in which we consider functional form for continuous predictors, overlooking marked nonlinearity would be to miss some important replicable structure.

In a paper on the use of resampling methods in model building, Sauerbrei (1999) argues that the original analysis of the data should be complemented by a stability investigation. The latter helps to identify the instability in the selected model, especially when a complex model including several weak factors is chosen. Based on work by Chen and George (1985) and Altman and Andersen (1989), Sauerbrei and Schumacher (1992) suggested that inspection of bootstrap inclusion frequencies may lead to a more informed interpretation of the importance of a selected variable than does the usual standardized parameter estimate. For a detailed discussion with an extension for assessing the stability of functions, see Chapter 8.

Bootstrap resampling and cross-validation can offer some insight into problems caused by variable selection. Many variants of these methods exist (e.g. k-fold cross-validation, parametric bootstrap, double bootstrap, balanced resampling). Much remains to be done to explore the possibilities and the limitations of resampling (LePage and Billard, 1992; Davison and Hinkley, 1997; Chernick, 1999). Use of the bootstrap helps a practitioner to recognize the instability of the selected model, especially when a complex model including several weak factors is chosen. A weak factor has low power to enter the final model and is included in only a proportion of the replications. Usually, only one among a set of correlated weak factors is selected to represent the effect of the variables from a 'correlated cluster' of several variables (Sauerbrei and Schumacher, 1992). If the effects are of a similar size, then the 'representative' that is selected in a given bootstrap replication depends on chance. Uninfluential variables are selected with a probability depending on the nominal significance level and the correlation structure. If several such variables are considered, then the probability that at least one of them is included in the final model is high. The stability of the selected model decreases with an increase in the number of candidate variables. A further consideration is whether a complex model with many variables which fits the data well is useful for prediction, or whether it is preferable to have a simpler model which does not provide an optimal predictor in the sense of minimizing the MSE or a similar criterion. If the strong factors are included, however, performance is typically only slightly inferior (Sauerbrei et al., 2007a). Here, we have considered complexity only in terms of the inclusion or exclusion of variables. Later, we consider the complexity of a functional relationship for a continuous predictor.

2.9.5   Conclusions and Outlook

The full model is a sensible starting point for scenarios considered in our book (see Section 1.6), but data-dependent model selection should determine the final model. Checks of the main assumptions by sensitivity analysis should become standard practice. Stability analyses should complement the main analysis (see Chapter 8). Since instability of a selected model often occurs, particularly with weaker factors, simpler models including only some strong factors may be preferred. Simple models are also less prone than complex models to bias in the estimated regression coefficient due to variable selection. Combining variable selection with shrunken estimators is becoming more popular, mainly in investigations of high-dimensional data, e.g. gene expression data (Zou and Hastie, 2005; Meinshausen and Bühlmann, 2006). Although the combination is an important step towards improving model building, much remains to be done. The advantages and weaknesses of several recent proposals for simultaneous variable selection and shrinkage require investigation. Some of these have been developed only for the normal-errors model. Transfer to GLMs and models for survival data may cause further difficulties. Parameterwise shrinkage, a heuristic approach based on cross-validation calibration, aims to reduce parameter estimates following selection of a model. Its properties remain to be explored.

Model averaging has been discussed as a possible way of improving predictors and making estimates of their variance more realistic. Variances are underestimated when the model-building process is ignored (Chatfield, 1995; Draper, 1995). Most researchers who incorporate model uncertainty in the model-building process work within the Bayesian framework (Raftery et al., 1997). Recently, a proposal by Buckland et al. (1997) using bootstrap resampling was extended by Augustin et al. (2005) to cope with situations considered in this chapter. Augustin et al. (2005) use model averaging, but try to improve practical usefulness by eliminating variables with no effect or a weak effect in a screening step. Their bootstrap approach for model averaging has its roots in the papers by Buckland et al. (1997) and Sauerbrei and Schumacher (1992). It lacks formal justification and requires some *ad hoc* decisions. Simulation studies of predictive performance show promising results (Holländer et al., 2006; Buchholz et al., 2008). They also indicate the usefulness of the screening step (Sauerbrei et al., 2008), a further indication of our preference for less complex models.

CHAPTER 3

Handling Categorical and Continuous Predictors

Summary

1. Observational studies comprise a mixture of binary, ordinal, nominal and continuous covariates. Preliminary to the analysis one must decide whether data manipulations are required.
2. For categorical variables, sparse categories may be combined, guided by the distribution of the variable and sometimes by subject-matter knowledge. Sparse categories may also indicate a small group of subjects who should have been excluded.
3. Reference categories should not be too small and should make scientific sense as a comparator for the other categories.
4. Having chosen the categorization, a satisfactory coding scheme is needed when variable selection is to be done.
5. Analysis of continuous variables using cutpoints should be avoided.
6. Investigation of nonlinearity is required.
7. Interpretability and transportability of functions for continuous predictors are important. Generally, simple functions are preferable to complex ones.
8. Global-influence functions, which include polynomials and FPs, are generally simpler than functions with local features.

3.1 INTRODUCTION

Nearly all datasets contain a mixture of types of variable. The way these are handled can strongly affect the results of an analysis. In this chapter, the different types of variable are described. We describe how to handle each of them in a univariate analysis, where the aim is to assess their effect on the outcome. For ordinal variables represented by dummy variables, the chosen coding scheme should be appropriate, particularly when selection of dummy variables is intended. For continuous variables, we argue against the still-popular use of categorization, and discuss how to model them while keeping their full information.

Multivariable Model-Building   Patrick Royston, Willi Sauerbrei
© 2008 John Wiley & Sons, Ltd

3.2   TYPES OF PREDICTOR

Studies requiring multivariable modelling contain a mixture of five basic types of covariate and a derived type (see Table 3.1). Derived variables, such as body mass index (i.e. `weight(kg)/height(m)`^{2}), are calculated from other variables and can be any of the five types. The main distinction between an ordinal variable and a count is that the latter contains values from the open-ended set of integers $\{0, 1, 2, \ldots\}$, whereas the former takes values from a fixed, ordered list. Ordinal variables are nonmetric and usually have a small number of categories. Continuous variables may, in principle, take any real value; in practice, however, almost all of them are either positive or nonnegative. Real data, of course, invariably have *rounding* of continuous variables – e.g. age is often given to the nearest completed year, and a biochemical measurement may be rounded automatically to reflect the sensitivity of the measurement method. Unless rounding is deliberately imposed by the analyst on the raw data to 'categorize' it (see Section 3.2.4), we regard rounded continuous variables as continuous.

3.2.1   Binary

Of these five types of variable, only the first presents no problems in analysis. A binary variable is represented by a dummy variable taking arbitrary values, preferably $\{0, 1\}$ or sometimes $\{-1, 1\}$ or $\{1, 2\}$, and is either included in a model or not. Its β-coefficient represents the difference in response (on the relevant scale) between the two classes.

3.2.2   Nominal

An issue with nominal variables, which falls under the heading of preliminaries for a multivariable analysis (see Section 2.3), is whether and how to combine categories. If a given category has few observations, then its regression estimate is imprecise, and in extreme cases may not even be computable. Judicious combining of categories may then be done but may be criticized, since one may be lumping together 'apples and pears'. An example from cancer is the classification of tumour histology into many categories, too many for practical data analysis. Subject-matter knowledge is required to ensure that categories are combined sensibly when combining cannot be avoided. An alternative for very sparse categories is simply to discard the relevant cases from the analysis, e.g. because the category represents a valid but 'forgotten' exclusion criterion. Justification would be needed.

Categorizations having been chosen, further issues remain in handling nominal variables in models. There are two important decisions: which category to use as the referent or base category, and whether to test all the dummy variables jointly or separately. The base category

Table 3.1   Five types of covariates found in datasets for multivariable modelling.

Type	Meaning	Example
Binary	Two classes	Gender (M/F)
Nominal	Categorical with no natural ordering	Continent (Europe, Asia, etc.)
Ordinal	Categorical with ordering	Tumour grade in cancer
Count	Number of instances of something	Number of children
Continuous	Can in principle take any real value	Serum cholesterol concentration

should make scientific sense. Its sample size should not be too small, otherwise standard errors of estimates comparing the base category with others are inflated. Depending on the results, tests of dummy variables separately may lead to further combining of categories with the base category. It is very different to a distributional approach to combining (e.g. avoiding very small categories or aiming at roughly equal sample sizes), which ignores the outcome. The issue is discussed for ordinal variables in Section 3.3.

3.2.3   Ordinal, Counting, Continuous

Increasingly many options are available for handling ordinal, counting and continuous variables. The options are considered in more detail in Sections 3.3, 3.4 and 3.6 respectively.

3.2.4   Derived

Derived variables may be of any of the preceding five (primary) types and are calculated by transforming and/or combining other variable(s). Composite derived variables are a combination of others, well-known examples being body mass index and mean arterial pressure, or MAP = (2 × diastolic pressure + systolic pressure)/3 mmHg. A variable created from a nominal variable by collapsing categories is a derived nominal variable.

Derived variables may depend only on the distribution of the variable or on the effect of the variable on the outcome, or a combination of the two. For example, a continuous variable may be collapsed to an ordinal variable with three to five categories. In epidemiology, groups based on quartile or quintile cutpoints are often used. For such a variable, parameter estimates for some of the groups may be roughly equal, leading to a decision to amalgamate the relevant categories. For correlated predictors, clustering variables and using a 'representative' of each cluster is also a possibility – see Harrell (2001) for details and Blettner and Sauerbrei (1993) for an example.

3.3   HANDLING ORDINAL PREDICTORS

3.3.1   Coding Schemes

When working with ordinal predictors, the coding scheme that is chosen may considerably affect the results of an analysis. Several possibilities exist. Two types of dummy-variable coding and two 'metric codings', in which a metric for ordinal variables is assumed, are considered. The four coding schemes are illustrated in Table 3.2 using a small, hypothetical dataset including an ordinal predictor (dose) with three categories and a binary outcome variable (alive, dead).

Half of the 400 patients are alive. The death rates for dose levels 2 and 3 are similar, but much lower for dose level 1 than the other two. Two coding schemes (M-1, M-2) assume a metric and represent dose by a single variable. The other two (categorical dummy-variable coding DC and ordinal dummy-variable coding DO) each represent the three levels by two dummy variables. Categorical coding requires a reference category with which the other categories are compared. 'Reference coding' is another term found in the literature. Note that if actual dose values were available, then a regression analysis on dose would be more appropriate.

Table 3.2   Metric, categorical and ordinal coding schemes and hypothetical outcomes for an ordinal variable dose with three levels.

dose	Metric coding		Dummy-variable coding				Dead/Total	(%)
			Categorical		Ordinal			
	M-1	M-2	DC1	DC2	DO1	DO2		
1, low	1	1	0	0	0	0	40/100	(40)
2, medium	2	2	1	0	1	0	70/130	(54)
3, high	3	2.5	0	1	1	1	90/170	(53)
Total							200/400	(50)

3.3.2   Effect of Coding Schemes on Variable Selection

The results of an analysis may depend on the coding scheme and model selection procedure used. Table 3.3 illustrates this point using stepwise FS and BE (see Table 2.1) on the very simple example data in Table 3.3.

With coding M-1, the t-value for the (linear) trend on dose is not significant at the 5% level ($P = 0.062$), suggesting that dose has no relation with the outcome. The OR for each dose comparison (levels 2 versus 1, 3 versus 2) is taken as 1.0. Ignoring the nonsignificant result of the test and estimating the regression slope for M-1 yields an OR of 1.26 for each comparison. For M-2, $P = 0.037$ and ORs of 1.43 and 1.19 result. For categorical coding, neither dummy variable is significant with FS, whereas both are significant with BE. For the latter, estimated ORs are identical to the crude ORs. For ordinal coding, only DO1 enters the model with both procedures. ORs are given for models with the relevant dummy variables selected. For example, according to the model with ordinal coding selected by BE, dose levels 2 and 3 do not differ significantly and, therefore, may be combined, which would imply an OR of 1.0 for 2 versus 3. The estimated OR for levels 2 and 3 versus level 1 is 1.71.

Table 3.3   Odds ratios (ORs) comparing dose level 2 with 1 and dose level 3 with 2 for different coding schemes and selection strategies for the data in Table 3.2. For details of variables selected, see the text. A nominal P-value of 0.05 was used to select variables.

dose	Crude OR	Model-based OR						
		Metric coding		Dummy-variable coding				
				Categorical		Ordinal		
		M-1[a]	M-2	FS	BE	FS	BE	
2 vs 1	1.75	1.0 (1.26)	1.43	1.0	1.75	1.71	1.71	
3 vs 2	0.96	1.0 (1.26)	1.19	1.0	0.96	1.0	1.0	

[a] The test is nonsignificant; values in parentheses give the estimated ORs (see text).

With only one covariate, variable selection does not make much sense. However, a similar situation can occur in a multivariable context where selection is required. The example is the simplest case and is used to illustrate the importance of coding schemes in data-dependent model building. In real studies with several variables, further difficulties, e.g. correlation between variables, complicates model building. This point is illustrated in Section 3.5.2.

With metric coding M-1, the effect is estimated assuming that the ORs for dose levels 1 versus 2 and 2 versus 3 are identical. This assumption contradicts the study results. The covariate would be nonsignificant, implying that all ORs are estimated as 1.0. The actual estimated effect with this coding is an OR of 1.26, which is smaller than the observed OR between levels 2 and 3 and not correctly reflecting the difference in outcomes. The coding 1, 2, 3 in M-1 is popular but arbitrary. In M-2, the metric is altered slightly. The difference between levels 1 and 2 is still one unit, but that between levels 2 and 3 is only 0.5 units. This coding reflects the difference in the outcomes somewhat better, and the covariate is now significant. However, such a metric may be criticized as data dependent. The estimated log OR is 0.355 giving an OR of 1.43 for dose level 2 versus 1 and 1.19 ($= \exp(0.355/2)$) for dose level 3 versus 2.

When the predictor is ordinal, ordinal coding is more appropriate than categorical coding. If effects are estimated without selection, then both codings give the same result, but the parameterization is different. With variable selection, the results may differ. Although illustrated here for two stepwise procedures, the principle is the same with other variable selection strategies. Ordinal coding is an attempt to reflect the ordinal nature of the covariate, whereas categorical coding is appropriate for an unordered covariate. The question is to how choose a suitable referent or base category (see also Section 3.2.2). Another possibility is to test the dummy variables jointly, whereas we have considered them separately.

Even with this simplest of examples, differences between codings and selection strategies can be illustrated.

With categorical coding, FS begins by comparing dose levels 2 versus 1 and 3 versus 1. Both effects are nonsignificant and the procedure stops. By contrast, BE starts with both dummies in the model and finds that neither can be eliminated. The result is the same as estimating both effects without selection. With ordinal coding, stepwise selection collapses dose levels 2 and 3 and finds a significant effect of dummy DO1. Adding dummy DO2 to this model would represent a difference between dose levels 2 and 3. Because of the similarity of the outcome in these two groups, the difference is nonsignificant. Both stepwise procedures collapse dose levels 2 and 3 and indicate a difference from dose level 1. Although collapsing of categories based on stepwise procedures is often criticized, we consider it sensible if done appropriately, guided by subject-matter knowledge and the size of the study.

As discussed in Chapter 2, BE has some advantages (Mantel, 1970; Sauerbrei, 1999). In this simple example, it represented the differences between groups very well. In more complex model-building tasks, it can give rise to parsimonious and interpretable models (Sauerbrei, 1999). Byar (1984) took a similar line (but with more groups) to modelling the effect of a continuous variable by step functions.

Other ways of investigating the effect of an ordinal variable using dummy variables have been suggested. For example, different types of contrast are often used. Contrasts are more appropriate for the analysis of factorial experimental designs and are not considered here. For further reading, refer to Walter et al. (1987) or Cohen et al. (2003).

3.4   HANDLING COUNTING AND CONTINUOUS PREDICTORS: CATEGORIZATION

To avoid the assumption of a linear effect, continuous variables are often converted into categorical factors by grouping into two or more categories. This is mostly done to simplify the analysis and interpretation of results. Often in epidemiological studies and other fields, an ordinal variable with several categories is created by using quartiles or quintiles. In clinical studies it is still popular to choose one cutpoint and create a binary variable representing 'low' and 'high' values. The latter may have its origin in clinical decision-making, which often requires two classes, such as normal/abnormal, malignant/benign, treat/do not treat, and so on. Although necessary for decision making, such simplicity is gained at high cost in a research context, and may create more problems than it solves (Altman et al., 1994; Royston et al., 2006).

From a biological point of view, a cutpoint model is unrealistic, with individuals close to but on opposite sides of the cutpoint characterized as having very different rather than very similar outcomes. The underlying relationship with the outcome would be expected to be smooth but not necessarily linear, and usually but not necessarily monotonic. Use of two groups makes it impossible to detect a nonlinear relationship.

From a methodological point of view, the disadvantages of grouping a predictor have been considered by many workers, including Altman et al. (1994), MacCallum et al. (2002), Irwin and McClelland (2003), Austin and Brunner (2004) and Royston et al. (2006). Grouping introduces an extreme form of rounding, with an inevitable loss of information and power, e.g. up to about 50%. Discarding a high proportion of the data is regrettable when many research studies are too small and, hence, underpowered. Probably many who do this are unaware of the implications (MacCallum et al., 2002). Furthermore, the number of groups is an issue.

3.4.1   'Optimal' Cutpoints: A Dangerous Analysis

The common approach of creating two groups by using one cutpoint may increase the probability of false positive results (Altman et al., 1994; Austin and Brunner, 2004). It becomes extreme when 'optimal' cutpoints are used. Every possible cutpoint on x is considered and the value of x which minimizes the P-value is chosen. The cutpoint actually selected is to some extent due to chance.

As an example of such an analysis, Figure 3.1 shows the P-values from logrank tests for dependence on the cutpoint used for S-phase fraction (spf) in the subset of lymph node positive patients from the Freiburg DNA breast cancer study. One hundred and nine patients with 56 events for recurrence-free survival were included in the analysis; for further details, see Pfisterer et al. (1995) or Holländer et al. (2004). The data are published in Lausen and Schumacher (1996). The cutpoint considered lies in the range from the 10th to the 90th centile of the distribution of spf. The optimal cutpoint is in the range [10.7, 10.9] and the corresponding P-value is 0.007. A slightly higher cutpoint (e.g. 12.0) gives $P > 0.1$. The conclusion as to the prognostic value of spf depends critically on the cutpoint chosen.

Altman et al. (1994) call this procedure the 'minimum P-value' approach. Multiple testing increases the type I error probability from a nominal 0.05 to around 0.4. The chosen cutpoint has a wide CI and is not usually clinically meaningful. The difference in outcome between the two groups is overestimated and the CI is too narrow. Corrections for these biases have been

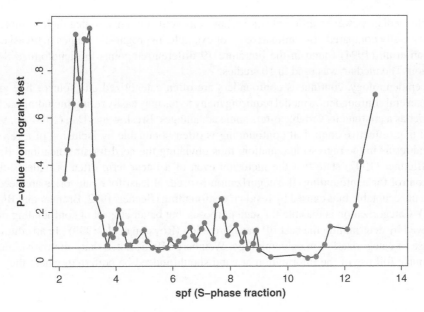

Figure 3.1 Freiburg DNA data. 'Optimal' cutpoint analysis of the effect of spf on recurrence-free survival. Horizontal line denotes $P = 0.05$.

suggested (Miller and Siegmund, 1982; Lausen and Schumacher, 1992; Altman et al., 1994; Hilsenbeck and Clark, 1996). Using a cutoff of 10.7 in the spf example gives a $P = 0.007$ and a rather impressive difference between the survival curves (Schumacher et al., 2006). The uncorrected hazard ratio is 2.37 with 95% CI (1.27, 4.43).

An approximate large-sample correction (Altman et al. 1994) for the multiple testing in the minimal P-value P_{min} is given by

$$P_{cor} = \begin{cases} -1.63 P_{min}(1 + 2.35 \ln P_{min}) & \text{if } \varepsilon = 0.1 \\ -3.13 P_{min}(1 + 1.65 \ln P_{max}) & \text{if } \varepsilon = 0.05 \end{cases}$$

where ε is the sample proportion of smallest and of largest values of the prognostic factor that are not considered as potential cutpoints. The approximation works well for P_{min} in the range (0.0001, 0.1). For example, to achieve a value $P_{cor} = 0.05$ requires $P_{min} = 0.002$ when $\varepsilon = 0.1$ and $P_{min} = 0.001$ when $\varepsilon = 0.05$. In the spf example with 10.7 as the cutpoint, a corrected $P_{cor} = 0.12$ and a corrected estimated effect with 95% CI is 2.10 (0.74, 5.97).

3.4.2   Other Ways of Choosing a Cutpoint

The methods for determining the cutpoint described below still incur information loss, but at least not an inflated type I error probability. For a few variables, recognized cutpoints are widely used (e.g. $> 25 \, \text{kg} \, \text{m}^{-2}$ to define 'overweight' based on body mass index). For some variables, such as age, it is usual to take a 'round number', an elusive concept which here usually means a multiple of 5 or 10. Another possibility is to use the upper limit of a reference interval in healthy individuals. Otherwise the cutpoint used in previous studies may be adopted. In the absence of a prior cutpoint, the most common choice is the sample median. With the sample median, however, different studies have different cutpoints, so that their results can be neither easily compared nor summarized. For example, in prognostic studies in breast cancer, Altman et al. (1994) found in the literature 19 different cutpoints to dichotomize S-phase fraction. The median was used in 10 studies.

In epidemiology, continuous confounders are often categorized into four or five groups. With several confounders, a model requiring many terms may easily result. Including each confounder as a continuous variable offers some advantages. Breslow and Day (1980, pp. 94–97) noted that 'effective control of confounding is often obtainable by inclusion of a few polynomial terms in the regression equation, thus obviating the need for stratification'. Brenner and Blettner (1997) state that the inclusion even of a linear term often provides satisfactory control for confounding. If categorization is used, at least four categories are needed to avoid unacceptable bias caused by residual confounding (Becher, 1992; Brenner and Blettner, 1997). Categorization is feasible for such purposes, but better control of confounding may be achieved by determining the best FP transformation (Royston et al., 1999) . In an editorial on the use of categorization in epidemiology, Weinberg (1995) noted that 'alternative methods that make full use of the information at hand should indeed be preferred, where they make sense'.

3.5   EXAMPLE: ISSUES IN MODEL BUILDING WITH CATEGORIZED VARIABLES

Continuous variables are often categorized, probably to simplify the analysis and presentation of results. We use an example from the literature to illustrate some problems of model building with categorized data, focusing on coding choices for ordinal variables (see also Section 3.3).

The following example on cancer of the cervix is given in Collett (2003a, Appendix C.9). The data were taken from the relevant website.

The aim of a hospital-based, multi-centre, unmatched case-control study, conducted under the auspices of the World Health Organization, was to determine whether oral contraceptive use affects the risk of invasive carcinoma of the cervix. The data relate to one of the participating centres, Siriraj in Thailand, and the database contains 899 records made up of 141 cases and 758 controls. The variables recorded are listed in Table 3.4 These data are used to examine

Table 3.4   Cervical cancer data. Variables and their coding.

Variable	Explanation	Coding					
		0	1	2	3	4	5
caco	Case-control status	Control	Case				
age	Age group		15–30	31–35	36–40	41–45	46–60
ocuse	Previous oral contraceptive use	No	Yes				
duroc	Months of oral contraceptive use	0	1–12	13–24	25–60	≥ 61	
sexrel	Number of sexual relations	0	1	≥ 2			
agesex	Age at first sexual relationship	None	8–16	17–18	19–20	21–23	≥ 24
npaps	No. of Papanicolaou smears	None	1–2	3–4	5–6	≥ 7	
vagdis	Doctor visited for abnormal vaginal discharge	No	Yes				
tpreg	Total number of pregnancies	None	1–2	3–4	5–6	≥ 7	

the relationship between oral contraceptive use and cervical cancer, paying attention to the possible confounding effect of other variables.

Univariate models for the effect of duration of oral contraceptive use are first considered, followed by multivariable models taking into account all the available risk factors. The coding schemes shown in Table 3.2 are compared.

3.5.1 One Ordinal Variable

Duration of oral contraceptive use (duroc), the main variable of interest, is coded as five categories. Table 3.6 gives cancer case/control status (caco) and duroc for all 899 cases and controls in the study. 'Never' is an obvious referent category. For other categories, estimated ORs and P-values from χ^2 tests are given in comparison with the referent category. Except for the sparse category '13–24', ORs increase with duration of oral contraceptive use.

As introduced in Section 3.3, the data are analysed by using a metric coding, a categorical dummy coding and an ordinal dummy coding (see Table 3.5). Table 3.7 gives the estimates for the three coding schemes. Only one variable is required for the metric coding, which implies a strong (and possibly incorrect) assumption of linearity. For the categorical and ordinal codings, the group 'never' is used as a referent and four dummy variables are needed for the other categories. The two types of dummy coding with four variables give the same results; only the parameterization and, therefore, the interpretation are different. For the categorical coding, estimates are the log OR compared with the referent. In principle, estimates for the metric coding can be interpreted in the same way. However it is assumed that an appropriate metric across the five categories from 'never' to '> 60' is the values 1, 2, 3, 4, 5. Effectively, duroc is handled as though it was a continuous variable with a presumed linear effect. The

Table 3.5 Cervical cancer data. Different ways of coding the ordinal variable duroc.

duroc	Metric	Categorical coding				Ordinal coding			
	M–1	DC1	DC2	DC3	DC4	DO1	DO2	DO3	DO4
Never	1	0	0	0	0	0	0	0	0
1–12	2	1	0	0	0	1	0	0	0
13–24	3	0	1	0	0	1	1	0	0
25–60	4	0	0	1	0	1	1	1	0
>60	5	0	0	0	1	1	1	1	1

Table 3.6 Cervical cancer data. Oral contraceptive use and cervical cancer in a case-control study. Estimates of ORs with P-values comparing with category 'never'.

duroc	Cases/n	(%)	OR	P
Never	88/629	(14)	—	
1–12	15/89	(17)	1.25	0.47
13–24	3/45	(7)	0.44	0.17
25–60	14/75	(19)	1.41	0.28
> 60	21/61	(34)	3.23	< 0.001

Table 3.7   Cervical cancer data. Parameter estimates for the coding schemes in Table 3.5. The estimated log OR for the metric coding is 0.209.

duroc	Categorical	Ordinal	Metric
Never	0	0	0
1–12	0.22	0.22	0.21
13–24	−0.82	−1.04	0.42
25–60	0.34	1.17	0.63
> 60	1.17	0.83	0.83

parameter estimate $\hat{\beta}$ for the metric variable M-1 is 0.21. ORs for the duroc categories are the exponential of $\hat{\beta}$, $2\hat{\beta}$, $3\hat{\beta}$ and $4\hat{\beta}$. The model estimates differ substantially from the observed data, especially for the category '13–24' (see Table 3.3), and from the estimates for the categorical coding.

The ordinal coding gives the estimated log(OR) comparing categories k and $k − 1$. For example, DO1 compares category 2 with category 1; as for the categorical coding, the estimated log OR is 0.22. The parameter estimate for DO3 is the log OR for duroc 25–60 versus 13–24. This estimate is identical to the log OR duroc 25–60 versus 'never' minus that for 13–24 versus 'never'. The latter two estimates are given for the categorical codings DC3 and DC2. Without elimination of a dummy variable, ORs for one coding scheme can be calculated from the ORs of the other.

3.5.2   Several Ordinal Variables

The same principle is applied to the variables age, sexrel, agesex, npaps and tpreg. The variable ocuse combines four categories of duroc and gives no additional information. sexrel is taken as a binary variable (0 or 1 versus 2), because only 1 of 197 women with no sexual relationship was a case. Taken together with the correlated variable agesex, a full model could not be fitted. None of the cases and controls had npaps > 4, so only three categories remain for this variable. This leaves seven potential risk factors. A 'full' model has seven variables with metric coding and linear effects, or 21 dummy variables with categorical or ordinal coding schemes. Results for the full model and three models obtained with BE are given in Table 3.8. In the full model, parameter estimates for some of the dummy variables are close to zero and P-values are > 0.5. For these categories, the OR is close to 1.0. Elimination of these dummy variables worsens the fit only slightly. The dummy variables representing agesex are another typical example of a problem with the full model. The five categories from '8–16' to '> 23' are compared with the referent 'never'. All parameter estimates are large and all P-values are significant. There is no monotonic trend with increasing age and parameter estimates are similar. This model indicates no trend on agesex but only a difference from 'never'. A new binary variable sex (yes or no) seems to be sufficient for this model. With ordinal categorization, an effect of sex is selected using BE(0.05). Only the dummy variable DO1 is included in the model. The parameter estimate is 2.96, which can be seen as a rough average of the estimates for the five dummy variables in the full model (range 2.39 to 3.06). With ordinal categorization, a simple model with six dummy variables is selected by BE(0.05) and sexrel is excluded from the model. The group '0' is identical to agesex equal to 'never'.

Table 3.8 Cervical cancer data. Estimated log OR for four multivariable models. For the full model, the P-value is also given.

Variable	Cat.	Full model		Selected models		
		$\hat{\beta}$	P	DC(0.05)	DO(0.05)	Final
age	15–30	0	–	0	0	0
	31–35	0.20	0.7	0	0	0
	36–40	0.77	0.06	0.51	0.68	0.79
	41–45	1.03	0.02	0.71	0.68	0.79
	46–60	0.46	0.31	0	0.68	0.42
duroc	Never	0	–	0	0	0
	1–12	−0.03	0.9	0	0	0
	13–24	−0.93	0.14	0	0	0
	25–60	0.11	0.8	0	0	0
	> 60	0.83	0.01	0.82	0.86	0.82
sexrel	0 or 1	0	–	0	0	0
	> 1	0.25	0.4	0	0	0
agesex	Never	0	–	0	0	0
	8–16	2.41	0.04	2.74	2.96	2.95
	17–18	2.75	0.02	3.10	2.96	2.95
	19–20	3.06	0.01	3.41	2.96	2.95
	21–23	2.75	0.02	3.04	2.96	2.95
	> 23	2.39	0.04	2.73	2.96	2.95
npaps	0	0	–	0	0	0
	1–2	−0.38	0.17	0	0	0
	3–4	−1.16	0.00	−1.07	−1.00	−1.04
vagdis	No	0	–	0	0	0
	Yes	0.76	0.00	0.71	0.73	0.70
tpreg	0	0	–	0	0	0
	1–2	0.39	0.5	0	0	0
	3–4	0.95	0.10	0.69	0.67	0.72
	5–6	0.84	0.17	0.70	0.67	0.72
	> 6	1.05	0.10	0.91	0.67	0.72

If categorical coding is used, then a model with 13 dummy variables is selected. It agrees better with the full model, but interpretation is more difficult. Again, agesex, but also tpreg, illustrates that a binary variable may be sufficient. The right column 'Final' gives our final model, which is a slight extension of the BE(0.05) model derived with ordinal coding. The only difference is the inclusion of an additional dummy variable (DO4) for age. This variable was the last to be excluded by BE(0.05), the P-to-remove value being 0.12. As indicated by the full model, the effect for category '46–60' seems weaker than for '36–40' and '41–45'. Provided sensible coding is used, BE is a suitable procedure to select a model. Sometimes, however, minor modifications can improve the model.

Table 3.9 gives three models using metric coding, and is presented for completeness and to illustrate the problems of using metric scaling with ordinal variables.

The full model indicates that age and sexrel have no significant effect at the 0.05 level. These variables and also duroc are excluded by BE(0.05). As duroc is the main variable

Table 3.9   Cervical cancer data. Models with linear ordinal coding for all predictors.

Variable	Full model		BE(0.05)	Duration[a]	
	$\widehat{\beta}$	P	$\widehat{\beta}$	$\widehat{\beta}$	SE
age	0.12	0.2	–	–	–
duroc	0.15	0.04	–	0.13	0.07
sexrel	0.38	0.2	–	–	–
agesex	0.13	0.04	0.15	0.14	0.06
npaps	−0.48	< 0.01	−0.43	−0.47	0.14
vagdis	0.91	< 0.01	0.90	0.89	0.21
tpreg	0.42	< 0.01	0.52	0.51	0.08

[a] Adjustment model selected by BE(0.157).

of interest, it is important to estimate its effect in a model adjusted for other variables. This is done in the last column of the table. The adjustment model was selected with BE(0.157).

The example shows how difficulties may arise when different coding schemes for ordinal variables are accompanied by selection of variables. From a subject-matter point of view, one may also consider creating subgroups, e.g. defined by sexrel. The group 'none' has such a low risk of cervical cancer that its exclusion from a main analysis may be appropriate.

3.6   HANDLING COUNTING AND CONTINUOUS PREDICTORS: FUNCTIONAL FORM

3.6.1   Beyond Linearity

Some 25 years ago, the automatic choice for coping with continuous predictors in almost all regression models was either the linear function or a step function (see Section 3.4). Sometimes, a function generally accepted according to subject-matter knowledge was available. In most cases, the use of one or other of these approaches was not questioned. Occasionally, when nonlinearity was strong and the linear function was obviously inadequate, a few 'popular' functions, such as the log, square root, quadratic, cubic or higher order polynomial, were considered. However, for various reasons, polynomials were rather unsatisfactory and were not often used. More important, depending on the interests and experience of the analyst and the amount of time available for the analysis, several functions might have been considered, but criteria for deciding between them were unclear.

Since then, the variety and power of models available to the analyst, capable of representing functional forms from the simplest to the most intricate, has increased greatly. We now have spline functions of different flavours and orders, local polynomial smoothers, kernel smoothers, wavelets, Fourier series, and so on. The list is long and perhaps baffling. Faced with such a multitude of choices, what should the analyst do? How much is the preferred strategy influenced by the problem at hand? Here, we restrict the discussion to some general issues (see also the Preface). A description of FP methodology for dealing with nonlinearity is given in Chapters 4–6.

3.6.2 Does Nonlinearity Matter?

In models with a single predictor, nonlinearity is easily seen and appreciated (if present – and if one bothers to look for it!). Even in generalized and Cox regression models, plots of smoothed residuals versus x allow approximate visualization of the function (see Section 1.4). Figure 3.2 shows martingale residuals (see Section 1.3.3) with a smoothing line plotted against nodes, the predictor in the GBSG breast cancer study with the largest standardized regression coefficient. The martingale residuals were calculated as the censoring indicator minus the cumulative hazard function. The Nelson–Aalen estimator was used for the latter (Hosmer and Lemeshow, 1999; Therneau and Grambsch, 2000). The plot reveals that the residuals clearly depend on nodes. Furthermore, the nonlinearity of the relationship is clear. If the aim was to identify and model this function, a straight line would not be a sensible choice. A function that is curved and flattening off at high values of the predictor is required. A straight line would predict an inappropriately high risk of an event for patients with a large number of nodes. The prediction accuracy of such a model would be poor. See also Section 6.5.4 for further discussion of this functional relationship.

In multivariable models, unrecognized nonlinearity may have more subtle effects. For example, Royston et al. (2006) considered adjustment of the treatment effect in a random-ized trial for a strongly prognostic covariate. They illustrated that, unless the covariate was modelled using a suitable nonlinear function, the treatment effect was underestimated and its P-value was too large. This is admittedly an extreme example, but it does show that correct modelling of nonlinearity may be critical. In general, when variables are selected, mismodel-ling the functional form of an influential covariate may cause correlated covariates to enter the model spuriously. This occurs in the prostate cancer data; if all variables are modelled linearly, then the model selected by BE(0.05) comprises cavol, weight, svi and pgg45, whereas

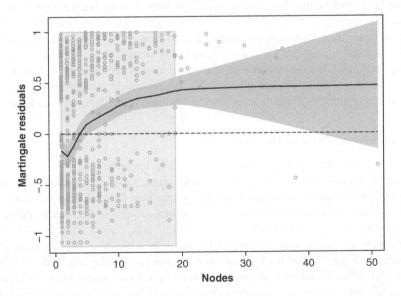

Figure 3.2   GBSG breast cancer data. Martingale residuals with a running line smooth and pointwise 95% CIs for the predictor nodes.

if log `cavol` is substituted for `cavol` then `pgg45` is excluded (and the fit is improved). The Spearman correlation between `cavol` and `pgg45` is 0.50.

Failing to search for nonlinearity may even cause the exclusion of an important variable from the model. For example, the variable `age` in the GBSG breast cancer dataset has a highly nonlinear effect (see Figure 1.5) and almost no linear effect at all. This is in close agreement with current medical thinking. `age` never enters the model unless nonlinearity is accounted for. Mismodelling a nonlinear effect of a covariate in a Cox model may induce a spurious time-varying effect of that variable (see Section 11.1).

3.6.3   Simple versus Complex Functions

A common belief is that complex functions give a more 'accurate' (less biased) representation of the unknown, underlying truth. Except in simulation studies where the analyst controls the game and knows the 'truth', this belief can never be tested. What is certainly true is that the fit of a sequence of increasingly complex models known to be good approximators, such as cubic smoothing splines (Green and Silverman, 1994), gets closer and closer to the *data* (as opposed to the *truth*). This introduces instability. An example is shown in Figure 3.3. Cubic regression splines with degrees of freedom between 1 (i.e. a linear function) and 10 were fitted to the relationship between `logpsa` and `cavol` in the prostate cancer data. Provided d.f. is greater than two, the overall shape is in every case similar. However, the fitted functions become more and more irregular as the complexity increases. It seems unlikely that the underlying function is so complicated.

It is clear that the fitted function in Figure 3.3 becomes more variable as the d.f. increase. This reflects the *bias/variance trade-off* (e.g. Hastie and Tibshirani, 1990, pp. 11–12, 40–42). The analyst may be willing to incur an increase in variance in return for a more accurate function. The dilemma is that if we choose too many d.f., then we fit the observed data very closely (i.e. 'overfit' it), but we may have problems interpreting the resulting function or reproducing it in another setting (external validation). However, if we choose too few d.f., then our estimate may be oversimplified. Every analyst must find a compromise between these two extremes. As already discussed in Section 1.7.3, we generally prefer simpler functions to more complicated ones, e.g. in Figure 3.3 the function with three d.f. seems to capture the shape of the function adequately without being overcomplicated. Our experience in clinical and epidemiological studies is that simple functional relationships are the rule rather than the exception.

3.6.4   Interpretability and Transportability

If there is specific interest in the form of, say, a dose–response relationship, then the function must be interpretable – the curve must make sense. Good examples of uninterpretable (nonmonotonic) functions are the plots for d.f. greater than four in Figure 3.3. The curve for 10 d.f. has three local maxima and two local minima; from a subject-matter point of view, such behaviour defies explanation.

Transportability (applicability of a model in different settings) is also important. A model that is so complicated that its fitted function is hard even to write down is hard to communicate to others and difficult for them to use with their own data. Smoothing splines and kernel-based estimates fall into this category, since they do not provide functional forms. Regression splines have more concise functional forms (see Section 9.3.1). Transportability is a requirement when seeking to validate a model externally. External validation is an important topic which

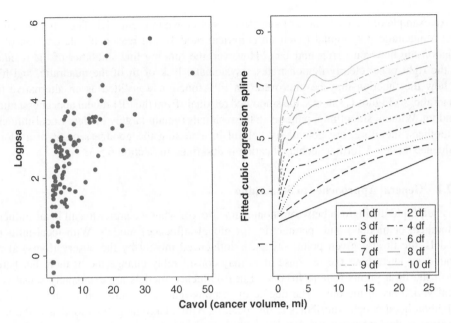

Figure 3.3 Prostate cancer data. Relationship between logpsa (log serum PSA concentration) and cavol (cancer volume in millilitres). Left panel: raw data. Right panel: fitted cubic regression splines with different d.f. (one d.f. means a straight-line model). For clarity, each fitted curve has been offset vertically from its neighbour by 0.5 units and the plot has been truncated horizontally at 26 ml.

receives insufficient attention, both theoretically and practically (see Altman and Royston, 2000). Validation is beyond the scope of our book. If a model is not transportable, then it will be 'quickly forgotten' (Wyatt and Altman, 1995). See also Section 1.7.2.

3.7 EMPIRICAL CURVE FITTING

The true model (or data-generating process) underlying the relationship between an outcome y and a predictor x is usually unknown and liable to depend on quantities not measured. In most applications, no prespecified mathematical function is known to be appropriate to describe such a relationship. Exceptions to this rule may arise from systems that can be described, at least to a first approximation, by fairly simple differential equations, e.g. in pharmacokinetics.

A statistical model is an attempt to describe reality empirically. Empirical curve fitting means smoothing the association between y and x. The principal aim is to visualize the relationship and thereby perhaps to provide or suggest a suitable functional form that can be used in more formal modelling, as described mainly in several later chapters.

Since the relationship is unknown, statisticians have tried to develop flexible smoothing tools which represent the function as accurately (unbiasedly) as possible. Examples of 'nonparametric' smoothers include kernel-based scatter-plot smoothers such as lowess (Cleveland and Devlin, 1988), regression splines (de Boer 2001), smoothing splines (Green and Silverman 1994), and wavelets (Vidakovic, 1999).

An example of the usefulness of smoothing was shown in Figure 1.1. The left panel shows that a quadratic polynomial (itself an empirical model) is a reasonable description of the relationship between `pbfm` and `bmi`. However, the running-line smoother of the residuals in the right panel clearly demonstrates a systematic lack of fit of the quadratic, and hints at how the function may be improved. An FP1 function was fitted as an alternative to a quadratic. As Figure 1.3 shows, the smoothed residuals from the FP1 model appear essentially random. Without smoothing, patterns in the residuals remain largely hidden. Local-influence smoothers, therefore, are potentially helpful in assessing the goodness of fit of a global-influence model. A further such application is described in Section 6.5.3.

3.7.1 General Approaches to Smoothing

It is helpful to distinguish between 'nonparametric' (or what we prefer to call local-influence smoothers or models) and 'parametric' (or global-influence) models. With local-influence models, the fit at a given point, say x_0, is influenced mostly by the observation(s) at and around x_0. Disturbing the response at x_0 may considerably change the fit there but hardly affect the fit at points remote from x_0. Local-influence models are exquisitely sensitive to local variations in the data. The degree of responsiveness is controlled by the size of the neighbourhood which contributes to the local fit. With a large neighbourhood size, the fitted function is less responsive to local disturbances, hence in general more smooth, but also more prone to bias; with a small neighbourhood size, the opposite holds, with a tendency to larger local fluctuations and a higher variance. We recommend Hastie and Tibshirani (1990, chapter 2) for a clear and detailed account of this type of smoothing. A thoughtful introduction to smoothing and nonparametric regression is given by Armitage et al. (2002, section 12.2).

By contrast, global-influence models may be less responsive locally to perturbations of the response at a given point but the fit at distant points may be affected, occasionally considerably. This property is sometimes regarded by proponents of local-influence models as a fatal flaw in global-influence models (e.g. see the discussion by Hastie and Tibshirani in Royston and Altman (1994)). A rigorous definition of a global-influence model is not, to our knowledge, available, but examples of such models include polynomials, nonlinear models such as exponential and logistic functions, and FPs.

3.7.2 Critique of Local and Global Influence Models

Flexibility to represent a wide range of functional forms is a desirable property, and local-influence models specialize in it. However, several disadvantages may be noted briefly (examples are given in Chapter 9). The tendency to reduce bias at the expense of variance means that artefacts (roughnesses in the curve) with no basis in reality may occur, hindering interpretation. Usually, no concise mathematical formula is available for the model, making it difficult to publish such that others can reproduce and use it precisely. Many types of model are available, and it is not obvious which one is best to use. There is no widely accepted approach to model selection, which in a univariate setting means determining the appropriate degree of smoothing to be applied. In a multivariable setting, i.e. the more relevant situation in analysing observational studies, the model selection issue is much more critical.

Among global-influence models, the disadvantages of polynomials are well-known: inflexibility in low-order polynomials, artefacts such as the Runge phenomenon (Runge, 1901) in high-order polynomials and others caused by the global influence property, as already noted. The Runge phenomenon occurs when a high-dimensional polynomial fits the observed data points exactly, but the estimated curve exhibits wild wave-like excursions between data points. The advantages of polynomials stem from familiarity, transparency and transportability (simple formula), and mathematically, from the classical Weierstrass theorem, which guarantees that any continuous function can be approximated arbitrarily closely by a polynomial.

Turning to FPs, although FP functions have global influence, they are much more flexible than polynomials. Indeed, low-dimensional FP curves may provide a satisfactory fit where high-order polynomials fail. FPs are intermediate between polynomials and nonlinear curves. They may be seen as a good compromise between ultra-flexible but unstable local-influence models, and less flexible conventional polynomials.

3.8   DISCUSSION

Observational studies usually contain a mixture of binary, categorical (with or without an ordering) and continuous covariates. Before analysis, it is necessary to decide whether any preliminary data manipulation is required.

3.8.1   Sparse Categories

For categorical variables, a sparse category is problematic. Combination with one or more other categories must be considered. The extent of combination depends on the distribution of a variable, and should be guided by subject-matter knowledge. Sometimes the intention is to create a reference category with which others are compared. The reference category should not be too small and must make sense from the subject-matter point of view.

A sparse category may indicate the presence of a few subjects who do not belong to the population of interest. For example, patients with a carcinoma *in situ* may have been included in a study of prognostic factors in invasive cancer. A decision on whether or not these patients should be excluded is required, and must be carefully justified.

3.8.2   Choice of Coding Scheme

Choosing an appropriate coding scheme is essential in multivariable modelling. Contrasts play an important role in designed experiments, but are less useful in observational studies. We illustrated a categorical and an ordinal coding scheme for a variable with $k > 2$ categories. If all $k - 1$ dummy variables remain in the model, then the two schemes differ only in their parameterizations. For an unordered variable, the obvious choice is the categorical scheme. For an ordered categorical variable, the ordinal scheme may make it easier to interpret the parameters of interest. For an illustration of the difference, see Table 3.7 and the related text.

If model building by variable selection is intended, then the ordinal coding scheme is mandatory for an ordered categorical variable. Eliminating one or more of the $k - 1$ dummy variables implies collapsing of categories, and is only reasonable for adjacent categories. Some researchers are not willing to collapse categories in a data-dependent fashion, preferring instead to test all $k - 1$ dummy variables jointly. In our view, however, the multivariable context demands that we develop parsimonious and interpretable models. The principle is impressively illustrated by comparing selected models with the full model in the cervical cancer example (see Table 3.8).

3.8.3   Categorizing Continuous Variables

Continuous variables are often modified by categorization, which makes them easier to use. For example, clinicians use categorized covariates in making treatment decisions (Mazumdar and Glassman, 2000). The functional form for the effect of a covariate is often unknown, and an analysis using a categorized variable is simpler than working with continuous variables directly. Categorization also allows graphical presentation, e.g. Kaplan–Meier curves can be produced in survival analysis. For a more detailed discussion of categorizing a continuous covariate, see Mazumdar and Glassman (2000) and Mazumdar et al. (2003).

Categorization into a small number of groups also raises the issue of how to code the resulting variable for the analysis. The ordinal coding scheme can be used. Alternatively, a coding scheme with a suitable choice of scores (e.g. $\{1, 2, 3\}$, $\{1, 2, 4\}$ or scores representing the median of each category), together with a test for linear trend, is possible.

None of these issues will be pursued further because, in our opinion, categorization has major disadvantages and should be avoided anyway (Altman et al., 1994; Royston et al., 2006).

3.8.4   Handling Continuous Variables

The main focus of our book is on methods which are intended to extract full information from continuous variables in a multivariable setting, resulting in models with plausible functional forms. Naturally, the selected model and the functional forms should be interpretable from a subject-matter point of view. Interpretability, transportability and general usability of a model demand simplicity; complex models, including complex functions of continuous covariates, are not useful when the aim is essentially explanatory (see Section 1.7.2).

Smoothers yielding only complex functional forms are helpful for representing the effects of single predictors, either on their own (i.e. univariately) or as components of a multivariable model. Such checks may reveal underfitting in parts of a model. The analyst should then try to assess whether the model requires some (slight) extension, or whether the lack of fit may be caused by peculiarities of the dataset. The finally chosen multivariable model should be simple and transportable. Therefore, we prefer simpler, global-influence models for continuous variables.

CHAPTER 4

Fractional Polynomials for One Variable

Summary

1. FP functions are introduced and defined mathematically.
2. Types of FP curve shape are discussed. First-degree FP (FP1) functions are monotonic. Second-degree FP (FP2) functions can represent a variety of curve shapes with at most one maximum or minimum.
3. The choice of powers is discussed. Powers are limited to eight integer or half-integer values between -2 and 3, with 0 denoting the log function. The class of FP2 functions consists of 36 pairs of powers (including eight repeated-power models).
4. A simple rule of thumb is available to calculate an origin when x includes nonpositive values.
5. Model fitting, parameter estimation, hypothesis testing and interval estimation for FP models are described. Parameters are estimated by maximum likelihood conditional on finding the best power(s), and also by maximizing the likelihood. Hypothesis testing is done by approximate likelihood ratio χ^2 tests.
6. A closed test procedure for selecting an FP function, called the function selection procedure (FSP), is introduced. The results are controlled by a tuning parameter α, which is the significance level of a test. This procedure is also of central importance to multivariable model-building with FP functions for continuous covariates.
7. Details of computational aspects are given, including how to calculate fitted values and their confidence intervals.
8. Presentation of FP functions is described. Function plots with pointwise CIs are recommended. If presentation in categories is desired, then tables giving model-based effect sizes for sub-ranges of x can be constructed.
9. An alternative way of modelling continuous variables with a substantial proportion of zero values is proposed.

Multivariable Model-Building   Patrick Royston, Willi Sauerbrei
© 2008 John Wiley & Sons, Ltd

4.1 INTRODUCTION

In this chapter, FP functions and models for a single predictor are introduced and studied. A 'closed test' procedure for selecting an FP function is described. The materials presented here are the building blocks for multivariable model-building with continuous predictors, the main topic of our book (see Chapter 6).

4.2 BACKGROUND

4.2.1 Genesis

It is immediately clear in Figure 1.1 that the relation between pbfm and bmi in the research body-fat data is nonlinear. The plot of smoothed residuals (right-hand panel of Figure 1.1) tells us that a quadratic curve, although a much better fit, is not ideal. In the past, when faced with such a problem, statisticians have often tried simple *ad hoc* transformations of the covariate x to improve the fit. Popular choices have been $\log x$, $\sqrt{x}$, or less commonly $1/x$. If we define x^0 as $\log x$, then transformations like this are power functions x^p for different values of p. Figure 4.1 shows what happens to the plot of smoothed residuals (see Section 1.4.2) for different choices of p from the set $S = \{-2, -1, -0.5, 0, 0.5, 1, 2, 3\}$. This set S includes the 'usual' transformations and the 'no transformation' case, $p = 1$. Figure 4.1 suggests that

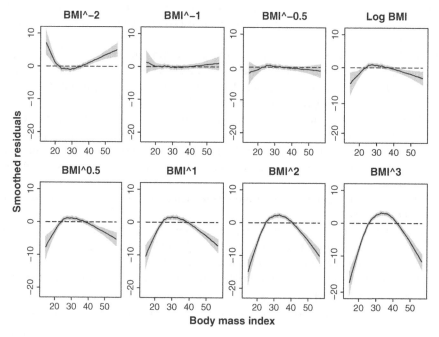

Figure 4.1   Research body-fat data. Smoothed residuals plot with 95% CIs for different power transformations of bmi.

the fit is best for power -1, with -0.5 a close second. The 'comparative anatomy' of such transformations was studied in some detail by Tukey (1957).

Royston and Altman (1994) devised a framework for these cases by defining functions of the form $\beta_0 + \beta_1 x^p$ with p in S to be *FP models of degree 1* (abbreviated to FP1 models). For example, with $p = 0.5$ the function is $\beta_0 + \beta_1\sqrt{x}$. Importantly, the linear function is the special case $p = 1$. The linear function occupies a special place, since it is usually chosen as the default when selecting FP models for a given dataset (more on this later).

It was natural for Royston and Altman (1994) to extend the definition to FPm functions, where m is an integer ≥ 1. FP2 functions, for example, are generalizations of quadratics. A remarkable range of curve shapes is available with FP1 and FP2 functions (see Section 4.5). In our book we are little concerned with FP3 and higher order functions. Except in some special situations, $m \leq 2$ provides enough flexibility for modelling many of the types of continuous function we encounter in the health sciences and elsewhere. Use of higher order FPs may be sensible in univariate modelling when the amount of noise is zero (e.g. when approximating smooth mathematical functions (Royston and Altman, 1997) (see Section 11.3.4)) or low (e.g. a complex but well-defined relationship must be approximated). Some examples of the latter case are given in Section 5.7. In multivariable modelling, use of FP3 or higher order functions is rarely sensible, since instability is much increased. Some of the available FP software (but not that in Stata) is restricted to analyses with FP1 or FP2 functions.

4.2.2 Types of Model

In the rest of this chapter, we develop FP modelling implicitly in the framework of the normal-errors model. However, the methodology transfers with only minor modifications to GLMs, survival models, etc. For example, in the Cox model, the intercept β_0 is replaced by an unspecified baseline hazard function $\lambda_0(t)$, and parameter estimation is by maximum partial likelihood rather than maximum likelihood. The basic features of the FP approach are (1) parameter estimation is by maximizing a likelihood function or equivalent criterion (e.g. penalized likelihood), (2) the model has at least one continuous covariate, and (3) the index is linear conditional on FP transformation of continuous x functions. Although there are, of course, major theoretical and practical differences between types of model, *the use of FPs* within these model types introduces very few new complications or problems.

4.2.3 Relation to Box–Tidwell and Exponential Functions

To our knowledge, Box and Tidwell (1962) were the first to propose the systematic use in data analysis of power functions of the form $\beta_0 + \beta_1 x^\pi$, where π is *any* real number (with $\pi = 0$ meaning log, as above). Mosteller and Tukey (1977) made extensive use of power transformations in data analysis, describing them as 're-expression' of the (scale of) x. Box and Tidwell (1962) also discussed using quadratic functions in x^π of the form $\beta_0 + \beta_1 x^\pi + \beta_2 x^{2\pi}$. The Box–Tidwell and FP2 classes of model have some members in common (e.g. for $\pi = \pm 0.5, \pm 1$), but in general they differ. Multi-exponential functions $\beta_0 + \beta_1 \exp(\pi_1 x) + \beta_2 \exp(\pi_2 x) + \ldots$ provide another link with FP functions. If we write $z = \exp x$, then a multi-exponential function with m terms is seen to resemble the FPm function $\beta_0 + \beta_1 z^{\pi_1} + \beta_2 z^{\pi_2} + \ldots$. The critical difference between Box–Tidwell, multi-exponential functions and FPs is that with FPs the powers are restricted to belong to the set S. This restriction has important practical implications, as discussed in Section 4.6.

4.3   DEFINITION AND NOTATION

4.3.1   Fractional Polynomials

An integer suffix to FP denotes the degree of FP (e.g. FP1, FP2, etc.). We distinguish between an FP transformation and an FP function or model. An FP1 *transformation* of a positive argument $x > 0$ with power p is defined as x^p, where p belongs to the set of powers $S = \{-2, -1, -0.5, 0, 0.5, 1, 2, 3\}$ proposed by Royston and Altman (1994). The set S could be changed (see Section 4.6). By convention, x^0 (i.e. with power $p = 0$) equals the natural log of x rather than 1. An FP1 *function* or *model* is defined as $\varphi_1^*(x; p) = \beta_0 + \beta_1 x^p = \beta_0 + \varphi_1(x; p)$. An FP2 transformation of x with powers $\mathbf{p} = (p_1, p_2)$, or for $p_1 = p_2$ (called 'repeated powers') (p_1, p_1), is the vector $x^{\mathbf{p}}$ with

$$x^{\mathbf{p}} = x^{(p_1, p_2)} = \begin{cases} (x^{p_1}, x^{p_2}), & p_1 \neq p_2 \\ (x^{p_1}, x^{p_1} \log x), & p_1 = p_2 \end{cases}$$

An FP2 function (or model) with parameter vector $\beta = (\beta_1, \beta_2)^{\mathrm{T}}$ and powers $\mathbf{p}$ is $\varphi_2^*(x; \mathbf{p}) = \beta_0 + \beta_1 x^{p_1} + \beta_2 x^{p_2} = \beta_0 + x^{\mathbf{p}}\beta = \beta_0 + \varphi_2(x; p)$. The constant β_0 is optional and depends on the context. For example, β_0 is usually included in a normal-errors model, but not in the Cox regression model for survival data.

When the meaning is clear, we also write FP1 and FP2 functions respectively as $\varphi_1(x)$ and $\varphi_2(x)$, or FP1 (p) and FP2 (p_1, p_2) when necessary to indicate only the powers, or simply FP1 and FP2. With the set S of powers as just given, there are eight FP1 transformations, 28 FP2 transformations with distinct powers $(p_1 \neq p_2)$ and eight FP2 transformations with equal powers $(p_1 = p_2)$.

The general definition of an FPm function with powers $\mathbf{p} = (p_1 \leq \ldots \leq p_m)$ is most easily written as a recurrence relation. Let $h_0(x) = 1$ and $p_0 = 0$. Then

$$\varphi_m^*(x; \mathbf{p}) = \beta_0 + \varphi_m(x; \mathbf{p}) = \sum_{j=0}^{m} \beta_j h_j(x)$$

where

$$h_j(x) = \begin{cases} x^{p_j} & \text{if } p_j \neq p_{j-1} \\ h_{j-1}(x) \log x & \text{if } p_j = p_{j-1} \end{cases}$$

for $j = 1, \ldots, m$. The φ^* functions always include β_0, whereas the φ functions always exclude it.

For example, for $m = 2$ and $\mathbf{p} = (-1, 2)$ we have $h_1(x) = x^{-1}, h_2(x) = x^2$. For $\mathbf{p} = (2, 2)$ we have $h_1(x) = x^2, h_2(x) = x^2 \log x$.

4.3.2   First Derivative

First derivatives are useful when summarizing the effect of a nonlinear function of a variable at different values (see Section 6.4.3). The first derivative (slope) of an FP function is given by

$$\varphi_m'(x; \mathbf{p}) = \sum_{j=1}^{m} \beta_j h_j'(x)$$

where

$$
h'_j(x) = \begin{cases}
1 & \text{if } p_j = 1 \text{ and } p_j \neq p_{j-1} \\
x^{-1} & \text{if } p_j = 0 \text{ and } p_j \neq p_{j-1} \\
p_j x^{p_j - 1} & \text{if } p_j \neq 0, 1 \text{ and } p_j \neq p_{j-1} \\
h'_{j-1}(x) \log x + x^{-1} h_{j-1}(x) & \text{if } p_j = p_{j-1}
\end{cases}
$$

for $j = 1, \ldots, m$.

For example, for $m = 2$ and $\mathbf{p} = (2, 2)$ we have $h'_1(x) = 2x$, $h'_2(x) = 2x \log x + x$.

4.4 CHARACTERISTICS

4.4.1 FP1 and FP2 Functions

FP1 functions are always monotonic (i.e. always have a positive or a negative slope) and those with power $p < 0$ have an asymptote as $x \to \infty$. FP2 functions may be monotonic or unimodal (i.e. have a maximum or a minimum for some positive value of x), and they have an asymptote as $x \to \infty$ when both p_1 and p_2 are negative.

An FP2 function $\beta_0 + \beta_1 x^{p_1} + \beta_2 x^{p_2}$ with different powers $p_1 \neq p_2$ is monotonic when sign $(\beta_1 \beta_2)$ sign (p_2)=sign (p_1), and unimodal otherwise. sign (0) is taken as $+1$. For example, $-4x^{-2} + 3 \log x$ is monotonic, since $(p_1, p_2) = (-2, 0)$ and

$$
\text{sign}(\beta_1 \beta_2) \, \text{sign}(p_2) = \text{sign}(-4 \times 3) \, \text{sign}(0) = (-1) \times (+1) = -1,
$$
$$
\text{sign}(p_1) = \text{sign}(-2) = -1
$$

One should be aware that a unimodal FP2 function may have its maximum or minimum outside the range of the observed data; then it is monotonic within the observed range but not outside it. As with any function, extrapolation beyond the observed range of x should be done with caution. Figure 4.2 presents a constructed example in which x is patients' age and y is the log relative hazard of an event in a time-to-event study. Imagine that functions 1 and 2 were fit from data for patients aged between 25 and 50 years old. We consider the behaviour of the functions extrapolated for patients older than 50 years. Both functions are monotonic on the interval 25–50 years and indicate that the hazard of an event consistently diminishes with age. Function 1 is monotonic for all ages, confirming this interpretation. However, function 2 actually has a minimum at age 56.65 years and increases thereafter; it is unimodal.

Function 1 is a rescaled version of $-4x^{-2}+3 \log x$, which we have already noted to be monotonic. Function 2 is a rescaled version of $-4x^{-2}-3 \log x$, for which sign$(\beta_1 \beta_2)$ sign$(p_2) = +1$ and sign$(p_1) = -1$, showing it to be unimodal.

4.4.2 Maximum or Minimum of a FP2 Function

An FP2 function with repeated powers $(p_1 = p_2)$ is always unimodal. When a general FP2 function is unimodal, the value of x at which the maximum or minimum occurs depends on the powers p_1 and p_2 and the ratio $r = -\beta_1/\beta_2$, as shown in Table 4.1. For example the function $-4x^{-2} - 3 \log x$ has $r = -4/3$ and a maximum at $x = [(-4/3)(-2)]^{-1/(-2)} = \sqrt{8/3} = 1.633$.

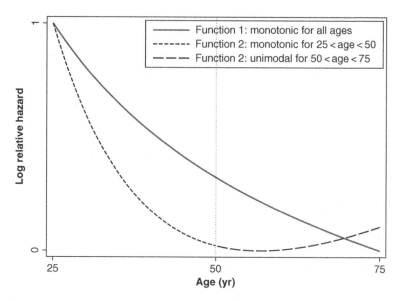

Figure 4.2 Constructed example showing a globally monotonic function (function 1) and a function (function 2) which is locally monotonic on (25, 50) but globally unimodal.

Table 4.1 The position (x) of the maximum or minimum of an unimodel FP2 function, $\beta_0 + \beta_1 x^{p_1} + \beta_2 x^{p_2}$ or $\beta_0 + \beta_1 x^{p_1} + \beta_2 x^{p_1} \log x$, in terms of its powers and coefficients. The quantity r is defined as $-\beta_1/\beta_2$.

Condition on p_2	$p_1 = 0$	$p_1 \neq 0$
$p_2 = 0$ ($p_2 \neq p_1$)	–	$(rp_1)^{-1/p_1}$
$p_2 \neq 0$ ($p_2 \neq p_1$)	$(r/p_2)^{1/p_2}$	$(rp_1/p_2)^{1/(p_2-p_1)}$
$p_2 = p_1$	$\exp(r/2)$	$\exp(r - 1/p_1)$

4.5   EXAMPLES OF CURVE SHAPES WITH FP1 AND FP2 FUNCTIONS

It is worth considering the FP1 and FP2 families in detail because, as already mentioned, we have found that models with degree higher than 2 are rarely required in multivariable analysis. Fractional polynomials with $m \leq 2$ offer many potential improvements in fit compared with conventional polynomials of the same degree. In only a few examples with a single predictor do higher degree FP functions improve the fit significantly (see Sections 5.7, 11.3.3, 11.3.4).

Although rather simple (a set of straight lines in x^p or $\log x$), the FP1 family, an approximation to the Box–Tidwell transformation, is often useful. Figure 4.3 shows schematically the eight FP1 curve shapes available with the powers in S. The curves were constructed on the interval [0.05,1.05] for x. If the lower extreme of the interval on x is reduced, then the curves with negative powers, which are rectangular hyperbolae, become steeper.

The FP2 family is much richer. Figure 4.4 shows examples of four shapes of FP2 functions, chosen to give some idea of the variety available with a few combinations of p_1 and p_2. The curves illustrate the region around the minimum or maximum and/or the turning point and represent only a portion of the whole curve for $x > 0$. In Figure 4.4, $\varphi_2(x; -0.5, 0)$

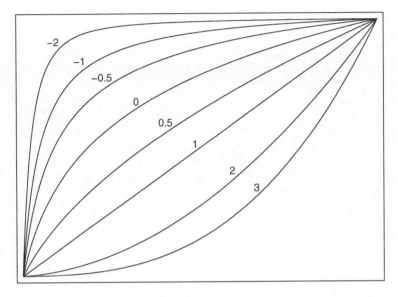

Figure 4.3   Schematic diagram of the eight FP1 curve shapes. Numbers indicate the power p.

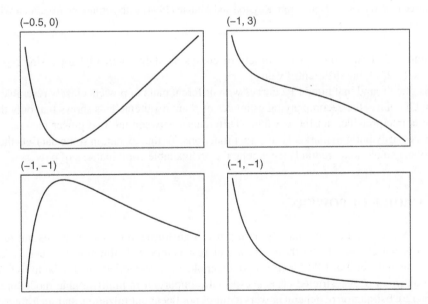

Figure 4.4   Schematic examples of FP2 curves. (Adapted from Royston and Altman (1994) with permission from John Wiley & Sons Ltd.)

resembles an asymmetric quadratic and $\varphi_2(x; -1, 3)$ an asymmetric cubic; both examples of $\varphi_2(x; -1, -1)$ have an asymptote at $x = \infty$, with the first curve rising steeply to a maximum and the second resembling a rectangular hyperbola. The ability to generate a variety of curves, some of which have asymptotes or which have both a sharply rising or falling portion and a nearly flat portion, is a particularly useful feature of FP2 functions.

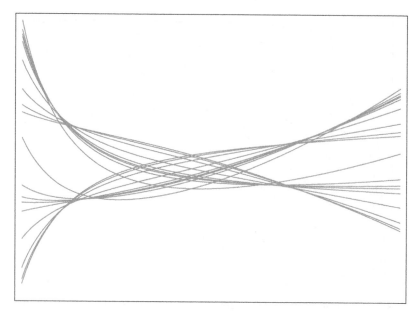

Figure 4.5   Illustration of some of the curve shapes available with FP2 functions with powers $(-2, 2)$ and different β values. (Adapted from Royston and Altman (1994) with permission from John Wiley & Sons Ltd.)

Figure 4.5 shows some of the variety of curves available with FP2 for a single pair of powers, $(-2, 2)$, but different β values.

It is often found that fitted FP2 curves with different pairs of powers closely resemble each other. Of course, the accompanying estimates of β are different. This shows that only the FP curve is interpretable, not the β values. Differences between the two power parameters do not necessarily point to markedly different functions. To that extent, in some datasets the FP2 family may have some redundancy among the 36 available combinations of powers.

4.6   CHOICE OF POWERS

The set S includes most commonly used power transformations. There are several reasons why S is not allowed to be the entire set of real numbers. Estimation is much eased by using a smaller set. For FP2 functions, for example, only 36 models need to be fit to find the best powers in S. A restricted class is especially important in multivariable modelling (see Chapter 6). Estimation of general powers requires nonlinear optimization, and such algorithms may not always converge. The powers selected from S may be seen as a rough approximation to the 'best possible' choice of powers in the interval $[-2, 3]$, both for one-term FP1 and two-term FP2 models. The FP restriction to S may have a small cost of a suboptimal fit of a function; but, besides the computational issues, it has the more important advantages of stabilizing a function and of transportability to other settings. Restricting permitted functions to the FP class substantially increases their general usefulness. Further, models with extreme positive or negative powers are very susceptible even to mild outliers or values close to zero in the covariate x, so such FPs would lack robustness (see Section 5.5).

Sometimes there are reasons to supplement S, e.g. to include $\frac{1}{3}$ if x has the dimensions of a volume. However, experience gained since 1994 has confirmed S as an excellent general choice. An extended set was used by Shkedy et al. (2006). Sometimes one may wish to restrict S, e.g. to negative powers if a function guaranteed to have an asymptote is required.

4.7 CHOICE OF ORIGIN

If nonpositive values of x can occur, then a preliminary transformation of x to ensure positivity is needed. Common cases are where x is a count, when $\log(x+1)$ is traditionally used; where x is a positive random variable such as a physical quantity for which recorded values can be zero, due to imprecise measurement and/or rounding of observations; or where x is a difference or log ratio between two quantities. A simple solution is to choose a nonzero origin γ and work with $x' = x - x_{\min} + \gamma$, where $x_{\min}$ is the smallest observed (or smallest possible) value of x. A possible choice of γ is the rounding interval of the sample values of x, or the minimum increment between successive ordered sample values of x (Royston and Altman, 1994). If x is a count (e.g. number of cigarettes smoked per day), then a natural choice is $\gamma = 1$. For more on this issue, see Section 5.4.

4.8 MODEL FITTING AND ESTIMATION

Typically, FP models are fitted by maximum likelihood. Since an FP model is linear in transformed x for any power(s) $\mathbf{p}$, maximum likelihood estimation amounts to finding the β which maximizes the likelihood of models with linear predictor $\beta_0 + x^{\mathbf{p}}\beta$. For a given class (FP1, FP2, etc.) this is done for each possible $\mathbf{p}$ with powers in S. The best-fitting model is the one whose $\mathbf{p}$ gives the highest likelihood. For the FP1 class, eight models must be fit, whereas 36 models are examined for FP2. The computational 'cost' of fitting an FP2 model is a little more than $36/8 = 4.5$ times as large as for an FP1 model. With modern computer technology, the cost issue can largely be ignored. Only in extreme situations (e.g. with datasets with millions of observations, simulation studies with many replications, or complex models) is it really relevant.

4.9 INFERENCE

4.9.1 Hypothesis Testing

Suppose that each of the m elements of $\mathbf{p}$ in an FPm model are allowed to vary continuously on $(-\infty, +\infty)$, rather than being restricted to S. Then the FP $\beta_0 + \varphi_m(x; \mathbf{p})$ is a nonlinear model with $2m+1$ parameters $\mathbf{p}, \beta_0, \beta$. Let $D(m, \mathbf{p})$ denote the deviance (minus twice the maximized log likelihood) of an FPm model with powers $\mathbf{p}$ and let $D(0)$ be the deviance of the null model β_0. Let $\widehat{\mathbf{p}}$ be the maximum likelihood estimate (MLE) of $\mathbf{p}$. Under the null hypothesis that $\beta = \mathbf{0}$, the distribution of $D(0) - D(m, \widehat{\mathbf{p}})$ is approximately χ^2 on $2m$ d.f. (Royston and Altman, 1994). The result applies asymptotically to all models of the type considered here.

Let $\mathbf{p}_{\text{FP}}$ denote the MLE of $\mathbf{p}$ with powers restricted to S. Then $D(m, \widehat{\mathbf{p}}) \leq D(m, \mathbf{p}_{\text{FP}})$. The deviance difference (or model χ^2 or likelihood ratio statistic) $\Delta D_{\text{FP}} = D(0) - D(m, \mathbf{p}_{\text{FP}})$

Table 4.2 Comparing FP1 and FP2 models. An argument to show that FP1 models are nested within FP2 models. See text for details.

Model class	FP1	FP2-p_1^*	FP2
Powers	p_1^*	(p_1^*, p_2)	(p_1, p_2)
Deviance	D_1^*	$D_2 \leq D_1^*$	$D_2^* \leq D_2$

is also distributed approximately as χ^2 on $2m$ d.f. Since $\Delta D_{FP} \leq \Delta D_p$, where $\Delta D_p = D(0) - D(m, \widehat{p})$, this second approximation is conservative in the sense that P-values derived from it overestimate the correct values (Ambler and Royston, 2001). Put another way, the 'true' d.f. of an FPm model (ignoring β_0) are somewhat lower than $2m$. However, the approximation is good enough for practical application. See Section 4.12 for an example.

By a similar argument, the deviance difference between FPm and FP($m - 1$) models is distributed approximately as χ^2 on two d.f. under the null hypothesis that the additional β is zero. The comparison between FP2 and FP1 models, therefore, requires two d.f., and that between FP1 and a linear model requires one d.f.

An FP1 model is nested in an FP2 model in the following sense. For every FP1 model with power p_1^*, there are eight FP2 models with powers (p_1^*, p_2), with p_2 in S, in which the FP1 model is nested. Furthermore, the deviances D_2 of all these (p_1^*, p_2) FP2 models are less than or equal to the deviance D_2^* of the best FP2 model (see Table 4.2). Therefore, the deviance difference between the best FP2 and the best FP1 model, i.e. $D_1^* - D_2^*$, must always be nonnegative.

With normal-errors regression models, the F distribution rather than the χ^2 distribution is used in hypothesis testing. This may affect the results in very small samples. The residual d.f. are calculated in the usual way by subtracting the model d.f. from the sample size n, also subtracting 1 for models with an intercept. The model d.f. are calculated as described above.

In Section 4.10 we discuss how to derive an FP model which is intended to fit the data 'best'. The criteria used to judge what is 'best' depend on the aims of the study.

Example

The sample size for the research body-fat dataset analysed in Section 1.1.3 is 326. The d.f. for several models are shown in Table 1.1. For an FP2 model, the residual d.f. are $326 - 4 = 322$. The F statistic for testing the two FP2 terms would be compared with an F distribution on $(4, 322)$ d.f.

4.9.2 Interval Estimation

Once the vector β for powers $\mathbf{p}_{FP}$ in an FP model has been estimated as just described, a pointwise CI for the fitted curve can be calculated. The usual pointwise CI is obtained from the predicted curve $\widehat{\eta} = \widehat{\beta}_0 + \varphi_m(x; \mathbf{p}_{FP})$ as $\widehat{\eta}$ plus and minus t times the SE of $\widehat{\eta}$, where t is the appropriate percentile of the t or normal distribution. For example, for a 95% CI using the normal distribution, $t = 1.96$. If SE($\widehat{\eta}$) is calculated 'naively' without allowing for the estimation (i.e. selection from S) of $\mathbf{p}$, it tends to be too small. This is an example of the general problem of underestimating uncertainty by ignoring the fact that the chosen model is merely one of many plausible candidates, each with their own uncertainty: the problem of model uncertainty (Chatfield, 1995).

Although by no means overcoming the problem of model uncertainty, a simple, general way to get a more realistic CI within the FP class is by using the bootstrap. B samples, consisting of selections on n rows of the data matrix randomly sampled with replacement, are drawn. The FP model of degree m is re-estimated (the m powers and the corresponding β values) in each sample, and the fitted values $\widehat{\eta}$ are predicted. The pointwise bootstrap $SE(\widehat{\eta})$ is calculated as the standard deviation (SD) of $\widehat{\eta}$ for each observation across the B bootstrap samples. $B = 100$ may be sufficient. This method allows variation in $\mathbf{p}_{FP}$ to affect the fitted values and hence the CI. Alternative approaches are available, e.g. computing the empirical centiles of the bootstrap distribution of $\widehat{\eta}$ at each observation, but then B must be much larger.

Figure 4.6 shows an example for the research body fat data. The left panel shows that the bootstrap SE is much larger than the naive SE outside the BMI interval $(20, 45)$. The right panel shows the additional uncertainty in the fitted values at the extremes of body mass index. In this example, the FP1 model with power -0.5 is selected in about 25% of bootstrap samples, the remaining samples giving power -1. The bootstrap SE (left panel of Figure 4.6), therefore, arises from a mixture of FP1 models with powers -0.5 and -1. The additional uncertainty, which is ignored by the naive SE estimator, arises because the data support power either -0.5 or -1 (but no other power, as it happens). Generally, there is more variation between the powers selected than in the present example, leading to larger differences between the bootstrap and naive SEs.

When bootstrapping data from a skew distribution with only a few high or low observations, some of the bootstrap samples do not include these observations. If the function is estimated over the whole range of the original data (as is done here), parts of the estimated function are the result of extrapolation. Further, some of the extreme values may be repeated in some

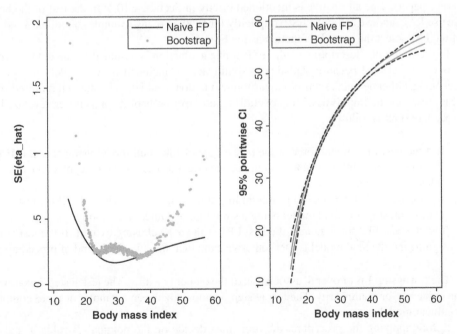

Figure 4.6   Research body-fat data. SEs and CIs for the fitted values from an FP1 model. Comparison of naive and bootstrap estimates, using $B = 100$ bootstrap replications.

bootstrap samples, which may affect the selected function (see Section 5.3). Both types of effect may inflate the CIs at the extremes. More realistic intervals may be obtained through consideration of model uncertainty (see Section 11.3.2).

4.10   FUNCTION SELECTION PROCEDURE

4.10.1   Choice of Default Function

Choosing the best FP1 or FP2 function according to the deviance criterion (see Section 4.9.1) would be straightforward. However, having a sensible default function is important for increasing the parsimony, stability and general usefulness of selected functions. In most of the algorithms implementing FP modelling, the default function is linear – to us, a natural choice. Therefore, unless the data support a more complex FP function, a straight-line model is chosen. There are occasional exceptions; for example, in modelling time-varying regression coefficients in the Cox model, our default time transformation is not t but $\log t$ (see Section 11.1).

4.10.2   Closed Test Procedure for Function Selection

An obvious question to ask with FP modelling is 'how do I select a suitable function for my data?' Our approach that has been developed to answer the question has the flavour of a closed test procedure (Marcus et al., 1976). It is called 'RA2' in Ambler and Royston (2001) and Sauerbrei and Royston (2002). Similar procedures are discussed in the cited papers and in earlier papers. One alternative is mentioned briefly in Section 4.10.4. In the rest of the book we use RA2, and we call it the FSP (implicitly for FPs). In many examples, modified versions of the FSP select the same FP function as the FSP.

In the following description of the FSP, a linear function is assumed as the default when x is either selected because statistically significant, or included in the model on *a priori* reasoning. Other defaults, such as a logarithmic function, could be substituted if required. The FSP preserves the 'familywise' (i.e. overall) type I error probability at a chosen level α. The procedure runs as follows:

1. Test the best FP2 model for x at the α level against the null model using four d.f. If the test is not significant, stop, concluding that the effect of x is 'not significant' at the α level. Otherwise continue.
2. Test the best FP2 for x against a straight line at the α level using three d.f. If the test is not significant, stop, the final model being a straight line. Otherwise continue.
3. Test the best FP2 for x against the best FP1 at the α level using two d.f. If the test is not significant, the final model is FP1; otherwise the final model is FP2. End of procedure.

The test at step 1 is of overall association of the outcome with x. The test at step 2 examines the evidence for nonlinearity. The test at step 3 chooses between a simpler or more complex nonlinear model.

Before applying the procedure, the user must decide on the nominal P-value α and on the degree m of the most complex FP model allowed. Typical choices are $\alpha = 5\%$ and FP2 ($m = 2$).

Note that it is possible within a single application of the procedure to choose different nominal significance levels for testing inclusion of x at step 1 ($\alpha = \alpha_1$, say) and for testing between functions at steps 2 and 3 ($\alpha = \alpha_2$, say). Often, x must be included in a model because of the study design, background knowledge or some other aim. Setting $\alpha_1 = 1$ is equivalent to forcing x into the model. Consequently, the type I error probability is preserved at α_2 for testing for nonlinearity in x, rather than relating to the overall influence of x. See also Section 6.6.2.

The FSP may easily be extended to $m = 1$ or $m > 2$. Starting with $m = 1$, the best FP1 function is first tested against the null model on two d.f. If the test is significant at the α level, then the FP1 function is tested against the linear model and the final model is either FP1 or linear. Starting with $m > 2$, the best FPm function is first tested against the null model on $2m$ d.f. Depending on the result, the FPm function is tested against the linear model, then if necessary against the FP1 model, and so on up the ladder of increasingly complex FPs.

4.10.3   Example

We apply the FSP with $m = 2$ and $\alpha = 0.05$ to the variable $x = $ bmi in the research body-fat data. The results are shown in Table 4.3. The comparison of FP2 versus null shows that x is highly significant, and the test of FP2 versus linear gives strong evidence that the function is nonlinear. The final comparison of FP2 versus FP1 is not significant ($P = 0.43$), so with any choice of α below 43% the procedure would select the FP1 model. The fitted FP1 and linear models are shown in Figure 1.3.

4.10.4   Sequential Procedure

Royston and Altman's (1994) original proposal for FP function selection was a sequential procedure: compare FP2 with FP1, then FP1 with linear, and finally linear with null. This approach increases the power when the 'true' function of x is linear. However, the cost is a greatly increased type I error probability when x is uninfluential – as much as a doubling or trebling of the nominal size (Ambler and Royston, 2001). Furthermore, the procedure can sometimes give inconsistent results. For example, tests of FP1 versus linear (one d.f.) and linear versus null (one d.f.) could each be nonsignificant, whereas a test of FP1 versus null (two d.f.) could be significant, creating doubt as to what is a reasonable model. For these reasons, we do not recommend the sequential procedure, but note it here for completeness.

Table 4.3   Research body fat data. Application of the closed test procedure to body mass index. The selected model is FP1.

Model	Deviance D	**P**	Step	Comparison	Dev. diff.	P
FP2	1631.30	$-2, -1$	1	FP2 vs null	649.32	< 0.001
FP1	1633.02	-1	2	FP2 vs linear	150.76	< 0.001
Linear	1782.06	1	3	FP2 vs FP1	1.72	0.43
Null	2280.62	–				

4.10.5   Type I Error and Power of the Function Selection Procedure

Ambler and Royston (2001) studied by simulation the type I error of the FSP. They considered normally distributed and binary responses y for the cases when the true model is null (y unrelated to x, $\eta = 0$) and linear ($\eta = \beta x$). Two different distributions of x were simulated: uniform and lognormal, the latter being positively skew. They looked at the FSP starting with either $m = 2$ or $m = 1$.

For practical reasons, the nominal size α was taken as 10% (note: not 5%) in all simulations. In the null case, Ambler and Royston (2001) found that the empirical size of the closed test procedure lay between 5% and 11% (to be compared with a nominal 10%). It was slightly larger for the case of y binary (using logistic regression) than for y normal. When the true model was linear, the empirical size of tests of FP2 versus linear and FP1 versus linear ranged between 5% and 7%. In all cases, therefore, the procedure was somewhat conservative, more so when the true model was linear, but never grossly so. The cost of the conservatism is a small loss of power, but the loss has not so far been quantified accurately.

Holländer (2002) performed a small simulation study of FP2 versus linear in the Cox proportional hazards model with no censored observations. When the nominal size was 5% the actual size was found to be 5.7% (95% CI: 4.3%, 7.1%). Here, the size is approximately correct, but slightly larger than might be inferred from the findings of Ambler and Royston (2001).

To our knowledge, the power of the FSP in the usual case of detecting linear and nonlinear functions of a covariate x has not been studied. The required simulations are quite complex, since many possibilities must be considered. This is an area for further research. Berger et al. (2003) obtained results on power favourable to FP methodology in the modelling of time-varying effects in the Cox model (see Section 11.1).

4.11   SCALING AND CENTERING

4.11.1   Computational Aspects

For completeness, we describe the preliminary scaling and centring automatically applied to a variable x by Stata when using the `fracpoly` and `mfp` routines to fit an FP model, making the Stata implementation numerically quite robust. However, `fracpoly` has `adjust (no)` and `noscaling` options to cancel scaling and centring if required. The SAS and R implementations work directly with the original values of x. Since x must be greater than zero, we assume that the origin for a nonpositive predictor has already been shifted to ensure positivity.

There are computational and conceptual advantages to ensuring that the transformed values x^p are sensibly scaled and centred. Scaling reduces the chance of numerical underflow or overflow in extreme cases, which would cause inaccuracies and difficulties in estimating the model. Centring ensures that the intercept β_0 or the baseline hazard function in a Cox model retains a meaningful interpretation. In Stata, the following preliminary scaling transformation is applied to x, independent of $\mathbf{p}$:

$$\texttt{lrange} = \log_{10}[\max(x) - \min(x)]$$

$$\texttt{scale} = 10^{\text{sign(lrange) int(|lrange|)}}$$

$$x^* = \frac{x}{\texttt{scale}}$$

where int(x) denotes the nearest integer below x. The quantity `scale` is an integer power of 10. The FP transformation of x^* is then centred on the mean of the observed values of x^*. For example, for the FP1 model $\beta_0 + \beta_1 x^p$, the actual model fitted by the software would be $\beta_0^* + \beta_1^*(x^{*p} - \overline{x^{*p}})$ where $\overline{x^*} = \frac{1}{n}\sum_{i=1}^{n} x_i^*$. This approach ensures that the revised constant β_0^* equals the fitted value of the FP function at $x^* = \overline{x^*}$ (also at the mean of x).

Note that the order of operations is: first (if necessary) scale x; second, estimate the FP powers; and finally, centre. Centring at an earlier stage may result in different powers, and should be avoided. Estimation of powers is unaffected by scaling, i.e. the same powers are found for $x/\texttt{scale}$ for any positive value of `scale`. In extreme cases, scaling is necessary to preserve accuracy (see example below).

4.11.2   Examples

If x were age observed in the range [25, 75] years, say, then $\texttt{lrange} = \log_{10}(75 - 25) = 1.6989$ and $\texttt{scale} = 10^{\text{int}(1.6989)} = 10^1 = 10$. Age would be divided by 10 and this would have little effect on the accuracy of the calculations.

However, suppose that another x had range $[1.76 \times 10^{-18}, 4.84 \times 10^{-18}]$ and mean 3.019×10^{-18}. Then $\texttt{lrange} = -17.51$ and $\texttt{scale} = 10^{-17}$. Now $x^* = x \times 10^{17}$, a rescaling which may greatly improve accuracy since the range of x^* is a much more reasonable [0.176, 0.484]. An experiment in Stata, suppressing the scaling, indeed showed that fitting an FP2 regression model to such an extreme x gave incorrect results.

(Correction: in the example in Royston et al. (1999), the factor $\texttt{scale} = 10^2$ was applied to systolic blood pressure, but the scaling was not reported in the text. Therefore, the definition of x in the legend of their table 5 is (systolic blood pressure)/100.)

4.12   FP POWERS AS APPROXIMATIONS TO CONTINUOUS POWERS

4.12.1   Box–Tidwell and Fractional Polynomial Models

We may be concerned that, by restricting S to a small set, we may lose something important by using an FP model compared with the 'best' model allowing all powers to be continuous. A way to check this for FP1 models is to estimate the parameter π of the Box–Tidwell transformation and compare the resulting value of $\widehat{\pi}$ with p for the chosen FP1 model. For FP2 models it is necessary to estimate π_1 and π_2 in the extended Box–Tidwell model $\beta_0 + \beta_1 x^{\pi_1} + \beta_2 x^{\pi_2}$, and compare $\widehat{\pi}_1$ and $\widehat{\pi}_2$ with p_1 and p_2. The fitted curves should also be compared.

For Stata users, Royston and Ambler (1999) provided a program `boxtid` to fit such models. Even without access to `boxtid`, for FP1 models a plot of the deviance against π (i.e. minus twice the profile likelihood function for π) also reveals whether a much better value of p is available. A much better value is occasionally available for FP1 models, and it often lies outside the interval $[-2, 3]$, but for FP2 models it is rare to find much better values.

4.12.2   Example

For the predictor `bmi` in the research body-fat data, we have seen that the FSP selects an FP1 model with power -1. Figure 4.1 suggests that power -0.5 provides nearly as good a fit (as is also suggested by the bootstrap analysis in Section 4.9.2). Perhaps the 'correct' power, and therefore a better model, lies between -0.5 and -1.

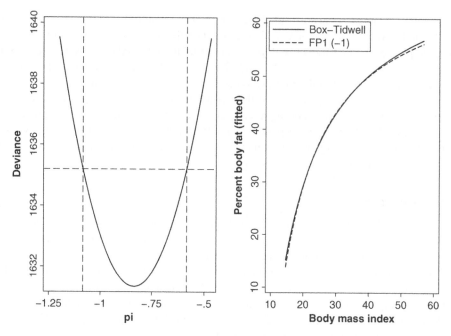

Figure 4.7   Research body-fat data. Left panel: deviance of Box–Tidwell model $\beta_0 + \beta_1 x^\pi$ versus π. Vertical lines show 95% CI for π. The MLE $\hat{\pi}$ of π is -0.84. Right panel: comparison of Box–Tidwell model $\beta_0 + \beta_1 x^{\hat{\pi}}$ with best FP1.

The MLE $\hat{\pi}$ is -0.84 (95% CI: -1.09, -0.59) (see left panel of Figure 4.7). Thus, the FP1 value of -1 is within the 95% CI for π. Inspection of the fitted curves for $\hat{\pi} = -0.84$ and $p = -1$ reveals them to be almost identical (see right panel of Figure 4.7). The deviances for the Box–Tidwell and FP1 models are 1631.34 and 1633.02 respectively (difference: 1.68). The values of explained variation R^2 between models with powers -0.84 and -1 differ only in the third decimal place. We conclude that using the MLE of π provides a negligible improvement in fit compared with the best FP1 model.

4.13   PRESENTATION OF FRACTIONAL POLYNOMIAL FUNCTIONS

4.13.1   Graphical

See Section 6.4.2 for a discussion of plotting FP functions and their CIs derived from a multivariable model. The principles are similar for a single predictor (univariate regression). Here, an example from a univariate regression analysis is given.

The best-fitting FP logistic regression model for predicting all-cause mortality from daily cigarette consumption (`cigs`) in the Whitehall I dataset is an FP1 with power $p = 0$, i.e. logit $\Pr(y = 1|\text{cigs}) = \beta_0 + \beta_1 \ln x$, where $x = \text{cigs} + 1$. (Adding 1 to `cigs` to avoid zeros is the usual approach for count variables; see Section 4.7).

Figure 4.8 displays the fitted function in four different ways. The upper panels show the probability of death, on the left as the logit (i.e. on the scale of the fitted function) and on the right on the probability scale. The lower panels show the log OR and the OR, compared with

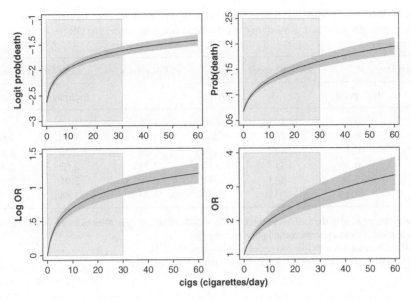

Figure 4.8   Whitehall I data. Graphical presentation of results from an FP analysis of cigarette smoking and all-cause mortality. The graphs depict the relationship in four different ways and are based on a logarithmic relationship fitted by logistic regression. Shaded boxes denote 2.5th to 97.5th centiles of the distribution of `cigs`.

a referent category of nonsmokers (`cigs` $= 0$.). In the lower panels, the width of the CIs at `cigs` $= 0$ is by definition zero. Which form of plot is preferred depends on the aim of the analysis, and to some extent on taste.

4.13.2   Tabular

Apart from assuming linearity, a conventional way to model and present results of an analysis involving a continuous covariate is in categories with corresponding dummy variables. In epidemiology, for example, use of three, four or five categories is common. Table 4.4 shows such an analysis of the relation between all-cause mortality and smoking for the Whitehall I dataset, using four categories of `cigs`; see Royston et al. (1999, Table 2).

Table 4.4   Whitehall I data. Conventional analysis in categories of the relationship between all-cause mortality and cigarette consumption.

Cigarettes/day	Number		OR of dying	
	At risk	Dying	Estimate	95% CI
0	10 103	690	1.00	–
1–10	2 254	243	1.65	1.41, 1.92
11–20	3 448	494	2.28	2.02, 2.58
> 20	1 455	243	2.74	2.34, 3.20

Adapted from Royston et al. (1999) by permission of Oxford University Press.

Table 4.5 Whitehall I data. Presentation of results of an FP1 model for mortality as a function of cigarette consumption.[a] The OR of dying is calculated from the model $\log \text{OR} = -2.62 + 0.293 \times \log(\text{cigs} + 1)$.

Cigarettes/day		Number of men		OR (obs.)	OR (model-based)	
Range	Ref. point[b]	At risk	Dying		Estimate	95% CI
0	0	10 103	690	1.00	1.00	–
1–10	6.5	2 254	243	1.65	1.80	1.68, 1.94
11–20	16.7	3 448	494	2.28	2.32	2.10, 2.57
21–30	26.3	1 117	185	2.71[c]	2.63	2.34, 2.96
31–40	37.4	283	48	2.79[c]	2.91	2.56, 3.31
> 40	49.0	55	10	3.03[c]	3.14	2.74, 3.61

[a] First three categories as in the analysis presented in Table 4.4; additional categories included for heavier smokers.
[b] Reference point is mean cigarette consumption in category.
[c] According to the analysis with four categories, all ORs would be 2.74.
Adapted from Royston et al. (1999) by permission of Oxford University Press.

The analysis is presented for illustration purposes only and is not adjusted for other covariates. Although the relationship between smoking and the probability of dying is statistically highly significant and may be roughly perceived from Table 4.4, more information can be extracted from the data by FP analysis and presented in an easily understood manner.

Table 4.5 displays results from the FP1 model mentioned in Section 4.13.1, presented in categories to preserve the familiar style of such tabulations. The category-based estimates have been supplemented with those from the FP model at relevant exposures, calculated as mean cigarette consumption in each category. The fit of the model appears to be good. For example, the relative risk for 11–20 cigarettes/day is estimated as 2.32, increasing to 3.14 for > 40 cigarettes/day. Corresponding crude ORs are 2.28 and 3.03. Use of a monotonic function makes interpretation of the relationship straightforward. Crude and model-based ORs differ more if the estimates of the latter are derived from a multivariable model.

The 95% CIs for ORs are calculated from the appropriate SEs as follows. Suppose x_{ref} is the referent, i.e. the value of x central to the preferred baseline category (here, 0 cigarettes/day). We wish to calculate the SE of the difference between the fitted log odds of dying given x relative to x_{ref}, according to the FP1 model $\widehat{\beta}_0 + \widehat{\beta}_1 x^p$. The difference in log odds ratios equals $\widehat{\beta}_1(x^p - x_{\text{ref}}^p)$ and its standard error is $\text{SE}(\widehat{\beta}_1)|x^p - x_{\text{ref}}^p|$. $\text{SE}(\widehat{\beta}_1)$ is obtained in standard fashion from the regression analysis (and neglects the uncertainty in estimation of p; see Section 4.9.2). A 95% CI for the OR is the antilog (i.e. exponential) of $(x^p - x_{\text{ref}}^p)[\widehat{\beta}_1 \pm 1.96\text{SE}(\widehat{\beta}_1)]$. For a worked example, see Section 4.14.5.

The width of the CI for the OR at a given value of x is strongly influenced by the type of model used and by the distance from the mean value of x, and much less by the number of observations in the region around x. With the traditional categorical analysis, the number of the observations in a category determines the width of the CI more directly. For example, according to the FP1 model the CI for > 40 cigarettes/day is only about 40% wider than for 21–30 cigarettes/day, whereas the latter is supported by about 30 times as many observations. The interval for the function is based on all the data, whereas that from the categorical model utilizes data only from the referent and the interval in question. The categorical approach has the major weakness of dependence on the categories chosen.

In general, the model-based intervals do not accommodate all the relevant uncertainty. The CI from the regression model may, therefore, be unrealistically narrow. As already mentioned in Section 4.9.2, model uncertainty methodology would be needed to address this issue more satisfactorily.

In addition, the equation for the regression model together with the covariance matrix of the estimated regression coefficients should be presented. For a model with many variables, this may be provided as additional material on a Web page. It must be admitted that we have not provided all such information in earlier papers. A graph of the fitted function and its pointwise CI should be given (see Figure 4.8). For FP models, this type of presentation transfers easily to the multivariable situation where parameter estimates are adjusted for other factors. With the categorical approach, it is difficult to present all the necessary details in the context of a multivariable analysis.

4.14   WORKED EXAMPLE

To help clarify the finer points of an FP analysis, we present an example in some detail. We derive a logistic regression model for the relationship between systolic blood pressure (sysbp) and all-cause mortality in the Whitehall I data.

4.14.1   Details of all Fractional Polynomial Models

Table 4.6 presents the deviance differences compared with a linear function for all eight FP1 and 36 FP2 models for sysbp. The best FP1 and FP2 models have powers 2 and $(-2, -2)$ respectively. Note that the deviances of models with powers $(-2, -1)$, $(-2, -0.5)$, $(-1, -1)$

Table 4.6   Whitehall I data. FP models for the relationship between systolic blood pressure and all-cause mortality. The deviance difference compares the fit with that of a straight line ($p = 1$). The maximum deviance difference and the best-fit FP1 and FP2 model powers are in italic.

FP1			FP2							
Power p	Dev. diff.	Power		Dev. diff.	Power		Dev. diff.	Power		Dev. diff.
		p_1	p_2		p_1	p_2		p_1	p_2	
−2	−74.19	−2	−2	26.22	−1	1	12.97	0	2	7.05
−1	−43.15	−2	−1	24.43	−1	2	7.80	0	3	3.74
−0.5	−29.40	−2	−0.5	22.80	−1	3	2.53	0.5	0.5	10.95
0	−17.37	−2	0	20.72	−0.5	−0.5	17.97	0.5	1	9.51
0.5	−7.45	−2	0.5	18.23	−0.5	0	16.00	0.5	2	6.80
1	0.00	−2	1	15.38	−0.5	0.5	13.93	0.5	3	4.41
2	*6.43*	−2	2	8.85	−0.5	1	11.77	1	1	8.46
3	0.98	−2	3	1.63	−0.5	2	7.39	1	2	6.61
		−1	−1	21.62	−0.5	3	3.10	1	3	5.11
		−1	−0.5	19.78	0	0	14.24	2	2	6.44
		−1	0	17.69	0	0.5	12.43	2	3	6.45
		−1	0.5	15.41	0	1	10.61	3	3	7.59

Adapted from Royston et al. (1999) by permission of Oxford University Press.

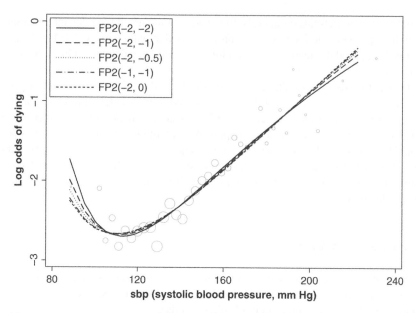

Figure 4.9   Whitehall I data. Fitted curves for the five best FP2 models for the log odds of dying as a function of sbp, ordered by deviance difference from linearity. Circles denote log odds of dying with sbp categorized into narrow intervals. Area of circles is proportional to the sample size in each sbp category. (Adapted from Royston et al. (1999) by permission of Oxford University Press.)

and $(-2, 0)$ are close to that of the best model. Despite differences in the powers, the functions exhibit a similar fit, as may be seen in Figure 4.9. The functions are almost identical in the range [100, 200] mmHg, where the observations are most plentiful: 97.8% of the data lie in this interval.

4.14.2   Function Selection

The steps of the FSP and the results of each test are shown in Table 4.7. It is clear that an FP2 model is needed, since (1) FP2 versus null is significant, (2) FP2 versus linear is significant, and (3) FP2 versus FP1 is also significant. For further general explanation, see Sections 4.10.2 and 4.10.3.

4.14.3   Details of the Fitted Model

Here, we include the preliminary scaling and centring automatically applied by Stata when fitting an FP model (see Section 4.11). With the current implementation of FP modelling in

Table 4.7   Whitehall I data. Application of the closed test procedure to the analysis of systolic blood pressure.

Model	Deviance	p	Comparison	Dev. diff.	P
FP2	10 641.17	$-2, -2$	FP2 vs null	332.57	< 0.001
FP1	10 660.96	2	FP2 vs linear	26.22	< 0.001
Linear	10 667.39	1	FP2 vs FP1	19.79	< 0.001
Null	10 973.74	–	–	–	–

Table 4.8 Whitehall I data. Estimates of regression coefficients and their variance–covariance matrix for the FP2 logistic model for systolic blood pressure.

Parameter	Estimate	SE	Variance–covariance matrix		
$\widehat{\beta}_1$	−5.433	0.321	0.103 3		
$\widehat{\beta}_2$	−14.300	1.317	0.374 8	1.734 2	
$\widehat{\beta}_0$	−2.388	0.032	0.005 597	0.024 7 94	0.001 036

other software (at present, only SAS and R), `lrange` is not relevant, `scale` $= 1$ and no centring constants are used. However, this does not change the principle of the computations given below, only minor details.

The range of `sysbp` is [85, 280], hence `lrange` $= \log_{10}(280 - 85) = 2.29$ and `scale` $= 10^{int(2.29)} = 100$ (see Section 4.11). The mean of `sysbp` is 136.1 mmHg. (Note that, to aid interpretation, it may be better to choose a different centring constant, e.g. here 140 mmHg.) For the FP2 transformation with powers $(-2, -2)$, two covariates are required to fit the logistic regression model: x^{*-2} and $x^{*-2} \log x^*$, where $x^* = $ `sysbp`$/100$. Centring constants of $(136.1/100)^{-2} = 0.5399$ and $(136.1/100)^{-2} \log(136.1/100) = 0.1664$ respectively are subtracted from these covariates, giving, say, $x_1 = x^{*-2} - 0.5399$ and $x_2 = x^{*-2} \log x^* - 0.1664$.

The logistic model actually fitted is $\beta_0 + \beta_1 x_1 + \beta_2 x_2$. The parameter estimates and their variance–covariate matrix are shown in Table 4.8. The constant $\widehat{\beta}_0 = -2.388$ gives the predicted log odds of dying for an individual having the mean blood pressure of the sample (136.1 mmHg), and is equivalent to a probability of $\exp(-2.388)/[1+\exp(-2.388)] = 0.084$.

4.14.4   Standard Error of a Fitted Value

We illustrate the calculation of the SE of a fitted value, $\widehat{\eta}$, from this model. Note that the SE ignores the fact that the FP powers are estimated. $\widehat{\eta}$ is the estimated log odds of death at a given value of the covariate. The variance of $\widehat{\eta}$ at (x_1, x_2) is given according to standard algebra by

$$\text{var}(\widehat{\eta}) = \text{var}(\widehat{\beta}_0 + \widehat{\beta}_1 x_1 + \widehat{\beta}_2 x_2) \tag{4.1}$$

$$= \text{var}(\widehat{\beta}_0) + \text{var}(\widehat{\beta}_1) x_1^2 + \text{var}(\widehat{\beta}_2) x_2^2$$

$$+ 2\text{cov}(\widehat{\beta}_0, \widehat{\beta}_1) x_1 + 2\text{cov}(\widehat{\beta}_0, \widehat{\beta}_2) x_2 + 2\text{cov}(\widehat{\beta}_1, \widehat{\beta}_2) x_1 x_2 \tag{4.2}$$

For example, with $x = 150$ mmHg then $x^* = 1.5$, $x_1 = -0.095\,46$, $x_2 = 0.013\,807$, $\widehat{\eta} = -2.0665$. From the above, $\text{var}(\widehat{\eta}) = 0.000\,936\,1$, $\text{SE}(\widehat{\eta}) = 0.0306$. The probability of dying is $\exp(\widehat{\eta})/[1 + \exp(\widehat{\eta})] = 0.1124$, and its 95% CI is $\exp(\widehat{\eta} \pm 1.96\text{SE}(\widehat{\eta}))/[1 + \exp(\widehat{\eta} \pm 1.96\text{SE}(\widehat{\eta}))] = (0.1066, 0.1185)$.

4.14.5   Fitted Odds Ratio and its Confidence Interval

For the FP2 model we are exemplifying here, the estimated log OR at x compared with a referent x_{ref} is given by

$$\log \text{OR} = (\widehat{\beta}_0 + \widehat{\beta}_1 x_1 + \widehat{\beta}_2 x_2) - (\widehat{\beta}_0 + \widehat{\beta}_1 x_1^{\text{ref}} + \widehat{\beta}_2 x_2^{\text{ref}})$$

$$= \widehat{\beta}_1(x_1 - x_1^{\text{ref}}) + \widehat{\beta}_2(x_2 - x_2^{\text{ref}})$$

where $x_1^{\text{ref}} = x_{\text{ref}}^{*-2} - 0.5399$, $x_2^{\text{ref}} = x_{\text{ref}}^{*-2} \log x_{\text{ref}}^* - 0.1664$, $x_{\text{ref}}^* = x_{\text{ref}}/100$. Its estimated variance is

$$\text{var}(\log \text{OR}) = \text{var}(\widehat{\beta}_1)(x_1 - x_1^{\text{ref}})^2 + \text{var}(\widehat{\beta}_2)(x_2 - x_2^{\text{ref}})^2$$
$$+ 2\text{cov}(\widehat{\beta}_1, \widehat{\beta}_2)(x_1 - x_1^{\text{ref}})(x_2 - x_2^{\text{ref}})$$

Using, for example, $x = 150$ mmHg and $x^{\text{ref}} = 105$ mmHg, we find $x_1^{\text{ref}} = 0.3671$, $x_2^{\text{ref}} = -0.1221$, $\log \text{OR} = 0.5691$, $\text{var}(\log \text{OR}) = 0.007017$, $\text{SE}(\log \text{OR}) = 0.0838$. The OR is $\exp(0.5691) = 1.77$ and a 95% CI for the OR is given by $\exp(\log \text{OR} \pm 1.96\text{SE}(\log \text{OR})) = (1.50, 2.08)$.

4.15   MODELLING COVARIATES WITH A SPIKE AT ZERO

In epidemiological and clinical studies, it may happen that the distribution of a continuous covariate x is composed of positive values and zeros. There is a probability mass or 'spike' at zero in an otherwise continuous, positive variable (Robertson et al., 1994). For example, the average amount of alcohol consumed per year is zero for teetotallers and some positive number for drinkers. When the individuals with $x = 0$ are seen as a distinct subpopulation, it may be necessary to model the outcome in the subpopulation explicitly with a dummy variable z, and the rest of the distribution as a positive continuous variable using a dose–response relationship. An approach using FP functions is as follows.

Assume that $x \geq 0$ for all individuals. In order for FP functions of x to be defined at $x = 0$, required also by the second stage of testing (see below), the origin of x is shifted by adding a small constant γ before analysis (see Section 4.7). Consider a model with index

$$\eta = \begin{cases} \beta, & x = 0 \\ \beta_0 + \varphi_2(x + \gamma; \mathbf{p}), & x > 0 \end{cases}$$

so that η is an FP2 function when $x > 0$ and a constant (β) when $x = 0$. Thus, η is a discontinuous function of x with a possible jump at $x = 0$. The expression for η is equivalent to $\eta = \beta_0 + \beta z + \varphi_2^+(x + \gamma; \mathbf{p})$, where

$$z = \begin{cases} 1, & x = 0 \\ 0, & x > 0 \end{cases} \tag{4.3}$$

$$\varphi_2^+(x + \gamma; \mathbf{p}) = \begin{cases} 0, & x = 0 \\ \varphi_2(x + \gamma; \mathbf{p}), & x > 0 \end{cases} \tag{4.4}$$

To select a model, we propose a procedure in two stages. In the first stage, the most complex model comprising z and $\varphi_2^+(x + \gamma; \mathbf{p})$ is compared with the null model on five d.f. If the test is significant, then the steps of the FSP for selecting an FP function are followed, but with z always included in the model. In the second stage, z and the remaining FP or linear component are each tested for removal from the model. If both parts are significant, then the final model includes both; if one or both parts are nonsignificant, then the one with the smaller deviance difference is removed. In the latter case, the final model comprises either the binary dummy variable or the selected FP function. If only an FP function is selected, then the spike at zero plays no further part. Since the selection of an FP function may be affected by the presence

Table 4.9 Prostate data. Analysis of pgg45 with a spike (probability mass) at zero. z is the dummy variable for zero values of pgg45.

Model for pgg45	Deviance	Dev. diff.[a]	d.f.	P	Power
First stage					
Null	302.1	29.8	5	< 0.001	–
Linear + z^b	273.7	1.4	3	0.73	1
FP1$^+$ + z^c	272.7	0.4	2	0.84	−0.5
FP2$^+$ + z^c	272.3	–	–	–	1, 3
Second stage					
Linear + z^b	273.7	–	2	–	
Linear [dropping z]	282.7	9.0	1	0.003	
z [dropping linear]	276.2	2.5	1	0.1	

[a] For first stage, deviance difference compared with FP2$^+$ + z model; for second stage, deviance difference compared with Linear + z model.
[b] Model chosen in the first stage.
[c] Modified FP1 and FP2 models, as defined by Equation (4.3).

of the binary dummy variable, the resulting model may differ from that from a standard FP analysis.

As an illustration, we consider applying the modified FSP to pgg45 in the prostate data (see Table 4.9). Some 36% of the values of pgg45 are zero. Because pgg45 increases in steps of 1, the small constant γ is taken as 1. The test of FP2$^+$ + z versus null is highly significant ($P < 0.001$). The next test, FP2$^+$ + z versus Linear + z, has $P = 0.73$. The selected model from the first stage, therefore, is Linear + z. At the second stage, dropping the selected linear term does not significantly worsen the fit ($P = 0.1$), whereas dropping z is highly significant ($P = 0.003$). The selected model, therefore, comprises only the binary variable z. Its regression coefficient is −1.16, the difference in mean lpsa between zero (1.74) and positive values (2.90) of pgg45.

In Figure 4.10, the three fitted functions corresponding to the first stage are plotted against pgg45, starting from the smallest positive value of 4 The second-stage model (z) is represented by the mean lpsa at zero and positive values of pgg45 (filled circle and horizontal line respectively). The shape of the FP1 function in Figure 4.10 suggests that an FP1 function may fit quite well as an alternative to z. Applying the standard FSP(0.05) to pgg45, an FP1 with power 0 is chosen. The selected model, log(pgg45 + 1), has deviance 273.0, very similar to that of the Linear + z model (273.7) and slightly lower than for the z-only model (276.2). Thus, an FP1 model is a viable alternative to the z-only model.

The decision to use a model including z as just described, or to work within the standard FP class, is best made on subject-matter grounds rather than by considering the fit of functions with or without z. Judgement of the relative merits of the two approaches is needed. For example, in some cases a discontinuous function may make no scientific sense. Since in an FP$^+$ + z model the FP function is not 'anchored' by the data at $x = 0$, the function may be unstable and poorly estimated for small positive x. When the proportion of zero values is small, it is preferable to ignore the special role of the zero subset and apply the standard FSP to x.

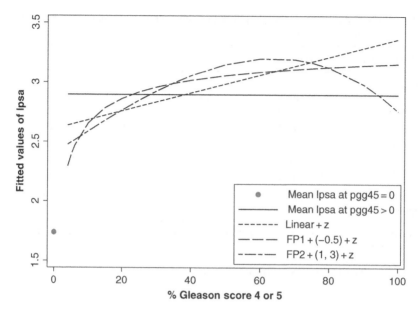

Figure 4.10 Prostate cancer data. Fitted linear and modified FP functions for pgg45 with a dummy variable z for zero values. FP1$^+$ and FP2$^+$ are FP functions estimated from positive values of pgg45.

Becher (2005) presented a similar approach when assessing smoking as a risk factor for lung cancer, but did not select the model for the continuous part of x in a systematic fashion. He assumed a logarithmic functional form for the association with daily cigarette consumption.

4.16   POWER OF FRACTIONAL POLYNOMIAL ANALYSIS

If there is a linear relationship between x and the outcome, then the FSP, which first tests FP2 versus null (or FP1 versus null), has less power than the usual test of x linear versus null. Some reduction in power when the underlying function is linear is a consequence of the search for a possible improvement in fit by nonlinear FP functions, while maintaining the type I error probability at a nominal level. For example, the deviance difference required for a test of FP2 versus null to be significant at $P < 0.05$ is 9.49. If the functional form was perfectly linear and a deviance difference of 9.49 was observed when testing linear versus null, then the P-value of the latter test would be 0.002; the corresponding P-value for testing FP1 versus null at the 5% level would be 0.014. It is clear from this example why power is lost when the underlying function is linear and best-fitting FP2 or FP1 functions (with their 'wasted' additional d.f.) are tested against the null model.

By performing some limited simulation studies, we illustrate the power of the FSP when testing FP2, FP1 and linear versus null in the three situations in which the underlying function is linear, FP1 or FP2. Clearly, much more extensive simulations are needed to explore a greater range of scenarios.

4.16.1 Underlying Function Linear

We illustrate the power of the FSP in a simple simulation scenario in which the underlying function is linear. We simulate $x \sim N(4, 1)$, $y \sim N(\beta x, 1)$, sample size 25, 50 or 100, and modest explained variation R^2 of 0.1, 0.2 or 0.3. β is calculated from R^2 via the relationship (Helland, 1987) $R^2 = \text{var}(x\beta)/[\sigma^2 + \text{var}(x\beta)] = \beta^2\text{var}(x)/[\sigma^2 + \beta^2\text{var}(x)]$. Since $\sigma^2 = 1$, we have $R^2 = \beta^2/(1+\beta^2)$ and so $\beta = [R^2/(1 - R^2)]^{0.5}$. The power to select x was estimated by counting the proportion of simulation replications in which x was declared significant at the $\alpha = 0.05$ level. Three tests were used: the FSP starting with most complex model FP2, the FSP starting with FP1, and a standard F-test of the significance of x as a linear term. The results for 1000 replications of each combination of n and R^2 are given in columns 4–6 of Table 4.10. The power loss (compared with assuming linearity) using the FSP starting with FP1 as the most complex function is moderate; but when FP2 is used, much more power is lost. For larger values of β and/or n, the power loss is small.

Columns 7–9 of Table 4.10 show the power when the significance level is relaxed to 0.157, equivalent to selection by AIC when testing a single parameter (see Section 2.6.3). The power of all the tests is much improved.

4.16.2 Underlying Function FP1 or FP2

Finally, we consider the power of the FSP when the underlying function is logarithmic or quadratic (examples of FP1 and FP2 functions respectively). The nine scenarios for n and R^2 used in the linear case were repeated. Significance testing was at the $\alpha = 0.05$ level only.

For the FP1 case, the simulation design was $\log x \sim N(0, 1)$, $y \sim N(\beta \log x, 1)$. Here, x has a skewed (lognormal) distribution, as is often seen in practice.

For the FP2 case, the simulation design was $x \sim N(4, 1)$, $y \sim N(\beta(x - 4)^2, 1)$. The underlying (quadratic) function is U-shaped and symmetric about $x = 4$, the mean of the distribution of x. In individual realizations (simulated datasets), the distribution of x is by

Table 4.10 Simulation study of the power to detect the influence of a variable x in univariate regression models assuming linearity, or using the FSP with maximum complexity FP1 or FP2 functions. Nominal significance levels of $\alpha = 0.05$ and $\alpha = 0.157$ were used when selecting models. See text for details.

R^2	β	n	$\alpha = 0.05$			$\alpha = 0.157$		
			Linear	FP1	FP2	Linear	FP1	FP2
0.1	0.33	25	34	21	9	56	42	21
		50	64	49	30	81	71	52
		100	89	82	67	97	93	83
0.2	0.5	25	63	51	28	82	70	49
		50	91	82	66	97	94	83
		100	100	99	97	100	100	100
0.3	0.65	25	85	75	48	95	89	73
		50	99	97	90	99	99	97
		100	100	100	100	100	100	100

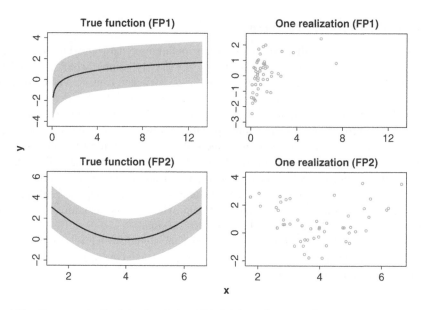

Figure 4.11   Power study. Explained variation R^2 is 0.3 in each case. Left-hand plots: underlying FP1 (logarithmic) and FP2 (quadratic) functions f of a single covariate x used in the simulations, plotted over the 1st to 99th centiles of the distribution of x. Bands show $f \pm 1.96\sigma$. Right-hand plots: one realization (simulated dataset) with $n = 50$ for each of the functions shown on the left.

chance not totally symmetric, so the test of a linear function still has some power to detect a relationship between y and x.

Figure 4.11 shows the underlying FP1 and FP2 functions, variation in the data and a single realization when $R^2 = 0.3$ and $n = 50$, plotted between the 1st and 99th population centiles of the distribution of x.

The results of the simulation study are given in Table 4.11. When the underlying function is FP1, the power of the FP1 test dominates that of the linear and FP2 tests, as would be expected. Increasing R^2 increases the power of all three tests, as does the increase in sample size. When the underlying function is FP2, the test assuming linearity has low power, which does not improve with increasing sample size. The FP1 test is better, and its power does improve with increasing R^2 and n. The FP2 test is easily the most powerful of the three.

4.16.3   Comment

Since the FSP is the building-block from which the multivariable model-building algorithm MFP is constructed, the results just presented should be borne in mind when performing multivariable analyses (see Chapter 6). If the effects of covariates are weak and/or the sample size is small, then variables whose linearity is in doubt are more likely to be selected by the FSP assuming FP1 as the most complex function than assuming FP2. Similarly, the power to detect weak nonlinear effects is enhanced if FP1 but not FP2 functions are allowed. The cost of using

Table 4.11   Simulation study of the power to detect the influence of a variable x in univariate regression models assuming linearity, or using the FSP with maximum complexity FP1 or FP2 functions. A nominal significance level of $\alpha = 0.05$ was used when selecting models. See text for further details.

R^2	β	n	True function FP1			True function FP2		
			Linear	FP1	FP2	Linear	FP1	FP2
0.1	0.33	25	27	26	13	10	9	10
		50	47	52	34	9	16	29
		100	75	82	69	10	32	64
0.2	0.5	25	50	52	36	10	15	24
		50	79	85	72	12	31	59
		100	97	99	97	13	58	94
0.3	0.65	25	69	73	56	16	28	46
		50	94	97	93	19	46	83
		100	100	100	100	17	77	99

FP1 but not FP2 functions is the risk of overlooking more complex or nonmonotonic relationships if they happen to be present. The third simulation demonstrates how a nonmonotonic function may be missed if FP2 functions are not considered.

One would hope that model checking (e.g. with smoothed residuals) would reveal such a lack of fit. Although unproblematic in a univariate setting, detection of anomalies is considerably more difficult in a multivariable setting. For example, unrecognized curvature in one variable may impact on the selection of other variables and/or functions.

4.17   DISCUSSION

FP functions of a single variable are the cornerstone of our approach to modelling continuous variables in regression. As seen in Chapter 6, they form the basis of multivariable FP models, the main topic of our book. In general terms, statistical models approximate relationships between variables. Usually, the coefficients lack a specific meaning, but the function is (or should be) interpretable. It is understood that a model is simply a convenient formula that gives a reasonable description of the relationship within the confines of the observed data.

There is no particular reason other than convention why regression models should include only positive integer powers of the covariate(s). As we have seen, FP models can give a good and yet parsimonious fit to different datasets. We have applied FP modelling to many datasets in addition to the ones considered in our book. We often find a model that is an improved fit in comparison with a linear function or even a conventional polynomial with the same number of terms. Because conventional polynomials are included in the FP family, we cannot actually obtain a worse fit with FPs.

Correcting for multiple testing always incurs a loss of power. Applying the FSP to a variable with a weak linear effect may lead to it being falsely declared as uninfluential. The possibility of an increased type I error rate is the cost of investigating nonlinearity in a systematic fashion. The cost is low with large sample sizes, but may be high for small samples. An option to consider in the latter case is to increase the significance level in the FSP, e.g. from 0.05 to 0.157 (the AIC level for one parameter).

Most, but not all, aspects of univariate FP models have been covered in this chapter. We address further issues in Chapter 5 and consider the use of FPs in modelling interactions of different types in Chapter 7.

CHAPTER 5

Some Issues with Univariate Fractional Polynomial Models

Summary

1. The chapter considers some additional aspects of FP modelling with a single covariate.
2. Fitted FP functions may contain artefacts due to a small number of covariate values; for example, one term in an FP model may depend on them. A diagnostic plot is described to reveal such a situation.
3. Because of the negative power and log transformations that may be used in FP functions, very small values of x may introduce artefacts into fitted FP functions. Selection of an origin for x may then be important. An approach to choosing a more suitable origin is described.
4. Improving fit, robustness or both by preliminary transformations of x is discussed. A general transformation for improving robustness, particularly useful in multivariable modelling, is described.
5. Occasionally, higher order FP functions (with $m = 3$ or $m = 4$) are useful in data analysis. Negative exponential pre-transformation may help to improve the fit of a complex function. Examples are given.
6. FPs are unsuitable for modelling some types of function. Examples of practical importance include sigmoid curves and complex spatial relationships.

5.1 INTRODUCTION

Additional issues affecting FP modelling of a single continuous covariate, including model diagnostics, robustness and choice of origin, are considered in this chapter.

Multivariable Model-Building   Patrick Royston, Willi Sauerbrei
© 2008 John Wiley & Sons, Ltd

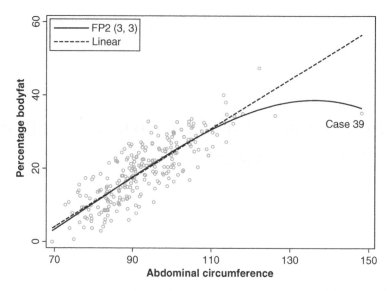

Figure 5.1   Educational body-fat data. Effect of case 39 on the FP model chosen for ab (abdominal circumference). Solid line: FP2 (3, 3) model; dashed line: linear model. (Reproduced from Royston and Sauerbrei (2007a) with permission from Elsevier Limited.)

5.2   SUSCEPTIBILITY TO INFLUENTIAL COVARIATE OBSERVATIONS

All statistical models are potentially disturbed by outliers in the response. That is a generic issue and is not considered here. FP models, moreover, may be affected by *covariate outliers*. A large value of x is much magnified in an FP1 model with power $p = 3$; similarly, a value near zero is enlarged by a transformation with $p \leq 0$. In each case, the spacing (relative distance) of such observations from their less extreme neighbours is increased. The leverage of such observations may be high. For example, an FP2 model may be made statistically significant compared with FP1 simply by the need to fit a single extreme observation of x.

By way of example, Figure 5.1 shows the relationship between percentage body fat (pcfat) and abdominal circumference (x_6 = ab) in the educational body-fat dataset. It is clear that ab has an extreme observation (case 39). With case 39 included, the FSP selects an FP2 model with powers (3, 3) ($P = 0.04$ for FP2 versus FP1). The fitted line nearly goes through case 39's response. With case 39 excluded, the selected model is linear ($P = 0.7$ for FP2 versus linear). The straight-line model fits the remaining observations well. The preferred model would, of course, depend on whether the outlier in ab was a credible observation or not. This cannot be determined from the data alone.

We consider methods to improve the robustness of FP models in Section 5.5. First, we describe a diagnostic plot to detect influential points.

5.3   A DIAGNOSTIC PLOT FOR INFLUENTIAL POINTS IN FP MODELS

A graphical tool to screen for observations influential on the selection of FP models is often helpful. A simple approach is the leave-one-out or jackknife method (Mosteller and Tukey, 1977) applied to FP models. The best FP2, FP1 and linear models are found after successively deleting each single observation (or covariate pattern, which may have more than one associated observation). The deviance of each of the three models at each omitted observation or pattern is stored. The deviance differences (linear minus FP2, or FP1 minus FP2, depending on which test is most relevant) are calculated and plotted against the 'case' or observation number (or covariate pattern number, or value of x at each excluded observation or covariate pattern). The χ^2 threshold for significance at a chosen α level may be shown on the plot. Single observations or covariate patterns which influence the choice of a more complex FP model stand out because, when they have been deleted, the deviance difference is reduced, sometimes dramatically. If this 'deleted' deviance difference is below the χ^2 threshold, then there is evidence that the more complex model selected using all the observations depends on this one observation or covariate pattern.

A minor elaboration when plotting the deviance difference against the covariate pattern or its corresponding unique value of x is to indicate the relative frequency by the size of the plotting symbol. It may happen that a low or high value of x is influential and also has several or even many associated observations. This potentially important feature can immediately be seen in such a plot.

Handling of influential points is a difficult topic. Problems are specific to each dataset considered. In general, we do not recommend deleting influential observations unless definite errors in the data have been identified. In the context of FP modelling, preliminary transformation of x (see Section 5.5) is an option to reduce undue influence of extreme observations.

5.3.1 Example 1: Educational Body-Fat Data

Diagnostic plots for the educational body-fat data are shown in Figure 5.2. The model selected on all the observations is FP2 (3, 3). The left-hand plot shows the deviance difference for FP2 versus linear against case number. The strong influence of case 39 is easily seen – it alone drives the selection of an FP2 model. Without case 39, the deviance difference is only 0.98, the test of FP2 versus linear is no longer significant and a linear model is chosen. Case 39 is included in all 251 other models. Deviance differences are larger than 17, clearly indicating that an FP2 model fits better if case 39 is included. Figure 5.1 shows that the FP2 function agrees very closely with the linear function for ab up to about 120 cm. Only three observations have higher ab values.

The right-hand plot of Figure 5.2 gives a similar message, additionally showing that case 39 is well separated from the bulk of the observations in both dimensions of the plot. In this example there are 185 covariate patterns or unique values of ab among the 252 cases. The highest frequency of any value is four.

5.3.2 Example 2: Primary Biliary Cirrhosis Data

The primary biliary cirrhosis (PBC) dataset is briefly described in Appendix A.2.7. We show that low values may also be influential in the selection of FP models. We apply the FSP at the $\alpha = 5\%$ level to the variable bil (bilirubin) in the full dataset (418 observations, 161 events). This is a survival-time dataset, and we use Cox regression. The test of FP2 versus linear is highly significant ($P < 0.001$), and that of FP2 versus FP1 is significant at the 5% level ($P = 0.04$). The fitted FP2 function is shown as the solid line in Figure 5.3.

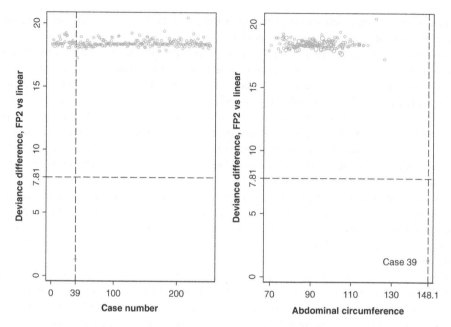

Figure 5.2   Educational body-fat data. Deviance difference between FP2 and linear models for the predictor ab (abdominal circumference). Left panel: excluding each of the 252 cases in turn. Right panel: excluding each of the 185 covariate patterns in turn (areas of circles are proportional to pattern frequencies). Horizontal line at 7.81 is the χ^2 threshold for significance of FP2 versus linear at the 5% level. Case 39 alone drives the selection of an FP2 model.

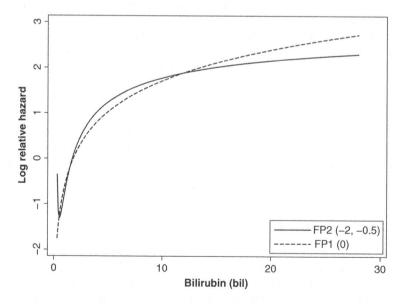

Figure 5.3   PBC data. Two fitted FP curves for the effect of bil on the log relative hazard of death.

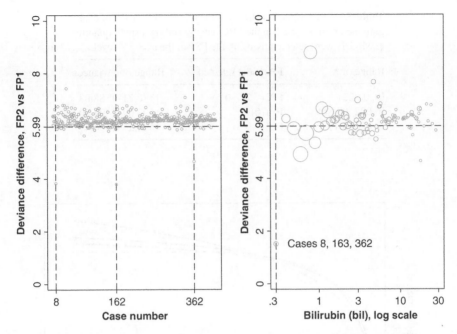

Figure 5.4 PBC data. Deviance difference between FP2 and FP1 models. Left panel: excluding each of the 418 cases in turn. Right panel: excluding each of the 98 covariate patterns in turn (areas of circles are proportional to pattern frequencies). Horizontal line at 5.99 is the χ^2 threshold for significance of FP2 versus FP1 at the 5% level. Cases 8, 163 and 362 mainly drive the selection of an FP2 model.

Note the 'hook' in the FP2 curve for low `bil`. Since bilirubin is known to be a strong prognostic factor in PBC with a monotonic effect, this feature is biologically implausible, similar to the hook in the risk curve for `nodes` in the GBSG breast cancer data; see Sauerbrei and Royston (1999).

The diagnostic plot for FP2 versus linear (not shown) does not suggest any influential points. We focus on the marginally significant test of FP2 versus FP1. Figure 5.4 shows the diagnostic plots for FP2 versus FP1 presented in a fashion similar to Figure 5.2. The left-hand plot suggests that cases 8, 163 and 362 are the most influential for selecting an FP2 rather than an FP1 model. The right-hand plot, based on covariate patterns, shows that these cases belong to the same covariate value, `bil = 0.3`.

5.4 DEPENDENCE ON CHOICE OF ORIGIN

We saw in Section 5.3.2 that the three lowest values of a covariate influenced the selection of an FP2 model with negative powers. Table 5.1 shows what happens to the selected model as the origin of `bil` is increased by adding a small constant ζ to `bil`. We see that at $\zeta = 0.022$ the model changes from FP2(-2, -0.5) to FP1(0), and the deviance shifts upwards. The fitted curve for the FP1 model with $\zeta = 0.022$ is shown by the dotted line in Figure 5.5. The hook is eliminated. As with the breast cancer example, the hook is a feature of the dataset,

Table 5.1   PBC data. The effect on the selected FP model of changing the origin for `bil` in the PBC data by adding a small quantity ζ. Models were selected by using the FSP at the $\alpha = 5\%$ level.

Range of ζ	FP model selected	Range of deviances
$[0, 0.022)$	FP2$(-2, -0.5)$	1587.71–1588.08
$[0.022, 0.086)$	FP1(0)	1594.09–1594.50
$[0.086, 0.1)$	FP1(-0.5)	1594.25–1594.49

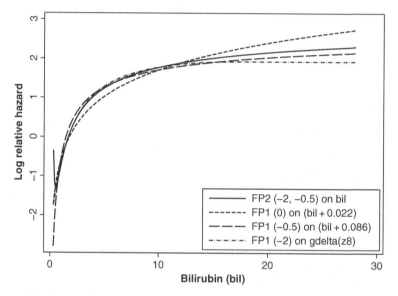

Figure 5.5   PBC data. Fitted FP curves for the effect of bilirubin (`bil`) on the log relative hazard of death. The dashed curves were derived after adding $\zeta = 0.022$ and $\zeta = 0.086$ to `bil`. The dot–dashed curve was derived after applying the pre-transformation g_δ(`bil`).

but the model is not sensible. A small amount of fit may be sacrificed in exchange for a more satisfactory estimate of the function.

As Royston and Altman (1994) noted, the origin is highly correlated with the FP power(s) needed; therefore, estimation of ζ is possible but not really sensible. It can be seen in Table 5.1 that, as ζ is increased from 0.085 to 0.086, the FP1 model power changes from 0 to -0.5 but the deviance is hardly affected. The function is nearly identical to FP1(0). All three functions are similar for bilirubin greater than about 0.5. We should not be concerned about changing the origin if there is a good reason to do so.

Ambler and Royston (2001) noted that the position of the origin of x affects the type I error probability for FP model selection. As the origin increases, the type I error is reduced. We consider the question of choice of origin further in Section 5.5.

5.5   IMPROVING ROBUSTNESS BY PRELIMINARY TRANSFORMATION

There is sometimes a need to shift the origin of x, e.g. when $x \leq 0$. A simple method was described in Section 4.7. An alternative approach was suggested by Royston and Sauerbrei (2007a) for dealing with the problems caused by undue influence of low and high extreme covariate values when FP transformations are applied. Let x_{min} and x_{max} denote the minimum and maximum observed sample values of x, $\bar{x}$ be the sample mean, s be the sample SD, and $\Phi(\cdot)$ be the normal cumulative distribution function. Let $q = \Phi[(x - \bar{x})/s]$ be the normal percentile corresponding to x, standardized to have zero mean and unit SD.

Royston and Sauerbrei (2007a) proposed the following pre-transformation (preliminary transformation) $g_\delta(x)$ of x:

$$g_\delta(x) = \delta + (1 - \delta)\frac{g(x) - g(x_{min})}{g(x_{max}) - g(x_{min})}$$

where

$$g(x) = \left[\ln\left(\frac{q + \varepsilon}{1 - q + \varepsilon}\right) + \varepsilon^*\right]/(2\varepsilon^*)$$

and

$$0 < \delta < 1, \quad \varepsilon = 0.01, \quad \varepsilon^* = -\ln[\varepsilon/(1 + \varepsilon)] = 4.615\,12$$

In practical terms, $g_\delta(x)$ would be used instead of x in FP models. For a general discussion of when to use $g_\delta(x)$, see Section 5.5.3.

At first sight, the function $g(x)$, which is closely related to a logistic function, appears very complicated, but in essence the idea is straightforward. The expression defining $g_\delta(x)$ linearly transforms $g(x)$ to the interval $[\delta, 1]$ with $0 < \delta < 1$. A suitable value of δ needs to be chosen such that the origin is not 'too small' (see Section 5.4). Royston and Sauerbrei (2007a) proposed the value $\delta = 0.2$ for practical use. They supported this value with results from simulation studies showing that the influence of small values of x was sufficiently reduced across a range of distributions of x when the FP1 model with $p = -2$ was used. The power $p = -2$ gives the most extreme transformation of small values of x in the FP1 class (see Figure 4.3).

The function $g(x)$ is constructed to be very nearly linear over the bulk of the observations (in fact, within about $\bar{x} \pm 2.8s$; see Figure 5.6). It smoothly tapers to zero as $x \to -\infty$ and to one as $x \to \infty$. The choice $\varepsilon = 0.01$ maximizes the linearity of $g(x)$ in the central region around $\bar{x}$. The value 2.8 arises as follows. $\kappa = 2.8$ minimizes over κ the integrated squared difference (i.e. the quadratic distance) between $g(x)$ and $\tilde{g}(x; \kappa)$, where $\tilde{g}(x; \kappa)$ represents double truncation at $\pm\kappa$. In other words, at $\kappa = 2.8$, $g(x)$ most closely resembles a linear transformation of x to $[0, 1]$ on the interval $[-\kappa, \kappa]$, the constant 0 for $x < -\kappa$ and the constant 1 for $x > \kappa$. For further details, see Royston and Sauerbrei (2007a, section 4.4).

The transformation $g(x)$ has the effect of smoothly pulling extreme low and extreme high values of x towards the centre of the distribution, reducing their influence in FP models. Figure 5.6 illustrates $g(x)$ when $\bar{x} = 0$, $s = 1$. In the central portion of the range $g_\delta(x)$ is almost perfectly linear. Effectively, the transformation prevents extrapolation of a function of $g_\delta(x)$ beyond $\bar{x} \pm 2.8s$.

We now revisit the examples of Sections 5.2 and 5.4 to examine the effect of $g_\delta(x)$.

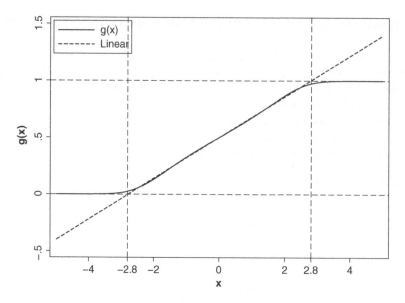

Figure 5.6   The pre-transformation $g(x)$ with $\bar{x} = 0$, $s = 1$.

5.5.1   Example 1: Educational Body-Fat Data

We first apply $g_\delta(x)$ to ab for all 252 observations and reapply the FSP. The test of FP2 versus linear on $g_\delta(x)$ has $P = 0.7$, so now a linear model is selected, rather than the previous FP2 model on x. The left panel of Figure 5.7 shows the fitted curve for this model. The right-hand vertical line represents ab $= 122.7$ cm, the effective upper truncation point $\bar{x} + 2.8s$ of the

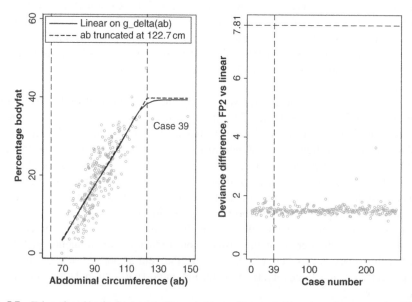

Figure 5.7   Educational body-fat data. Left panel: Linear fit on g_δ(ab) and on ab truncated at 122.7 cm. Right panel: diagnostic plot for FP2 versus linear on g_δ(ab).

pre-transformation. The lower effective truncation point $\bar{x} - 2.8s$ falls below the minimum of the distribution and does not come into play here. As shown in the right panel, case 39 no longer has a marked influence on the fitted line. The linear fit on ab truncated at 122.7 cm (long dashes) is similar to that on $g_\delta(\text{ab})$, but has a 'corner' at 122.7 cm.

The plot of the deviance differences (right panel) may now suggest that case 216 is another influential point. However, the deviance difference is only 3.65, well short of the value of 7.81 nominally required for significance of the χ^2 test (three d.f.) of FP2 versus linear at the 5% level (even ignoring the issue of repeated significance testing).

5.5.2   Example 2: PBC Data

The problem with this dataset was the influence of three cases with the lowest observed value of bil. Shifting the origin of bil to the right by at least 0.022 caused an FP1 function to be selected for the effect of bil on the log relative hazard of death (see Figure 5.5). Without the shift in origin, an FP2 function with an implausible hook was selected.

We now examine the effect on this relationship of applying $g_\delta(x)$ to bil. FP2 versus linear on $g_\delta(\text{bil})$ is still highly significant ($P < 0.001$), whereas FP2 versus FP1 is no longer significant ($P = 0.4$). Therefore, the FP1 model, which has power -2, is selected. This model does not have (indeed, being an FP1, cannot have) a hook. The fit of the FP1 model on $g_\delta(\text{bil})$, with the default $\delta = 0.2$, is compared with the FP2 model on untransformed bil and with the two FP1 models with the origin shifted in Figure 5.5. Apart from differences at the extreme low and extreme high values of bil, the fitted curves are all similar. The diagnostic plot for FP2 versus FP1 models on $g_\delta(x)$ no longer indicates any influential observations.

As a sensitivity analysis of δ, the values were varied from 0.01 to 0.3. With $\delta = 0.01$, the test of FP2 versus FP1 on $g_\delta(x)$ is significant ($P = 0.03$), whereas for $\delta = 0.05$, 0.1, 0.2 and 0.3 the P-values are 0.8, 0.6, 0.4 and 0.6 respectively. Too small a value of δ is unhelpful. When $\delta > 0.086$, an FP1 function of $g_\delta(x)$ is always selected.

5.5.3   Practical Use of the Pre-transformation $g_\delta(x)$

We have seen that the pre-transformation $g_\delta(x)$ may be effective in removing a disproportionate influence of low or high values of x when selecting FP models. Would we recommend its use routinely in data analysis? The answer is 'no'. The disadvantages are (1) the interpretation of the relationship between the response and the covariate is less natural, (2) the relationship is less easily communicated, (3) $g_\delta(x)$ changes the shape of the FP function at extreme values of x, which may induce unacceptable bias when the 'true' functional form is simple. Rather, we suggest the use of $g_\delta(x)$ as a sensitivity analysis of a selected FP model. This application is particularly important in multivariable models (see Chapter 6), where the presence of an extreme value of an x may affect the functions selected for itself and for other, correlated x variables.

When there is a need to shift the origin of x, either because influential low (but not necessarily outlying) observation(s) have been detected, or because x has zero or negative values, the origin-shift transformation incorporated in $g_\delta(x)$ may be used on its own. The origin-shift transformation with parameter δ is given by

$$\omega_\delta(x) = \delta + (1 - \delta)(x - x_{\min})/(x_{\max} - x_{\min})$$

and is a robust alternative to the heuristic origin-shift proposal described in Section 4.7. The values of $\omega_\delta(x)$ lie in the interval $[\delta, 1]$. As with $g_\delta(x)$, Royston and Sauerbrei's (2007a) suggested that a default value of $\delta = 0.2$ is recommended for practical use.

5.6 IMPROVING FIT BY PRELIMINARY TRANSFORMATION

5.6.1 Lack of Fit of Fractional Polynomial Models

Even in situations considered in our book, FP models may not always provide a satisfactory fit in some datasets, even when the relationship to be modelled is relatively simple (for an example, see Section 5.8.2). Occasionally, preliminary covariate transformation may substantially improve the fit of FP models. In one example (see Section 5.7.2) we show that the fit of an FP model can be improved by using a preliminary exponential transformation. In the previous examples in this chapter, FP2 modelling gave results that did not conform with subject-matter knowledge, motivating pre-transformation to produce models more in agreement with current thinking. However, the need for pre-transformation is, in our experience, the exception rather than the rule.

5.6.2 Negative Exponential Pre-transformation

Royston and Altman (1994) remarked that pre-transformations x^* of x of the form $x^* = \exp[-x/\text{SD}(x)]$, where $\text{SD}(x)$ denotes the sample SD of the observations of x, can sometimes help to improve the fit of an FP model, but did not pursue the theme. The pre-transformation pulls in large values of x and thereby changes the shape of the resulting FP functions. The transformation considerably reduces positive skewness and the influence of observations in the extreme upper tail of x. Furthermore, x^* always takes positive values, as required for FP functions, so is potentially useful when x has zero or negative values. Log transformation, a possible competitor, does not have the latter property.

Sauerbrei and Royston (1999) used the pre-transformation $x^{**} = \exp(-0.12 \times \text{nodes})$ for the number of positive lymph nodes in the breast cancer example (see also Section 6.5.4). The constant -0.12 was the MLE of γ in the nonlinear Cox regression model with predictor $\eta = \beta \exp(-\gamma \times \text{nodes})$. The pre-transformation was motivated by the implausible hook in the fitted FP2 model and the need for a monotonic function with an asymptote (see Figure 6.9).Using $x^* = \exp[-x/\text{SD}(x)]$, the multiplier $-1/\text{SD}(x)$ of x is found to be -0.18. When x^* and x^{**} are used (univariately) as covariates in an FP model, the FSP with $\alpha = 5\%$ chooses a linear model for each of them. The resulting Cox models have similar deviances (3500.1 and 3498.0 respectively), and the fitted functions are also similar.

When using $x^* = \exp[-x/\text{SD}(x)]$ as a covariate in an FP model, the effect of a power transformation x^{*p} with $p \neq 0$ is to multiply $-1/\text{SD}(x)$ by p. When $p = 0$ the linear model is recovered, since $\ln\{\exp[-x/\text{SD}(x)]\}$ is proportional to x. Thus, the family of FP models in x^* includes the straight line.

In practice, it is worth trying the negative exponential pre-transformation when FP models on untransformed x do not produce a good fit. The approach may be particularly useful when the function to be modelled appears to be approximately constant for larger values of x, as in an example presented in Section 5.7.2. Also, when the distribution of x is strongly positively skewed, the transformation makes the distribution much less skewed and may improve the model fit.

5.7 HIGHER ORDER FRACTIONAL POLYNOMIALS

FPs of order m may have up to $m - 1$ turning points (local maxima or minima). If, therefore, a response variable has, say, two turning points then one must expect that nothing lower than an FP3 model would be sufficient. The study by Royston and Altman (1997) uses high-order FP functions to approximate statistical functions. There was no risk of overfitting or other instability, since the functions to be modelled were completely smooth and the only issue was how close an approximation was needed – i.e. what degree of FP was adequate. In some cases the additional flexibility of FP3 functions allows a better fit, even for curves with only one turning point (see Figure 5.8, for example).

In this section we explore two datasets in which a high-order FP model improves the fit.

5.7.1 Example 1: Nerve Conduction Data

In this dataset, the response y is the speed of nerve impulses in human muscle and x is the age in years since conception (the data includes a number of fetuses). The sample size is 406. An analysis using the FSP starting with the usual default of $m = 2$ shows that FP2 is significantly better than FP1 and gives the results shown in the upper panels of Figure 5.8. The smoothed residuals indicate a lack of fit for ages < 2 years.

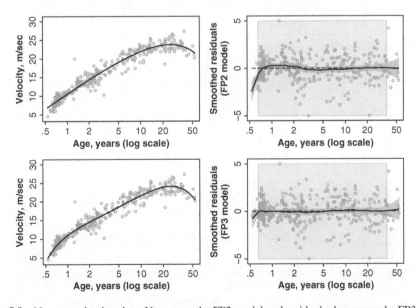

Figure 5.8 Nerve conduction data. Upper panels: FP2 model and residuals; lower panels: FP3 model and residuals. Raw residuals have been truncated to $[-5, 5]$ for plotting purposes.

Applying the FSP starting with $m=4$ shows that FP3 fits significantly better than FP2 ($P<0.001$), but that FP4 is little better than FP3 ($P=0.08$). At $\alpha=5\%$, therefore, the FP3 model is selected. The lower panels of Figure 5.8 show the improved fit of this model compared with FP2. Both models clearly indicate that nerve conduction velocity peaks at a little over 20 years of age in humans.

5.7.2   Example 2: Triceps Skinfold Thickness

The data are derived from an anthropometric study of 892 females under 50 years in three Gambian villages in West Africa (Cole and Green, 1992). We examine the relationship between triceps skinfold thickness (a crude measure of body fat) and age. The distribution of triceps skinfold thickness is positively skewed; therefore, we work with a log-transformed response variable (see Figure 5.9). The plot of smoothed values shows a local maximum in skinfold thickness at about 2 years and a local minimum at 10 years. With two turning points, we must expect to need at least an FP3 model. For individuals aged 25 to 50 years, the mean skinfold thickness is nearly constant.

The FSP with $m=4$ in fact suggests that an FP4 function is needed ($P<0.001$ for FP4 versus FP3). However, the smoothed residuals from this model indicate lack of fit (see Figure 5.10, upper panels). The local maximum at about 2 years is missed completely, and the behaviour of the function at the highest ages is implausible. This is an instability typical of high-order polynomial-type models.

The approximate constancy of the function for age > 25 years suggests trying the negative exponential pre-transformation, $\exp[-x/\mathrm{SD}(x)]$, which compresses the upper age range. The FSP still indicates that FP4 (with powers (2, 3, 3, 3)) is needed. The lower panels of Figure 5.10 show the greatly improved fit. The wavy fitted curve for ages > 25 years visible in the upper left

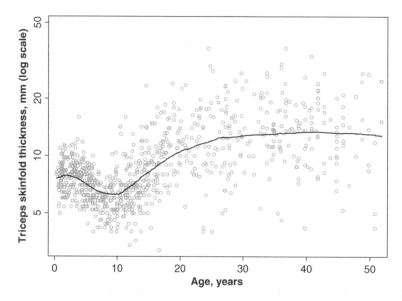

Figure 5.9   Triceps skinfold thickness data. Running line smooth for $E(\texttt{triceps}|\texttt{age})$.

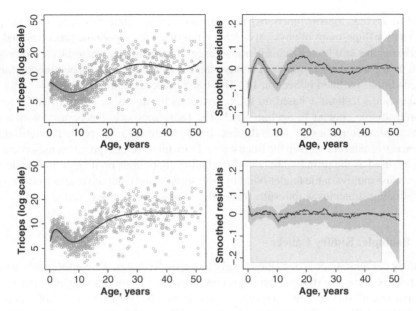

Figure 5.10   Triceps skinfold thickness data. Upper panels: FP4 model on `age` with smoothed residuals; lower panels: FP4 model on exponentially pre-transformed `age` with smoothed residuals.

panel is abolished. The fitted curve, although still not 'perfect', is altogether more satisfactory as a description of the relationship between geometric mean skinfold thickness and age. However, an FP4 function for y of the form $\beta_0 + \beta_1 x^2 + \beta_2 x^3 + \beta_3 x^3 \log x + \beta_4 x^3 (\log x)^2$, where $x = \exp[-\text{age}/\text{SD}(\text{age})]$, is unusually complex. Although acceptable in a univariate analysis, such a complicated function would be unlikely to be considered in multivariable model-building.

Despite its complexity, the FP4 model is interpretable and practical. Triceps skinfold thickness is a crude measure of body fat. Babies become 'chubbier' as they grow from an early age, and then begin to lose some of this fat as they develop into children. Around puberty, general body size increases rapidly. This is reflected in triceps skinfold, which increases quickly from the age of 10 years, eventually levelling off in adulthood.

The original aim of this study was to derive age-specific reference intervals for triceps skinfold thickness. This requires modelling of the whole age-specific distribution, not just the mean, as here. For simpler types of function, we consider the use of FPs in such applications in Section 11.2.

5.8   WHEN FRACTIONAL POLYNOMIAL MODELS ARE UNSUITABLE

5.8.1   Not all Curves are Fractional Polynomials

While we are obviously enthusiastic about FP modelling, we would not want the reader to get the impression that we believe FPs are suitable for all modelling tasks with continuous variables. That is patently not so. There are many types of relationship which demand more flexible and sometimes highly specialized types of model. Examples which come to mind are sigmoid curves (e.g. logistic functions), human growth curves, time-series, complex temporal–spatial relationships, maps of physical or other kinds of territory, periodic functions, and so on. Nevertheless, this still leaves a considerable range of applications where FPs are very useful. Perhaps their particular strength is in multivariable model-building with continuous variables under conditions outlined in Section 1.6. This topic is addressed in Chapter 6.

We describe a technique based on spline functions for assessing the fit of an FP function in a multivariable context in Section 6.5.3. The same approach can be used with a single predictor. Spline functions are more flexible than FPs, but are more prone to overfit the data and, hence, to create artefacts in the fitted curve. Difficulties of interpretation may occur when splines are used as the primary method of modelling continuous predictors. In Chapter 9, we briefly discuss multivariable model-building with splines, and compare results from two spline model-building strategies with results from MFP in several examples.

5.8.2   Example: Kidney Cancer

The kidney cancer dataset is described briefly in Appendix A.2.9. We consider only the effect on the relative hazard of dying of an important prognostic factor, haemoglobin (haem). The outcome is overall survival, and Cox regression is used. The FSP at the $\alpha = 0.05$ significance level selects an FP1(-1) function as the best model. The resulting curve is shown in the upper left-hand panel of Figure 5.11 The upper right-hand panel, showing the smoothed martingale residuals from the FP1 model, suggests a systematic lack of fit. The function is overestimated at low values of haem and underestimated at high values. The lower left-hand panel shows the estimated log relative hazard from a logistic growth-curve model (Richards, 1959),

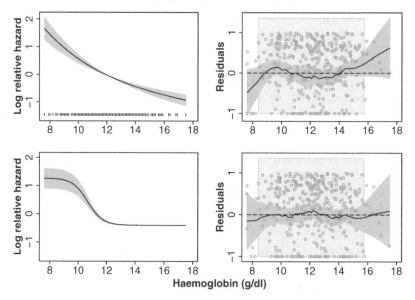

Figure 5.11   Kidney cancer data. Upper panels: FP1 model fit and smoothed martingale residuals versus haem; lower panels: logistic growth curve model and smoothed martingale residuals. Short vertical lines indicate the positions of observations of haem. Raw residuals (circles) have been truncated at -1 for clarity.

$\beta_1[1 + \exp(\beta_2 + \beta_3 \text{haem})]^{-1}$, fitted by maximum partial likelihood. The growth-curve model was chosen purely heuristically because it is a better fit to the data than FP1. Judging by the absence of a pattern among the smoothed martingale residuals, the model is indeed an excellent fit. The deviance of the growth-curve model is some 12 lower than the FP1. An FP2 model improves the fit, but not substantially (deviance difference 4.1 for FP2 versus FP1, $P = 0.13$).

Medically, the growth-curve model may make more sense, at least for higher values. The fitted curve suggests that for values of haem above about $12 \, \text{g} \, \text{dl}^{-1}$ the risk is constant, whereas the FP1 model indicates a diminishing risk as haem increases, which is not very likely. Very high values (above $\sim 17 \, \text{g} \, \text{dl}^{-1}$) are abnormal and could even increase the risk, although there is little evidence of that in the data. Values below about $12 \, \text{g} \, \text{dl}^{-1}$ are also abnormal; the growth-curve estimate nicely shows an increased risk below $12 \, \text{g} \, \text{dl}^{-1}$.

5.9 DISCUSSION

Although we have not illustrated it directly here, Royston and Sauerbrei (2007a) showed that the *leverage* of extreme observations after FP1 transformation was generally brought under control by using the robustness pre-transformation $g_\delta(x)$ (see Section 5.5). The primary aims of the pre-transformation are to provide a formal approach to the origin problem, and to reduce the influence of extreme observations on the selected model and its parameter estimates without removing any observations from the dataset. Depending on the distribution of x, the pre-transformation is usually not required, allowing one to work with the original data and to use the FP function selection procedure directly with x. For 'well-behaved' data this is our preferred approach, and one with which we have much experience.

In practical data analysis, there are no rigid criteria as to what constitutes 'heavy skewness', 'high leverage' or 'an influential covariate observation'. Detailed investigation of such properties by data analysts is certainly the exception rather than the rule. Therefore, we propose using the pre-transformation as a route to a systematic sensitivity analysis of the chosen model, as already mentioned. We suggest that a sensitivity analysis should include a graphical comparison of the fitted functions of x before and after pre-transformation, with the pseudo-truncation points $\bar{x} \pm 2.8s$ marked on the plot (e.g. see Figure 5.7). The graph shows how similar or different the functions are and how far into the tails of the distribution of x the pseudo-truncation points fall. It is also reasonable to compare the likelihoods of the two models. If desired, a formal comparison of likelihoods may be made by applying nonnested hypothesis tests (e.g. Royston and Thompson, 1995).

As always, the preferred model must take into account subject-matter knowledge. Implausible functions of continuous covariates should be rejected unless the evidence from the data favouring them is very strong. Finally, it may be wise not to rely on estimates of functions in the regions beyond the pseudo-truncation points.

We also saw that the negative exponential pre-transformation can greatly improve the fit in some cases. We see this approach as just another item in the FP toolkit, but one that should be used only when definitely needed. For example, when according to subject-matter knowledge the relationship should be modelled as monotonic with an asymptote, an FP1 model with a negative power may fit satisfactorily, but if not, the negative exponential transformation may be useful. This was the case for the variable nodes in the breast cancer dataset analysed by Sauerbrei and Royston (1999) (see also Section 6.5.4).

The two examples we have given of higher order FPs (Section 5.7) are by no means to be taken as typical of the sorts of function that are commonly encountered in applications in the health sciences. In fact, we have only come across about four such datasets over a period of more than a decade. We have included these examples for completeness, to demonstrate that FP3 or even FP4 functions may be useful on rare occasions.

CHAPTER 6

MFP: Multivariable Model-Building with Fractional Polynomials

Summary

1. Categorization or an unchecked assumption of linearity is a poor strategy for assessing the effects of continuous predictors in a multivariable context.
2. Combining BE with the FP function selection procedure defines the MFP, our preferred method of building a multivariable model when some predictors are continuous.
3. The complexity of the MFP model is controlled by two tuning parameters, α_1 and α_2, which are the nominal P-values for selecting a variable with BE and for determining an FP function using the function selection procedure respectively. Different α values may be chosen for selecting variables of primary interest and confounders.
4. An MFP model, once derived, must be critically assessed for lack of fit and other peculiarities. Function plots and graphical analysis of residuals are powerful tools. Since FPs are global functions, techniques are described to check for overlooked local features in a sensitivity analysis.
5. Techniques are available to detect and overcome the effects of influential observations and outliers on the chosen functions.
6. With adequate sample size, the MFP is shown to produce interpretable and transportable models in several studies with continuous predictors.

6.1 INTRODUCTION

The present chapter is the heart of the book. Here, FP modelling of continuous predictors (see Chapters 4 and 5) is extended by combining selection of variables with the determination of suitable functional forms from the FP class. In many studies, the aim is to derive an interpretable multivariable model which gets the 'big picture' right: the stronger predictors are included, and plausible functional forms are found for continuous variables. To build

Multivariable Model-Building   Patrick Royston, Willi Sauerbrei
© 2008 John Wiley & Sons, Ltd

such models, a systematic search for possible nonlinearity, provided by the function selection procedure (FSP), is added to BE. The extension is feasible with any type of regression model to which BE is applicable. Sauerbrei and Royston (1999) called it the MFP procedure, or simply MFP. MFP requires only general knowledge about building regression models. Detailed 'tuning' to select a 'good' model is not required. Checks of the selected model are discussed.

The nominal significance level is the main tuning parameter required by MFP. Actually, two significance levels are needed: α_1 for selecting variables with BE, and α_2 for comparing the fit of functions within the FSP. The notation MFP(α_1, α_2) denotes a model selected using α_1 for BE and α_2 for the FSP, with BE(α_1) and FSP(α_2) representing the individual aspects. The common case $\alpha_1 = \alpha_2 = \alpha$ is abbreviated to MFP(α).

In this chapter, particular points and features are illustrated through analyses of several example datasets. The prostate study is used in the primary presentation of MFP. Further studies are used to present other issues in MFP modelling. An important aspect that we try to convey is how to determine a useful model with MFP.

6.2   MOTIVATION

To motivate the simultaneous selection of variables and functional form, three models are derived to predict log(psa) in the prostate data. Table 6.1 gives parameter estimates and P-values for the following models: full linear, BE(0.05) linear, and MFP(0.05). The variables are sorted by P-values in the full linear model (most significant first), because this determines the order in which variables are considered by MFP. All variables except svi are continuous. Three of the seven variables in the full model have $P > 0.05$ and are

Table 6.1   Prostate data. Regression coefficients and P-values for three selected models: full (full linear model), BE(0.05) (linear model selected by backward elimination at 0.05 level), and MFP(0.05) (MFP model with variables and functions selected at the 0.05 level).

Variable	Model selection method								
	Full			BE(0.05)			MFP(0.05)		
	$\widehat{\beta}$	SE	P	$\widehat{\beta}$	SE	P	$\widehat{\beta}$	SE	P
cavol	0.075	0.014	< 0.001	0.063	0.012	< 0.001			
log cavol							0.54	0.07	< 0.001
svi	0.75	0.27	0.007	0.58	0.25	0.020	0.68	0.21	0.002
pgg45	0.0078	0.0034	0.022	0.0067	0.0031	0.035	–		–
weight	0.012	0.005	0.024	0.016	0.004	< 0.001	0.014	0.004	< 0.001
bph	0.058	0.034	0.094	–		–	–		–
cp	0.039	0.034	0.26	–		–	–		–
age	−0.0076	0.0120	0.53	–		–	–		–
Intercept	1.52	0.72	–	1.06	0.20	–	1.02	0.18	–
R^2		0.60			0.58			0.63	

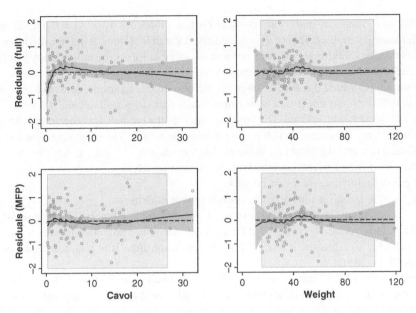

Figure 6.1   Prostate data. Residual plots comparing the fit of the full linear and MFP models to log(psa). Upper panels: full model; lower panels: MFP model. Residuals and smoothed residuals with 95% pointwise CIs are plotted against cavol (left) and weight (right).

all excluded by BE(0.05). The full and BE(0.05) models explain similar amounts of variation R^2. The parameter estimates for variables in the BE(0.05) model are similar to those in the full model. The largest difference is for weight whose estimated regression coefficient increases and whose P-value is reduced, because bph, cp and age are eliminated from the model. Spearman rank-correlation coefficients of weight with bph, cp and age are 0.49, 0.18 and 0.40 respectively, suggesting that confounding may occur with these variables.

MFP selects a model with only three variables, but the fit is improved because log(cavol) is a better fit than untransformed cavol. R^2 is increased to 0.63 and the better fit is visible in Figure 6.1.

The pattern of smoothed residuals in Figure 6.1 clearly indicates that cavol is mismodelled in the full model, whereas residuals from the MFP model appear random. The residual plots support a linear relationship for weight in both models. Linearity was assumed in the full model and MFP found no strong evidence against it. Residual plots of other variables in the model (not shown) gave no indication of mismodelling.

6.3   THE MFP ALGORITHM

Here, we precisely describe the steps of the MFP(α_1, α_2) procedure and exemplify the algorithm with the prostate data. All significance tests are from likelihood ratio (χ^2) statistics. Output from Stata's mfp command is used to clarify the steps.

1. Nominal P-values α_1 and α_2 are chosen. Typical values are $\alpha_1 = \alpha_2 = 0.05$. Values may differ among variables. Taking $\alpha_1 = 1$ for a given variable 'forces' it into the model (no variable selection). Taking $\alpha_2 = 1$ for a continuous variable forces the most complex permitted FP function to be fitted for it (no function selection).
2. Maximum permitted d.f. for FP functions are chosen; for example, four, two and one d.f. mean FP2, FP1 and linear functions respectively. Our suggested default is four d.f.
3. The full linear model is fitted. The 'visiting order' of the predictors is determined according to the P-value for omitting each predictor from the model. The most significant predictor is visited first and the least last. Assume that the variables $x_1, \ldots, x_k$ have been arranged in this order, which is retained in all cycles of the procedure.
4. Let $c = 0$, to initialize the cycle counter.
5. Let $j = 1$, to initialize the variables counter within each cycle.
6. If x_j is continuous, go to step 7. Otherwise, x_j is categorical or binary. The joint significance of its dummy variable(s) is tested at the α_1 level. All other variables currently in the model are included as adjustment terms. If x_j is significant, then it is retained; otherwise it is dropped. Go to step 8.
7. Step 1 of the FSP is applied to x_j at the $\alpha = \alpha_1$ level (see Section 4.10.2). If x_j is not significant it is dropped. Otherwise, steps 2 and 3 of the FSP are applied at the $\alpha = \alpha_2$ level, to choose an FP or linear function. All other variables currently in the model are included as adjustment terms. In the case of an FP with powers $\mathbf{p}$ being chosen, x_j is represented by transformed variables $x_j^{\mathbf{p}}$ in subsequent steps in which other variables are considered.
8. Including or dropping x_j applies until x_j is reconsidered in the next cycle.
9. Let $j = j + 1$. If $j \leq k$, return to step 6 to process the next predictor. Otherwise, continue to step 10.
10. Let $c = c + 1$. The cth cycle is complete. If $\dot{c} > c_{max}$, stop (a practical value is $c_{max} = 5$). Report that the algorithm has failed to converge in c_{max} cycles. Otherwise, check whether included variables and FP transformations have changed from cycle $c - 1$ to cycle c. If so, return to step 5 to start a new cycle. If not, stop and report the current model estimates. End of procedure.

6.3.1 Remarks

1. Typically, MFP requires two, three or occasionally four cycles for convergence. Lack of convergence involves oscillation between two or more models, and is extremely rare.
2. In step 6, a joint test of several dummies for a categorical variable is performed. One may prefer to test dummies separately, e.g. for ordinal variables as in the cervical cancer example in Section 3.5.2. Dummy variables are then defined before MFP starts and are treated as separate covariates. We often use this approach in examples.

6.3.2 Example

The first cycle of MFP with the prostate data is shown in Box 6.1. The order of the variables is determined by the P-values from the full linear model given in Table 6.1. Settings $\alpha_1 = \alpha_2 = 0.05$ were used, with four d.f. (i.e. FP2) for each continuous predictor. Since four

---

Box 6.1   Prostate data. Stata output for first cycle of the MFP algorithm.

```
. mfp regress lpsa age svi pgg45 cavol bph cp weight, select(0.05)
Deviance   for   model   with   all   terms   untransformed = 214.267,97
observations
```

Variable	Model	(vs.)	Deviance	Dev diff.	P	Powers	(vs.)
cavol	null	FP2	240.057	43.782	0.000*	.	-.5 1
	lin.		214.267	17.992	0.001+	1	
	FP1		199.664	3.389	0.215	0	
	Final		199.664			0	
svi	null	lin.	208.646	8.982	0.004*	.	1
	Final		199.664			1	
pgg45	null	FP2	202.042	5.346	0.309	.	-2 -2
	Final		202.042			.	
weight	null	FP2	209.680	10.352	0.052	.	-2 -2
	Final		209.680			.	
bph	null	FP2	217.669	10.010	0.057	.	-1 3
	Final		217.669			.	
cp	null	FP2	217.871	4.647	0.365	.	2 3
	Final		217.871			.	
age	null	FP2	217.877	1.241	0.884	.	-1 -1
	Final		217.877			.	

```
End of Cycle 1: deviance = 217.877
```

---

d.f. is the default, it does not have to be specified in the mfp routine. Box 6.1 gives the shortest version of the mfp command. The outcome is lpsa. The predictors can be given in any order. The command name regress denotes a normal-errors model; logistic or Cox models would be specified by logistic or stcox respectively.

The variable cavol has the strongest effect in the full linear model and so is considered first. Adjusting for all other variables currently in the model, the comparison of FP2 with null shows cavol to be highly significant ($P < 0.001$). The test of FP2 versus linear is also highly significant ($P = 0.001$), whereas FP2 versus FP1 is not significant ($P = 0.215$). Therefore, an FP1 function (with power 0, i.e. a log transformation) is selected for cavol; log(cavol) is included in all subsequent models in cycle 1. In the Stata output, '*' denotes that the test for inclusion of a variable (whether binary, categorical or continuous) is significant at the α_1 level, and '+' that the test of FP2 versus FP1 or FP2 versus linear for a continuous variable is significant at the α_2 level.

The next variable, svi, is binary and is significant ($P = 0.004$) in a model with log(cavol) and the rest of the variables linear. Adjusting for log(cavol), svi and variables considered later in the list, pgg45 is not significant at $P < 0.05$ and is dropped. The same happens with the remaining four variables, two of them (weight and bph) falling just short of the 5% significance level. The model at the end of cycle 1, therefore, comprises only log(cavol) and svi.

Box 6.2   Prostate data. Stata output for second cycle of the MFP algorithm.

Variable	Model	(vs.)	Deviance	Dev diff.	P	Powers	(vs.)
cavol	null	FP2	264.616	49.574	0.000*	.	-.5 1
	lin.		238.091	23.048	0.000+	1	
	FP1		217.877	2.835	0.257	0	
	Final		217.877			0	
svi	null	lin.	226.908	9.031	0.003*	.	1
	Final		217.877			1	
pgg45	null	FP2	217.877	3.597	0.497	.	.5 3
	Final		217.877			.	
weight	null	FP2	217.877	15.749	0.005*	.	-2 -2
	lin.		205.000	2.872	0.439	1	
	Final		205.000			1	
bph	null	FP2	205.000	5.852	0.246	.	.5 .5
	Final		205.000			.	
cp	null	FP2	205.000	4.680	0.362	.	2 3
	Final		205.000			.	
age	null	FP2	205.000	1.821	0.793	.	-.5 0
	Final		205.000			.	

End of Cycle 2: deviance = 205.000

Box 6.2 shows the second cycle. The second cycle starts with the model from the first cycle (log(cavol) and svi). Each variable is considered afresh for inclusion or exclusion. First, cavol is considered, now adjusting only for svi, and an FP1 (log transformation) is again selected. svi is still significant ($P = 0.003$); the P-value is different from the first cycle because five other variables have been eliminated. pgg45 is still dropped, but weight now enters the model ($P = 0.005$) with a linear function. The inclusion of weight in the second cycle is a result of eliminating bph, cp and age in the first cycle, and of the correlation with these variables (see Section 6.2). None of the remaining variables enters. The model at the end of the second cycle is log(cavol), svi and weight. In the third cycle (not shown) no further change in the selected variables or functions occurs, so MFP terminates with the three variables shown in Table 6.1.

The P-values in Table 6.1 are from LRTs for excluding each (possibly transformed) variable and ignore the model-building process. Stata gives the parameter estimates and other details in a separate table after completion of the final cycle.

6.4   PRESENTING THE MODEL

6.4.1   Parameter Estimates

The parameter estimates for the MFP$(0.05, 0.05)$ model for the prostate data are shown in Table 6.1. Binary variables or continuous variables with a linear effect are presented conventionally, showing $\widehat{\beta}$ with its SE or CI and perhaps P-value.

6.4.2   Function Plots

Values of $\widehat{\beta}$ for FP terms are incomplete as a report of results since they tell us little about the fitted function for an x. A *function plot* is more informative. Suppose an x is modelled as an FP with power(s) $\mathbf{p}$ and parameter estimate $\widehat{\beta}$. The *partial predictor* for x is defined as $\widehat{\eta}_x = \widehat{\beta}_0 + x^{\mathbf{p}}\widehat{\beta}$ (see Section 4.3). Its SE (conditional on selection of $\mathbf{p}$) is given by the square root of Equation (4.1). In a multivariable model, the interpretation of $\widehat{\eta}_x$ is an adjusted estimate of the functional form for the effect of x. With the recommended centring of FP functions around the mean $\overline{x}$ of x (see Section 4.11), $\widehat{\eta}_x$ equals the intercept $\widehat{\beta}_0$ at $\overline{x}$. In a model without an intercept (e.g. the Cox model), $\widehat{\eta}_x$ and its SE equal zero at $\overline{x}$.

A *function plot* (if residuals are plotted, sometimes called a *component-plus-residual* plot) for x is a plot of $\widehat{\eta}_x$ plus or minus twice its SE against x, overlaid with the sum of $\widehat{\eta}_x$ and the residual for each individual observation. Function plots for `cavol` and `weight` are shown in Figure 6.2.

In a model with a single predictor, the points would exactly equal the y values. They show the amount of residual variation at each x and may indicate lack of fit and outliers in y. They also serve to show the positions of the covariate observations. There are no obviously disturbing features to note about the model in this example. See Section 6.5.1 for further discussion of the role of function plots in model criticism.

6.4.3   Effect Estimates

Although function plots are helpful, they do not nicely summarize the magnitude of a non-linear effect of a continuous x on the outcome. When x is modelled linearly, its effect is summarized through its regression coefficient $\widehat{\beta}$. The latter can be regarded as the slope of the partial predictor on x, i.e. $d\widehat{\eta}_x/dx$, which is constant. With a non-linear function, the slope changes with x. By focusing on a few relevant values of x, the effect can be summarized through $d\widehat{\eta}_x/dx$ and its CI at those values. Differences in $d\widehat{\eta}_x/dx$ also help to quantify the

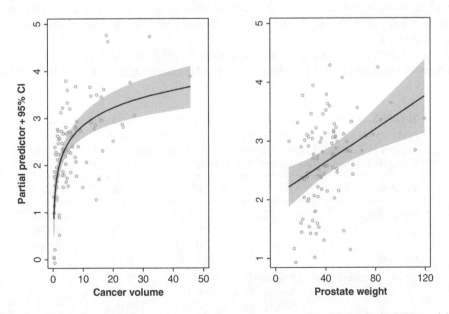

Figure 6.2   Prostate data. Function plots for predictors `cavol` and `weight` in the MFP model.

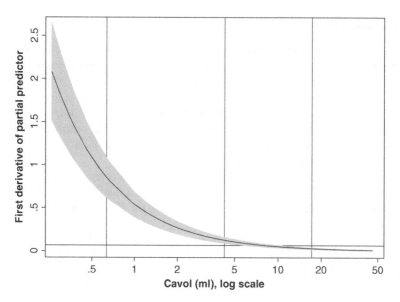

Figure 6.3 Prostate data. First derivative of the partial predictor for `cavol`, with 95% pointwise CI. Vertical lines represent the 10th, 50th and 90th centiles of the distribution of `cavol`. Horizontal line is $\widehat{\beta}$ for `cavol` in a linear model.

extent of non-linearity of the function, also enabling comparisons between different types of model for x.

First derivatives of FP functions are described in Section 4.3.2. Figure 6.3 shows $d\widehat{\eta}_x/dx$ ($=0.54 \times$ `cavol`$^{-1}$) and its 95% pointwise CI of the selected FP1 function of `cavol` in the prostate data. The effect of a 1 ml change in `cavol` on the outcome decreases rapidly up to about 8 ml and is small thereafter. The vertical lines show the 10th, 50th and 90th centiles of `cavol`. The horizontal line shows $\widehat{\beta} = 0.066$ for `cavol` when fitted as linear in a model including `weight` and `svi` (the other variables selected by `mfp`). Comparing this constant with the whole curve for $d\widehat{\eta}_x/dx$ shows how poor an approximation a linear function is (see also Figure 6.2).

Table 6.2 shows a possible table of estimates for the prostate model. The effect of `cavol` has been summarized through $\widehat{\beta}$ for log(`cavol`) and the values of $d\widehat{\eta}_x/dx$ at the 10th, 50th and

Table 6.2 Prostate data. Suggested summary of the MFP model, including effect estimates for `cavol` using first derivatives of the partial predictor. Top part is the selected MFP model. To assist interpretation, the bottom part shows the effect (first derivative) of the function for `cavol` at particular centiles of the distribution.

Variable		Effect	SE	P
log(cavol)		0.54	0.07	< 0.001
svi		0.68	0.21	0.002
weight		0.014	0.004	< 0.001
Intercept		1.02	0.18	–
cavol (centiles)	10th (0.63)	0.85	0.12	
	50th (4.25)	0.13	0.02	
	90th (17.1)	0.032	0.004	

90th centiles. The effect of a change of 1 ml in `cavol` is large for low values (0.63 ml, 10th centile), whereas such a change has hardly any effect at large values (17.1 ml, 90th centile). For FP1 functions, the standardized effect estimates (i.e. effect/SE) for every point on the derivative curve are the same as for log `cavol`. Differences in Table 6.2 are due to rounding.

6.5   MODEL CRITICISM

Having built a model, it is good statistical practice to check it for weaknesses and features that depend strongly on a small number of observations. Such problems include: lack of fit (particularly of functions), observations (e.g. some types of outlier) that are influential on the variables and/or functions selected, and inappropriate behaviour of functions (e.g. artefacts contravening subject-matter knowledge). Many of the issues are not specific to the analysis of continuous functions or to the use of FP methodology. We concentrate on aspects that are particularly relevant to FP models, and that may be investigated using techniques that are not unduly restricted by the type of regression model employed. Plots play an important role; e.g. see Ryan (1997, chapter 5). With adaptations, some of the methods can also be used in GLMs or models for survival data.

We consider FP function plots, goodness of fit of functions through graphical analysis of residuals and more formally by extending the model, identifying observations influential on the choice of function, and incorporating subject-matter knowledge.

6.5.1   Function Plots

The importance of a function plot in identifying outliers and other anomalies in the data is illustrated by a reanalysis of the prostate data. The outlier in `weight` was retained and a model was built using MFP(0.05). The model selected was `svi`, `log(cavol)` and `weight`$^{-0.5}$. Figure 6.4 shows the function plot for `weight`. The large outlier in `weight` is obvious. With

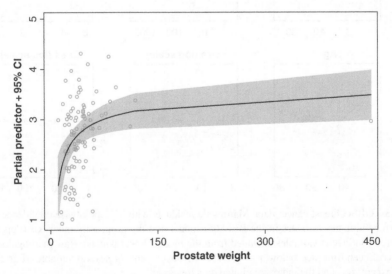

Figure 6.4   Prostate data (with outlier). Function plot for `weight`. Note the way the FP1 function accommodates the outlier.

the observation at 449 g corrected to 44.9 g (see Appendix A.2.5), a straight line is chosen for weight ($P < 0.01$) (see Figure 6.2), but the rest of the MFP model is unchanged. The need for power -0.5 is driven entirely by this single observation.

6.5.2 Graphical Analysis of Residuals

Our suggested approach to plotting residuals was described in Section 1.4.2. As an example, the GBSG breast cancer dataset, first considered in Section 1.1.4, is reanalysed with a focus on the graphical analysis of residuals. A linear Cox regression model selected by BE(0.05) has the four predictors gradd1, nodes, pgr and hormon. Five further predictors are omitted as not significant at $P < 0.05$: age, gradd2, meno, size and er. Smoothed martingale residuals are used to help assess the fit of the Cox model to the continuous predictors. Figure 6.5 shows the results for the two selected continuous covariates and the index (upper panels) and the three excluded ones (lower panels). For legibility, the plots have been truncated vertically to the interval $[-1, +1]$. Lack of fit is seen for nodes, pgr and the index. The apparently nonrandom patterns suggest that age, size and to a lesser extent er, although excluded by BE(0.05), might improve the model fit.

By applying MFP(0.05), Sauerbrei and Royston (1999) found that age was required, with nonlinear functions for age, nodes and pgr. Their model II comprised FP2($-2, -0.5$), FP2($-2, -1$) and FP1(0.5) functions for these three variables respectively. Binary variables

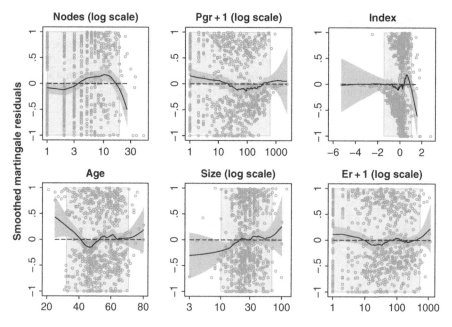

Figure 6.5 GBSG breast cancer data. Martingale residuals with 95% pointwise confidence intervals for the two continuous variables in the BE(0.05) model and the prognostic index $\mathbf{x}\widehat{\beta}$ (upper panels) and the three continuous variables excluded from the model (lower panels). Raw martingale residuals (circles) have been truncated below at -1. Lack of fit is evident for several variables. For clarity, all variables except age and the index are plotted on a log scale.

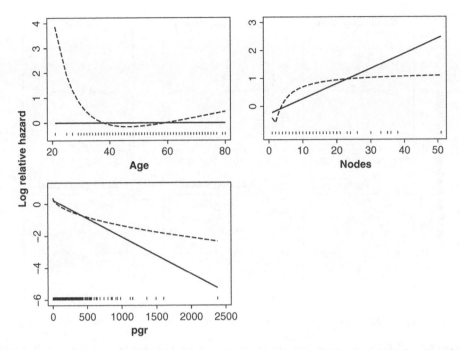

Figure 6.6   GBSG breast cancer data. Plots of linear functions (solid lines) and FP functions from model II (dashes). Short vertical lines indicate positions of observations (but not their frequencies).

gradd1 and hormon were also selected. The fitted linear and FP functions are compared in Figure 6.6. Since age did not enter the linear model, its estimate is zero. There is a considerable difference between the linear and FP functions, confirming that the linear model fits poorly.

Figure 6.7 shows the plot of residuals from model II for all continuous variables, whether selected or not. Lack of fit is much less obvious now. Minor effects at large values of nodes, size and er are driven by isolated values, as is clear from the wide CIs at these points. Some 'signal' may remain for age and size. The smoothed residuals hint at a more complex function for age. Regarding size, Royston and Sauerbrei (2003) noted that the power to detect a nonlinear effect was low. In 5000 bootstrap replications, they found that the data supported three types of model, of which the present model was one type. Another was the present model plus a nonlinear function of size. This example is continued in Section 6.5.3.

6.5.3   Assessing Fit by Adding More Complex Functions

As suggested by Royston (2000), one way to check the goodness of fit of an estimated function for a continuous covariate is to increase the complexity of the model and check whether the additional terms are required. If they are statistically significant, then the original function may be a poor fit and require improvement.

Several examples have been presented in the book showing how smoothed residuals are helpful in examining specific features of a model, such as the adequacy (e.g. Figure 6.7) or inadequacy (e.g. Figure 6.5) of the index or of individual predictors in a multivariable setting. Here, it is shown how such appraisals may be made in a more formal manner. The approach also suggests how the model may be improved when required.

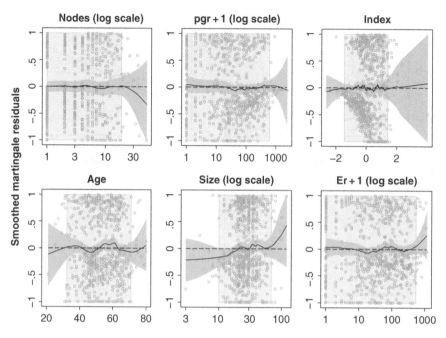

Figure 6.7 GBSG breast cancer data. Martingale residuals for three continuous variables and the index in the MFP(0.05) model II (upper panels and left-hand lower panel), and the two continuous variables excluded from model II (centre and right-hand lower panels). To be compared with Figure 6.5.

One possibility for improving the fit of a particular function is to relax the significance level α_2 of the FSP part of the MFP procedure. However, the resulting function still belongs to the FP class, and this may not be rich enough to represent the data adequately. Another possibility is to add to the selected MFP model (summarized through its index) functions not from the FP class. In a normal-errors model, a search for model improvement would be made by using the residuals. For each continuous predictor, a flexible modelling techniques would be applied to the residuals to check for lack of fit (i.e. non-random patterns). For logistic and Cox models (etc.), the equivalent analysis would involve 'offsetting' the index from the MFP model in a model that included additional terms. The added functions may be tested for statistical significance in the usual way.

Many types of function are available (see Chapter 3 for some examples); but, for simplicity and generality, 'natural' or restricted cubic regression splines are used. See, for example, de Boer (2001) and Harrell (2001, pp. 20–23). Restricted cubic splines are linear beyond boundary knots placed at or near the extremes of the predictor. Provided sufficient degrees of freedom are allowed, this class of functions is flexible enough for present purposes. Furthermore, it is parametric, straightforward to compute and applicable to all model types in which use of FPs is also an option. Interpretability and overfitting of the resulting models are certainly potential problems with this approach; therefore, the technique is presented here as a way of checking the MFP model for serious lack of fit. Comparisons between MFP and spline models are made in Chapter 9; see also Sauerbrei et al. (2007a).

How complex should the extended model be? It is hard to answer that question in absolute terms, but a reasonable compromise is to allow three or four d.f. for the extension of the function. With restricted cubic splines for a single predictor, three internal knots and two boundary knots may be used to achieve four d.f. This gives four basis functions, one of which is linear. If the original function to be extended is linear, the linear basis function is redundant and is not included in the model. Otherwise it is included. Rather than search for optimized knot positions (which raises difficult issues), the internal knots are placed at fixed centiles of the observed distribution of the covariate, and the boundary knots at the observed minimum and maximum. The suggestion of Harrell (2001, table 2.3) to place three knots at the 10th, 50th and 90th centiles of x is adopted. Although in general Harrell recommends more than three knots for modelling a continuous covariate, the smaller number of three is used here, since an existing function, which already accounts for much of the structure in the data, is being extended.

Example 1
Using the BE(0.05) and MFP(0.05) models for the prostate data as an example (see Table 6.1), the adequacy of the postulated functions for cavol, weight and pgg45 is checked. Table 6.3 gives the results of testing the addition of regression spline functions of each continuous predictor when the response variable is the residuals from the respective model. The message is quite clear: cavol is mismodelled in the BE(0.05) model but not in the MFP(0.05) model. Apart possibly from bph, there is no clear evidence of lack of fit for any other covariate in either model. Mismodelling of cavol in the BE(0.05) model may have increased residual confounding for bph and cp.

Figure 6.8 shows the estimated function and pointwise 95% CI for cavol from the spline model, added to the original linear function of cavol. This 'reconstruction' of the estimator gives an approximate idea of what the function may look like. The mismodelling of log(psa) for cancer volumes up to about 10 ml is clearly visible. The 'reconstructed' function resembles

Table 6.3 Prostate data. Spline-based tests of function misspecification for continuous predictors in the BE(0.05) and MFP(0.05) models. The binary variable svi was included in both models. F-tests and P-values for joint testing of the additional spline terms are given. Dashed lines separate predictors that were selected from those not selected by the given model-building procedure.

Variable	Model selection method			
	BE(0.05)		MFP(0.05)	
	F	P	F	P
cavol	5.92	0.001	0.89	0.47
weight	0.64	0.59	0.57	0.64
pgg45	1.47	0.24	0.71	0.55
age	0.13	0.97	0.76	0.55
bph	2.56	0.06	2.18	0.10
cp	2.01	0.12	1.47	0.23

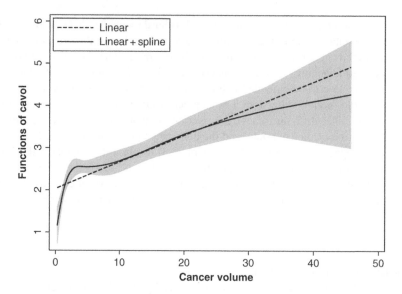

Figure 6.8   Prostate data. Partial predictors for `cavol`. Dashed line, from BE(0.05) model; solid line: linear function plus spline function, with 95% pointwise CI.

the FP functions for this variable seen in Figure 6.2, but exhibits some implausible local fluctuations.

Example 2

The adequacy of the postulated functions for continuous variables in the GBSG breast cancer data is further checked by spline analysis. The question of whether any of the omitted continuous variables should have been included is also addressed. The results are given in Table 6.4.

Table 6.4   GBSG Breast cancer data. Spline-based tests of function misspecification for continuous predictors in the BE(0.05) and MFP(0.05) models. χ^2-tests and P-values for joint testing of the additional spline terms are given. Dashed lines separate predictors that were selected from those not selected by the given model-building procedure.

Variable	Model selection method			
	BE(0.05)		MFP(0.05)	
	χ^2	P	χ^2	P
nodes	31.6	< 0.001	2.5	0.5
pgr	14.0	0.003	2.5	0.5
age	18.6	0.001	1.0	0.9
size	10.8	0.03	9.4	0.052
er	1.7	0.6	2.3	0.5

Adding splines significantly improves the fit of the linear model (BE(0.05)) for nodes, pgr and age. In contrast, MFP(0.05) has already selected nonlinear functions for these variables and the splines do not suggest any improvement is possible. There is some evidence from both the BE(0.05) and MFP(0.05) analyses that size could enter the model, although it appears to be only weakly influential (see also Section 6.5.2).

A point to remember is that the knots chosen for the spline model affect the results of tests for more complex functions. Depending on the knot positions, local features may or may not be incorporated in the fitted functions. We recommend, therefore, that the knot positions should be predefined, as has been done here. Choosing knots in a data-dependent fashion increases the risk of overfitting.

6.5.4   Consistency with Subject-Matter Knowledge

Although in most cases FP modelling generates functional forms for the effects of continuous predictors that are sensible and consistent with subject-matter knowledge and understanding, this is not always the case. A good example of what may happen arises in the GBSG breast cancer study. When MFP(0.05) is used to select a model, the variables age, grad1, nodes, pgr and hormon are chosen (see also Section 1.1.4). The FP2 function φ_2(nodes; $-2, -1$) is determined for nodes, and is plotted in Figure 6.9. The general shape of this function makes good sense, with larger numbers of positive nodes having little additional effect on the risk of an event. However, the minimum at two nodes is unreasonable, since on biological grounds the underlying risk must increase monotonically with nodes. The selected function fits the data but contradicts medical knowledge.

If such a contradictory function is obtained, then it must be rejected and a better one sought, even at the cost of a slightly worse fit to the data. With the GBSG data, Sauerbrei and Royston (1999) noted the resemblance of the FP2 nodes function to a negative exponential. They replaced nodes with the variable enodes $= \exp(-0.12 \times$ nodes) in the MFP analysis. The factor -0.12 was determined univariately by maximum likelihood. The model, including enodes instead of nodes, was reselected by MFP, resulting in a linear function for enodes. This function is shown as a dashed line in Figure 6.9. It agrees quite closely with the FP2 function, but the unsatisfactory hook is eliminated.

In general, the most likely issue with FP models is selecting a nonmonotonic FP2 function when a monotonic function is required. Since all FP1 functions are monotonic, a simple way to prevent this is to specify maximum complexity FP1 for the variable, but the fit may be slightly compromised. A second possibility is pre-transformation of the variable; see Sections 5.5 and 5.6.

6.6   FURTHER TOPICS

6.6.1   Interval Estimation

As discussed in Section 4.9.2, uncertainty in estimating the powers of an FP model in a univariate setting is neglected when calculating CIs for fitted values, e.g. in a function plot. The same is true in a multivariable setting. However, the bootstrap can again be used to obtain

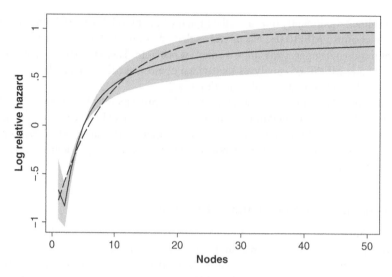

Figure 6.9 GBSG breast cancer data. Solid line: plot of fitted function φ_2(nodes; $-2, -1$) for the number of positive lymph nodes. The 'hook' at two nodes is medically implausible. Dashed line: fitted function from negative exponentially transformed variable, enodes.

more realistic SEs. Rather than considering the most complicated case of selecting an MFP model for all variables in each bootstrap replication, we limit the investigation to uncertainty in functions of selected variables only. The following approach, therefore, does not take into account all the uncertainty arising from the model-building process.

A large number B of bootstrap samples is drawn. In each sample, the MFP($1, \alpha_2$) procedure is applied to the variables selected by MFP(α_1, α_2). Different powers may be selected in different bootstrap samples. The partial predictor for each continuous x is calculated, always adjusted to the same value of x (typically the sample mean). A pointwise CI is finally calculated using the original estimate, $\widehat{\eta}_x = \widehat{\beta}_0 + x^p \widehat{\beta}$, plus or minus t (e.g. 1.96) times the bootstrap SE. The latter is the SD of the B function estimates for each individual in the dataset. The same approach may be used to get a bootstrap estimate of uncertainty in the overall index $\widehat{\eta}$.

Figure 6.10 shows an example using the prostate data. The selected model is log cavol, weight and svi. Since maximum complexity FP2 was allowed for the continuous predictors (cavol and weight), FP2 models are again permitted for them in the bootstrap analysis. The bootstrap CIs are similar to those for the selected model over most of the data, but much greater uncertainty is seen in the upper extreme (for cavol) or lower extreme (for weight) of the distribution. This corresponds to the selection of FP1 or FP2 models in a proportion of bootstrap samples.

6.6.2 Importance of the Nominal Significance Level

In Chapter 2, the importance of the nominal significance level in model building was discussed. Depending on the aim of a study, different significance levels may be chosen: α_1 for BE and α_2 in the FSP. With larger significance levels, variables with weaker effects are included, and

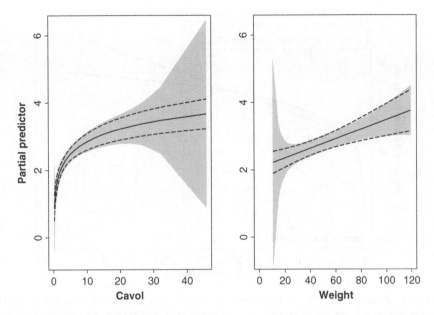

Figure 6.10 Prostate data. Bootstrapped function plots ($B = 200$ replications). Continuous lines show partial predictors for the model selected in the original data, and dashed lines 95% pointwise CIs ignoring uncertainty due to selection of FP functions. The greater uncertainty in the bootstrapped 95% pointwise CIs is shown by the grey bands.

the models become more complex and also more unstable (Royston and Sauerbrei, 2003). Nevertheless, MFP usually selects similar functions for strong factors, although the powers chosen may vary.

On reanalysing the prostate data with MFP, but using a nominal significance level of 0.25 for both BE and FSP, an FP1 function of bph with power -2 is added to the variables selected by MFP(0.05). Furthermore, MFP(0.25) selects the FP2 powers $(-1, 1)$ for the strongest factor, cavol. As illustrated by Figure 6.11, the function for cavol closely resembles the log function derived with MFP(0.05). Major differences are seen only for large cancer volumes where the information is sparse.

R^2 is increased slightly from 0.63 for MFP(0.05) to 0.66 for the more complex MFP(0.25) model. This analysis illustrates the effect of different significance levels. Inclusion or exclusion of bph is the principal difference between the two models. Preference between the models may depend on whether the function selected for bph is plausible or not. However, from a subject-matter viewpoint, the small sample size seriously limits the conclusions that can reasonably be drawn from the data.

6.6.3   The Full MFP Model

With concern for a potentially nonlinear effect of any continuous predictor, the full MFP model introduced by Royston and Sauerbrei (2003) is the natural extension of the full linear model. In the full MFP model, written MFP(1, 1) or MFP(1), all continuous variables are included with their best-fitting FP2 functions, and other binary, categorical and ordinal variables are included as dummy variables. No variable selection and no function selection are done; only

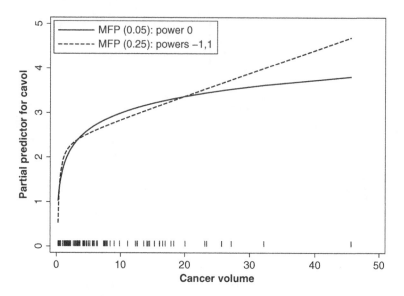

Figure 6.11   Prostate data. The estimated effect of `cavol` on log(psa) in MFP models selected with nominal significance levels of 0.05 and 0.25. Short vertical lines show the positions (but not frequencies) of values of `cavol`.

the best of 36 FP2 transformations are chosen for each continuous variable. In the prostate data, with one binary and six continuous variables, the selection is made from a maximum of $36^6 \simeq 2 \times 10^9$ different models – the binary variable `svi` is always included. Of course, because BE is used, the MFP algorithm 'sees' only a tiny fraction of these possible models.

Figure 6.12 shows the functions selected in the full MFP model for the prostate data. Another power transformation is chosen for `cavol`, but for this, the strongest predictor, the function is similar to the ones presented in Figure 6.11. Some of the other functions are barely interpretable, so in this respect the full MFP model is quite useless and selection of variables and functions is required. R^2 is increased from 0.633 for the MFP(0.05) model to 0.721 for the full MFP model, but the increase is likely to be due to substantial overfitting.

6.6.4   A Single Predictor of Interest

In many analyses, one variable is of primary interest; other variables may act as confounders. In epidemiology, a variable may be investigated as a potential new risk factor in addition to those already known. In a prognostic study, the predictive ability of a new variable, taking 'standard' factors into account, may be of concern. Interest here lies in the potential effect of a new factor, adjusting for other variables. The model-building procedure may be modified to allow for this situation. A common approach is to build a confounder model by a mixture of *a priori* knowledge and more or less formal variable categorization and selection. The effect of the new factor is then assessed in a model adjusting for the confounders. When the factor is continuous, this is done by categorizing the variable, or assuming a linear effect, or modelling its effect using splines (e.g. Hastie and Tibshirani, 1990; Rosenberg et al., 2003) or fractional polynomials (e.g. Royston et al., 1999), or by some other method.

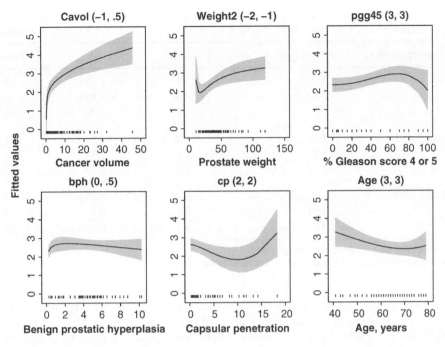

Figure 6.12   Prostate data. Functions selected in the full MFP model (MFP(1)), with 95% pointwise CI for each partial predictor. Short vertical lines show the positions (but not frequencies) of the predictor values.

If subject-matter knowledge is limited, then decisions on the selection and modelling of confounders and the main factor of interest are required. Within the MFP framework, Royston et al. (1999) suggested using certain options of the MFP procedure to build the confounder model and to model the effect of the main variable. The first key element is the choice of nominal significance levels α_1 and α_2, and the second is the development of the confounder model without using the main variable. The aim of a good confounder model is the reduction of residual confounding. Interpretation of the effects of variables in the confounder model is less important and a more complicated model is acceptable. This can be done by choosing relatively large values of α_1 and/or α_2, e.g. 0.20 (Dales and Ury, 1978; Greenland, 1989). Choosing the confounder model independently of the variable of interest ensures that the incremental effect of the variable is estimated, and it reduces the selection bias.

In the following, such an analysis is illustrated with the prostate data. The main aim of the study is to assess whether cp has any additional value in predicting log(psa). This question is considered only to illustrate our approach and the importance of the confounder model. However, such investigations are the main aim in many studies.

The first step is to select the confounder model. Initially assuming linear effects, a confounder model C_{lin} was selected using BE(0.20) for all candidate variables except cp. A second model, C_{mfp}, allowing for possible nonlinear effects of continuous confounders, was chosen using MFP(0.20, 0.05).

Model C_{lin} comprised cavol, pgg45, weight, bph and the binary variable svi. MFP(0.20, 0.05) selected the same confounder model C_{mfp} as MFP(0.05, 0.05) with the three

Table 6.5 Prostate data. Models for cp. With no adjustment for other variables, cp is a highly significant predictor of log(PSA). With adjustment by linear or MFP models, it is no longer significant.

Adjustment model	Model for cp	Dev. diff.[a]	P	Power
None	Linear	29.7	< 0.001	1
	FP1[b]	34.8	< 0.001	0
	FP2	37.4	< 0.001	−0.5, 3
Linear BE(0.2)	Linear	1.4	0.26	1
	FP1	1.4	0.54	1
	FP2	5.5	0.29	−0.5, 0
MFP(0.2, 0.05)	Linear	0.3	0.62	1
	FP1	0.5	0.80	3
	FP2	4.7	0.36	2, 3

[a] Deviance difference compared with model omitting cp.
[b] Function for cp selected with FSP(0.1) (not significant in adjusted cases).

variables log(cavol), svi and weight (see Section 6.2). Table 6.5 gives results for the effect of cp estimated in an unadjusted model and in models adjusted for C_{lin} and C_{mfp}.

Application of FSP(0.10) to cp – a rather liberal significance level may be acceptable for the main effect of interest – results in a log function in the unadjusted model. According to the deviance difference, the FP2 function with powers (−0.5, 3) is highly significant compared with the model excluding cp. This FP2 function fits significantly better at the 10% level than a linear function (deviance difference is $37.4 − 29.7 = 7.7$; $P = 0.059$ for $F_{3,92}$), but is little better than log, the best FP1 function.

In the two adjusted cases, the best FP2 functions hardly improve the fit of the model. In a variable selection situation, cp would be eliminated (see also the result from MFP in Section 6.3). Here, with a specific interest in this variable, a linear function would be postulated. The P-value of 0.62 from the model adjusted by C_{mfp} clearly indicates that there is no additional effect of cp. The corresponding P-value from model adjusted by C_{lin} is 0.26. As shown in Figure 6.13, the effect of cp is negligible when adjusted for C_{mfp} and stronger when adjusted for C_{lin}. The stronger effect in the latter model is caused by the strong correlation between cp and cavol ($r = 0.66$) and the mismodelling of cavol by assuming a linear effect. However, the main message is that the effect of cp is nonsignificant in both adjusted models, whereas it is highly significant in the unadjusted model.

6.6.5   Contribution of Individual Variables to the Model Fit

Having obtained a multivariable model, it is of interest to ask how much each of the selected variables contributes to the model fit (Schemper, 1993). Statistical significance, the criterion used when choosing a variable, does not guarantee importance. In this section, the contribution of variables to a survival model is assessed by the change in an R^2-like measure called R_D^2. R_D^2 is a monotonic transformation of the D measure of prognostic separation proposed by Royston and Sauerbrei (2004b), itself a refinement of an earlier measure, SEP (Graf et al., 1999). The usual R^2 can be used in normal-errors regression models. Several other R^2-type measures

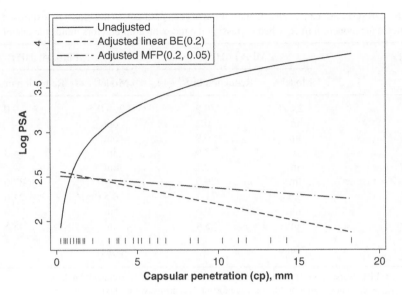

Figure 6.13   Prostate data. The effect of cp on log(psa) with different adjustment models.

have been proposed for GLMs (summarized by Hardin and Hilbe (2007, pp. 59–62)) and censored survival data.

An option is to remove from the model variables that contribute little to the fit. For example, their regression coefficients could be small. Harrell (2001) discussed simplifying a 'full' model by removing predictors that make only a small contribution to $\mathbf{x}\widehat{\beta}$ (e.g. because of small β values). This was done by doing OLS regression of the index $\widehat{\eta} = \mathbf{x}\widehat{\beta}$ on all variables except the x under consideration and inspecting the (conventional) explained variation, R_{η}^2 say. With all variables present, $R_{\eta}^2 = 100\%$ by definition. If x is unimportant, then R_{η}^2 after its removal should still be close to 100%. Harrell (2001) suggested removing all variables that reduced R_{η}^2 by less than a fixed amount, such as 1% or 5%. Such a strategy of model simplification was further studied by Ambler et al. (2002). The magnitude of the reduction in R_{η}^2 is a measure of the importance of x.

As an example, consider the 'full' model for the GBSG breast cancer data obtained with MFP(1, 0.05); that is, all variables are included, but functions are selected at the 0.05 significance level. The model is given in Table 6.6. It is clear that meno, size, gradd2 and er contribute little to the MFP (1, 0.05) model; the reduction in R_{η}^2 is < 1% for each. The major contributions of exp($-0.12 \times$ nodes) and pgr are also apparent. Their contribution to the MFP (0.05, 0.05) model is similar.

If desired, one can also examine what happens to R_D^2 when weak contributors are removed. For example, the MFP (0.05, 0.05) model has $R_D^2 = 0.275$, barely less than 0.278 for MFP (1, 0.05) (see Table 6.6). By retaining only exp($-0.12 \times$ nodes) and φ_1(pgr + 1, 0.5), R_D^2 is reduced to 0.22. Adding φ_1(age; -2, -0.5) increases it to 0.26; thus, although highly statistically significant, age makes a relatively minor contribution to explained variation, as may also be seen in Table 6.6. This may be explained by the distribution of age. Its prognostic effect is mainly in young patients (< 40 years), but most patients are older (91% are $\geq$ 40 years). R_D^2 for exp($-0.12 \times$ nodes) alone is 0.14.

Table 6.6 GBSG breast cancer data. Contribution of each predictor to the model fit, expressed in terms of the percentage reduction in R^2_η when regressing the index on all predictors minus the one of interest.

Predictor	MFP(1, 0.05)		MFP(0.05, 0.05)	
	Model[a]	Reduction in R^2_η (%)	Model[a]	Reduction in R^2_η (%)
age	$-2, -1$	8.9	$-2, -0.5$	9.0
meno	in	0.2	out	–
size	lin	0.9	out	–
gradd1	in	3.9	in	4.3
gradd2	in	0.1	out	–
$\exp(-0.12 \times \text{nodes})$	lin	29.9	lin	37.6
pgr	0.5	21.2	0.5	24.6
er	lin	0.6	out	–
hormon	in	5.4	in	5.5
R^2_D	0.278^b		0.275^b	

[a] Numbers are FP powers; in: binary variable included; out: binary variable excluded; lin: linear.
[b] Value of R^2_D, an explained-variation-like measure for survival analysis – see text.

As a second example, the model selected by BE(0.05) for the glioma data is tumour grade (x_3), age (x_5), Karnofsky index (x_6), type of surgery (x_8) and epilepsy (x_{12}), for which $R^2_D = 0.26$. The percentage reductions in R^2_η by omitting each variable in turn from this model are 21, 28, 4, 17 and 3 respectively. Clearly, x_3, x_5 and x_8 are the major contributors. R^2_D for a model including only x_3 and x_5 is 0.15; adding x_8 gives 0.23. The full model (15 variables) has $R^2_D = 0.30$, which is not appreciably larger than that of the BE(0.05) model with five variables.

6.6.6 Predictive Value of Additional Variables

It may happen that a multivariable model has been established in a given study and new measurements become available. The new data may include variables not investigated before whose additional predictive ability is of interest. The issue is nicely discussed by Kattan (2003). Two new measures applicable to models for binary outcomes are proposed by Pencina et al. (2008). As in Section 6.6.5, we use R^2_D to assess predictive ability.

A good example concerns primary operable breast cancer. A well-known risk score for disease recurrence or death, the Nottingham prognostic index (NPI), was developed some years ago (Haybittle et al., 1982; Galea et al., 1992). The NPI is defined as $0.2 \times$ tumour size (cm) $+$ lymph node stage $+$ tumour grade, where lymph node stage $= 1$ for node negative, 2 for one to three positive lymph nodes, and 3 for four or more nodes. Values of tumour grade are 1, 2 or 3. Higher values of the NPI correspond to a worse prognosis. Because NPI includes tumour size, it is a continuous variable, but generally it is used and presented in categorized form. Categorization reduces its predictive ability.

The NPI has been validated in several independent studies (e.g. Balslev et al., 1994), basically by categorizing the index into three or more groups and plotting Kaplan–Meier recurrence-free survival curves for the groups. Clear separation between the curves is usually seen, denoting definite differences in event rates. The GBSG breast cancer study, which was here first analysed in Section 1.1.4, is restricted to node-positive patients, which reduces the predictive ability. The results presented here are for illustrative purposes only. The study has several variables not featured in the NPI, including nodes (the exact number of lymph nodes), age and pgr. The GBSG variable grade is measured in a different way from the NPI's tumour grade. It is clinically and statistically relevant to ask how much predictive ability (in terms of R_D^2) these extra variables can add to that of the NPI.

To assess the additional prognostic value of nodes (as a continuous variable), age and pgr, the transformations $A = \exp(-0.12 \times \text{nodes})$, $B = \varphi_2(\text{age}; -2, -0.5)$ and $C = \varphi_1(\text{pgr} + 1; 0.5)$ are used. Table 6.7 gives R_D^2 for models including only NPI, then adding all seven combinations of the variables A, B and C to the Cox model. It is clear that all three factors can contribute to improving the predictive ability of the NPI. With A, B and C included, the model R_D^2 rises to 0.27, compared with 0.15 for the NPI alone (as a continuous score).

The third column of Table 6.7 shows what happens if a trichotomized version of NPI is combined with the continuous predictors, instead of taking NPI as continuous. Standard clinical cutpoints of 3.4 and 5.4 for NPI separating low-, medium- and high-risk patients were applied, dividing the patients into unequal groups of size 18 (2.6%), 367 (53.5%) and 201 (43.9%). This unbalanced distribution arises from the inclusion criteria of the GBSG study. Trichotomization reduces R_D^2 for NPI alone from 0.15 to 0.13, reflecting a loss of information due to categorization (see Royston et al. (2006) for further examples). All other values are reduced, but R_D^2 for $+A + B$, $+A + C$ and $+A + B + C$ changes little.

Note that in the analyses just described, the regression coefficient on the NPI has been estimated from the data. Ideally, Cox regression on the NPI should yield a $\widehat{\beta}$ near unity, whereas in fact $\widehat{\beta}$ (SE) is 0.553 (0.064). The value of 0.553 indicates that NPI overestimates the log relative hazard by a factor of about two, meaning that NPI is poorly calibrated in this independent dataset (van Houwelingen, 2000). One reason could be differences in measuring tumour grade.

Finally, Table 6.7 illustrates that R^2-like measures for nonnormal-errors models, such as Cox regression, can behave in a counter-intuitive fashion. In column 2, for example, $R_D^2 = 0.12$ for NPI $+ C$ is slightly smaller than 0.13 for NPI only, even though the model has one extra variable. Unlike in the normal-errors model, adding a variable is not guaranteed to increase R_D^2 (although usually it does).

Table 6.7 GBSG breast cancer data. R_D^2 for models including the NPI (continuous or trichotomized) and combinations of three continuous prognostic factors, A, B and C (for definitions, see text).

Model	NPI	
	Continuous	Trichotomised
NPI only	0.15	0.13
NPI $+ A$	0.19	0.15
NPI $+ A + B$	0.24	0.23
NPI $+ A + C$	0.22	0.20
NPI $+ B$	0.19	0.15
NPI $+ B + C$	0.22	0.17
NPI $+ C$	0.18	0.12
NPI $+ A + B + C$	0.27	0.26

6.7 FURTHER EXAMPLES

6.7.1 Example 1: Oral Cancer

The primary aim of the study was to establish a dose–response relationship in an African-American subpopulation between the risk of oral cancer and intake of ethanol. The main risk factor was drinks, the estimated average number of 1 oz alcoholic drink units ingested per week. The design was a case-control study matched on age and sex. A known confounder also measured was tobacco consumption (cigs, number of cigarettes smoked per day).

Rosenberg et al. (2003) used this study to raise and illustrate issues arising when modelling the effect of the continuous variable drinks on risk in a flexible way using splines and other types of function. Major problems Rosenberg et al. noted were obtaining unambiguous information on the functional form and deriving an interpretable model.

This example belongs to analysis class 4 (see Section 2.4). With class 4, the confounding variables are prespecified and no data-dependent selection of these variables is done. Typically, only the functional form for continuous confounders remains to be determined. Usually, residual confounding can be sufficiently controlled using a step function chosen by categorizing the continuous factors into four or five classes (Becher, 1992; Brenner and Blettner, 1997). Rosenberg et al. (2003) used four categories for age and cigarette smoking.

The drinks risk factor and all three confounders are included in all models without significance testing for selection of variables. There remains the question of the functional form for the continuous variables. Since there is less concern about overfitting the confounder part of the multivariable model, a nominal significance level of $\alpha_2 = 0.20$ is applied for selecting the functional form for age and cigs; that is, FSP(0.20) is applied to these variables within MFP. The selected confounder model comprises the matching factors sex and age, the latter with an FP2 function with powers $(0, 0)$, and a linear function of the confounder cigs.

For drinks we use the conventional level of $\alpha_2 = 0.05$. Using MFP with logistic regression on the case-control status as outcome, an FP2 function for drinks is found to be highly significant ($\chi^2 = 72.6$ on four d.f., $P < 0.001$). FP2 is a signficantly better fit than linear ($\chi^2 = 7.9$ on three d.f., $P = 0.048$), but little better than FP1 ($\chi^2 = 0.8$ on two d.f., $P = 0.7$). An FP1 model, therefore, is selected by MFP. It has power $p = 0.5$, i.e. a square-root transformation of drinks. In fact, to avoid zero values, the default preliminary transformation drinks $+ 0.0625$ is used in the modelling (see Section 4.7). Defining a separate category for nondrinkers would also be possible (see Section 4.15). Note that the test of FP2 versus linear is only just significant.

The left panel of Figure 6.14 shows the fitted partial predictor for drinks, adjusted for confounders. Some nonlinearity is seen in the relationship, with the risk increasing more rapidly (on the log odds scale) at low alcohol consumption than at high values. For up to 100 1 oz drinks the shape of the FP1 curve agrees well with the more complicated and generally less interpretable spline functions given in Rosenberg et al. (2003, figures 1 and 2). For larger values, different functional forms are presented with different spline analyses. The message is of a strong dose–response effect of alcohol consumption on the risk of oral cancer. (Note that, since this is a case-control design, the log odds of caseness cannot be converted directly into a risk estimate due to the oversampling of cases relative to controls.)

The plot of smoothed residuals against drinks (right panel of Figure 6.14) shows no sign of systematic lack of fit.

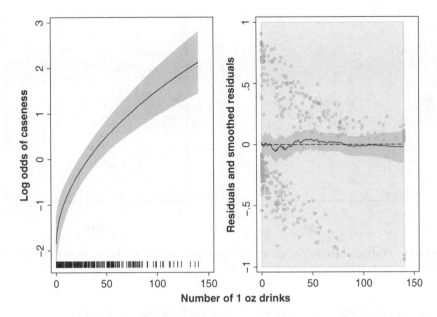

Figure 6.14   Oral cancer data. Left panel: fitted FP1 function for `drinks` with 95% pointwise CI, adjusted for `age`, `sex` and `cigs`. Vertical spikes show the positions of `drinks` values. Right panel: raw and smoothed residuals for caseness with 95% pointwise CI.

We postpone detailed comparisons between spline and FP functions for this dataset until Chapter 9.

6.7.2   Example 2: Diabetes

The diabetes data were provided by Hastie and Tibshirani (1990, p. 6). The observations arise from a study of the factors affecting patterns of insulin-dependent diabetes mellitus in 43 children (Sockett et al., 1987). The aim was to investigate the dependence of serum C-peptide on other factors, better to understand the patterns of residual insulin secretion. The response, `cpep`, is the log of C-peptide concentration at diagnosis, and the selected covariates are `age`, the child's age at diagnosis, and `base`, minus their base deficit. Base deficit is a measure of acidity. These covariates are a subset of those studied by Sockett et al. (1987). The data are investigated here as a relatively simple multivariable modelling problem.

Figure 6.15 shows scatter plots of the response, `cpep`, against each of the two predictors `age` and `base`. Each predictor is clearly associated with `cpep`. Although there is considerable scatter, the relationship between `cpep` and `age` looks nonlinear. The Spearman rank correlation values among the three variables are $r_S(\text{cpep}, \text{age}) = 0.35$, $r_S(\text{cpep}, \text{base}) = -0.55$ and $r_S(\text{age}, \text{base}) = -0.28$.

On applying MFP(0.05), both `age` ($P = 0.005$) and `base` ($P = 0.02$) are significant when FP2 is tested against the null model. However, no FP transformation appears to be required; tests of FP2 against linear have P-values of 0.052 and 0.29 respectively. The explained variation R^2 for the linear model is a modest 0.36.

FP2 might be regarded as unnecessarily complex for such a small dataset ($n = 43$). On reapplying MFP(0.05) with the most complex permitted function FP1 instead of FP2, the

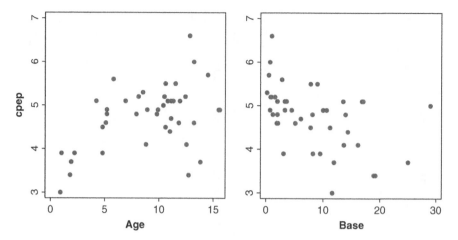

Figure 6.15   Diabetes data. Relationship between cpep and the predictors age and base in 43 children.

P-values for FP1 versus linear are 0.008 for age and 0.095 for base. An FP1 transformation with $p_1 = -1$ is selected for age and a linear function for base. The R^2 increases to 0.47.

There is clearly an issue of low power when testing main effects and nonlinear terms in this small dataset (see Sections 4.16.3 and 6.9.2). A third approach is just to estimate the best-fitting FP1 model with two predictors. This model has powers -1 for age and 0 for base. The R^2 increases slightly to 0.51.

The fit of these three models is studied by smoothing the residuals from each model as a function of the covariates. If the fit is good, then no systematic patterns should be visible. Figure 6.16 shows the results. Curvature is seen for both variables for the residuals from model 1 and additionally for base in model 2. The residuals from model 3 appear randomly scattered, with no obvious trend for either covariate.

Finally, Figure 6.17 shows the fitted FP1 functions for each predictor from model 3. The lower graph shows a good fit to the observed values of cpep.

6.7.3   Example 3: Whitehall I

One of the original aims of this famous epidemiological study was to evaluate the possible effects of social status on health. Social status was reflected in a specific British organization (the Civil Service in London) by several categories of job grade. It was found that, over a 10-year follow-up period, health outcomes, particularly all-cause mortality and death due to coronary heart disease (CHD), were significantly worse in junior job grades than senior job grades (Marmot et al., 1984). To do such analyses, the authors adjusted for established risk factors, including age, serum cholesterol, blood pressure, body weight and height, and cigarette consumption.

The Whitehall I dataset is reanalysed in a rather different way. CHD is the main outcome of interest, but time to death from CHD is considered as the response variable in Cox models. Use of CHD death includes all 2576 such events rather than only the 707 occurring in the first 10 years (for which a complete follow-up was available), greatly increasing the power of

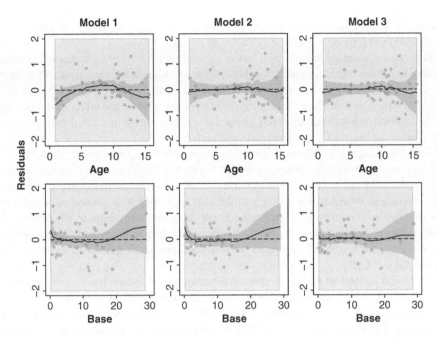

Figure 6.16 Diabetes data. Residuals and smoothed residuals with 95% pointwise CIs. Model 1: linear in age and base; model 2: FP1(-1) for age, linear for base; model 3: FP1(-1) for age, FP1(0) for base.

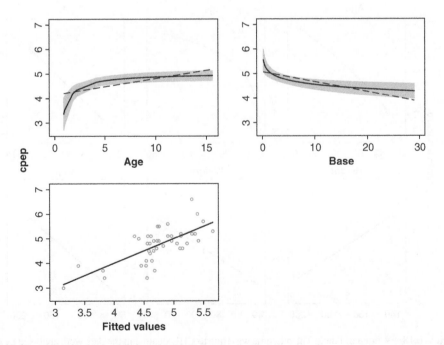

Figure 6.17 Diabetes data. Upper graphs show the partial predictors for age and base, fitted by FP1 functions, together with 95% CIs. Long dashes: linear function. Lower graph shows the observed and fitted values of cpep from the bivariate FP1 model. Diagonal line is $y = x$.

the analysis. The median follow-up time is 26.9 years and the longest is 27.4 years. Marmot et al. (1984) used logistic regression of the event status at 10 years.

Rather than evaluating the effect of job grade adjusted for other factors, each continuous covariate is considered in turn as a possible risk factor, adjusting for other factors in an MFP model. A similar analysis was reported by Royston et al. (1999) using MFP with logistic regression on ten-year all-cause mortality. As is sometimes done, the effect of the two highly correlated blood pressure variables (diasbp and sysbp) are encapsulated through the mean arterial pressure: map $= (2 \times \text{diasbp} + \text{sysbp})/3$.

MFP(0.05) selects the following model: $\varphi_1(\text{age}; -0.5)$, $\varphi_1(\text{cigs}; 0)$, chol, height, weight, map, jobgrade (three dummies). Estimated functions with 95% CIs are shown in Figure 6.18. Of the two variables requiring FP transformation, the effect of age is only mildly nonlinear, whereas that of cigs increases rapidly at low consumptions and much more slowly for heavy smokers. Increases in age, cigarette consumption, cholesterol, body weight and mean arterial pressure are all associated with increasing risk of CHD, whereas the opposite is true for height. These results are in accordance with established knowledge of the epidemiology of CHD.

A final point of interest for the modelling concerns cigs, which has a 'spike at zero' (see Section 4.15). A choice may be made between a model with a dummy variable for nonsmokers and a continuous function of cigs for smokers, or one with a continuous function of cigarette consumption for all individuals. Some 59% of the subjects were nonsmokers, of whom 6.8% died, compared with 13.7% of the smokers. Adding a dummy variable for nonsmokers to

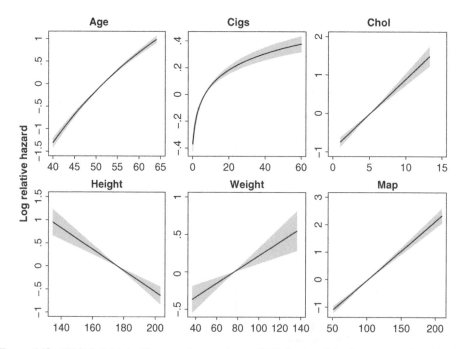

Figure 6.18 Whitehall I data. The outcome was time to CHD death and the data were analysed by Cox proportional hazards modelling. Partial predictor (log relative hazard) with 95% pointwise CI for each continuous predictor in the selected MFP model.

the MFP model does not improve the fit significantly ($P = 0.07$). If the analysis described in Section 4.15 is done, then the selected MFP model includes a dummy variable for nonsmokers and a FP1(-0.5) function for `cigs` $+ 1$. Both components are highly significant ($P < 0.001$). The deviance is 4.4 lower for this model than for the 'pure' FP1(0) model reported above. Overall, the evidence favouring one or the other model is not decisive. The final model should be chosen on subject-matter grounds.

Goodness of Fit

Figure 6.19 shows smoothed martingale residuals for each predictor in the selected MFP model. To make the graph more readable, the plotted range has been restricted by truncating lower confidence limits at -0.2 where necessary. There is no striking lack of fit for any predictor in the main body of the data (shaded boxes).

The fit was further assessed by extending the function for each covariate with splines, as described in Section 6.5.3. For `cigs`, knots were placed at the 10th, 50th and 90th centiles of the positive values. The additional spline terms were not significant at the 5% level for any variable except `map`, for which $P = 0.039$. In view of the large sample size and considerable power available, this suggests that lack of fit of the functions is at most minor. However, it is of interest to compare the spline function for `map` with the selected linear function (see Figure 6.20). The best-fitting FP2 function of `map` has been included in the plot. All estimates are adjusted for the other five variables in the MFP model. In the bulk of the data (75–140 mmHg – approximately 2nd–98th centiles), the three fitted functions are almost identical (see also lower right panel of Figure 6.19). Compared with the linear model, the

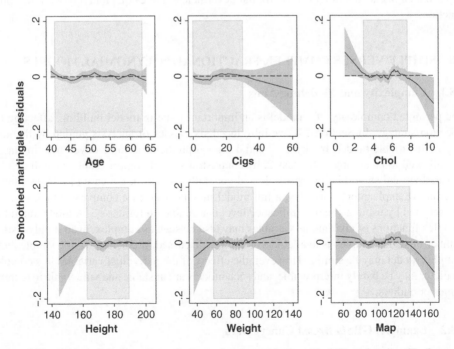

Figure 6.19 Whitehall I data. Smoothed martingale residuals with 95% pointwise CIs for each continuous predictor in the MFP model.

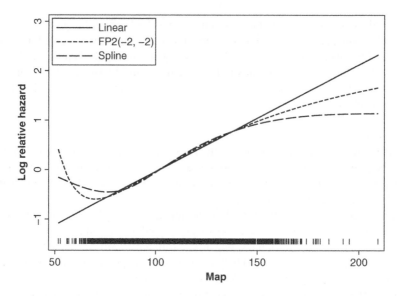

Figure 6.20   Whitehall I data. Three fitted functions for map. Short vertical lines indicate positions (but not frequencies) of observations. See text for further details.

FP2 and spline functions indicate an increased risk of CHD at low blood pressures and a reduction in the upward risk trend at high pressures. The FP2 function does not fit significantly better than a linear function ($P = 0.18$), but nevertheless agrees qualitatively with the spline function.

6.8   SIMPLE VERSUS COMPLEX FRACTIONAL POLYNOMIAL MODELS

6.8.1   Complexity and Modelling Aims

The permitted complexity of a model is an important issue in model building, affecting the selection strategy. In Section 2.4, we introduced six classes of analysis guided by the aims of a multivariable model. In classes 2 and 3 a good predictor is to be developed. In class 3 it is allowed to be complex. In class 2, by contrast, it should comprise only a small number of variables and should be interpretable and generally usable. In the case of a large number of available explanatory variables, a full model may be seen as a complex model, whereas one selected by using a small significance level is a candidate for class 2. A model including complex functions of continuous variables may also be seen as complex, even if only a small number of such variables is available. As already discussed, predictors from a simple and a complex model may be similar. However, the effects of the individual variables in a complex predictor may be barely interpretable, and such models are unstable and vulnerable to extreme values or outliers.

6.8.2   Example: GBSG Breast Cancer Data

We illustrate these issues with the GBSG breast cancer data. Because the number of variables is only eight, the full linear model is not complex. We introduced the full MFP model as a natural extension in Section 6.6.3. It gives a predictor from class 3, with the advantage that model building is limited to the selection of FP2 functions. By contrast, to derive a predictor consisting of a small number of variables and giving interpretable effects, we use MFP(0.05). This would be an analysis belonging to class 2.

With MFP(0.05), a fairly simple Cox model comprising $\varphi_2(\text{age}; -2, -0.5)$, gradd2 (binary dummy for grade 2/3), enodes, $\varphi_1(\text{pgr}; 0.5)$ and hormon (binary dummy for hormonal treatment) is selected. Here, enodes $= \exp(-0.12 \times \text{nodes})$ is a preliminary negative exponential transformation of nodes (see Section 5.6). We compare the resulting predictor η_1 with η_2, that from the full FP model MFP(1) derived using all available covariates, also replacing nodes by enodes. The full FP model consists of $\varphi_2(\text{age}; -2, -0.5)$, meno, $\varphi_2(\text{size}; -1, 3)$, gradd2, gradd3, $\varphi_2(\text{enodes}; -2, 1)$, $\varphi_2(\text{pgr}; 0, 3)$, $\varphi_2(\text{er}; 2, 2)$, and hormon.

The left panel of Figure 6.21 shows that η_1 and η_2 broadly agree except for four outliers defined by $\eta_2 < -4$. When the outliers are excluded from the plot, the agreement is seen more clearly (right panel). The intraclass correlation coefficients (see Section 2.5) are $r_1 = 0.237$ for all the data and $r_1 = 0.958$ if the outliers are excluded. With the outliers excluded the predictors are similar, but the models differ markedly. The variables meno, size, gradd3 and er do not enter the simple model. The chosen functions also differ. The functions from the simple model are interpretable and appear to be consistent with medical knowledge. However, the function for nodes in the complex model is nonmonotonic.

Clearly, a model that generates such extreme predictions as does MFP(1) here is unacceptable. An approach that can improve robustness by reducing the influence of extreme covariate

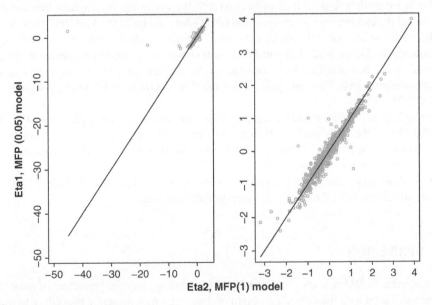

Figure 6.21 GBSG breast cancer data. Comparison of indexes (eta, η) from MFP(0.05) and MFP(1) models. Left panel: all 686 observations; right panel: main body of the data, excluding four outliers with $\eta_2 < -4$.

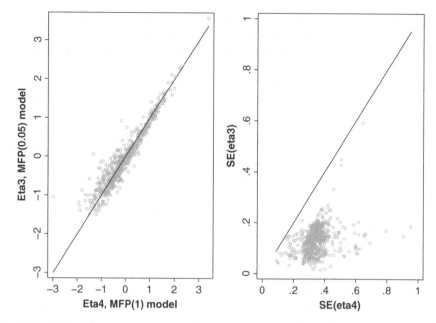

Figure 6.22   GBSG breast cancer data. Comparison of individual predictions from MFP(0.05) and MFP(1) models with all continuous covariates subjected to the preliminary robustness transformation $g_\delta(x)$. Left panel: individual prediction; right panel: estimated SE of predictions.

values in FP models is described in Section 5.5. We apply the preliminary transformation $g_\delta(x)$ to all the continuous covariates in the GBSG data and build the MFP(0.05) and MFP(1) models on these, including the binary covariates as before. Call the resulting predictors η_3 and η_4 respectively. Figure 6.22 (left panel) shows that no obvious outlier is present when using the preliminary transformation and that the simple and complex predictors η_3 and η_4 again agree reasonably well. Pairwise intraclass correlation coefficients between η_1, η_3 and η_4 all exceed 0.95.

The right panel of Figure 6.22 illustrates an additional cost of a complex model. It shows that the SEs of the individual predictions are very much larger for η_4 than η_3. In fact, the median ratio of the SEs is 2.5. Because the model selection process has been ignored, the SE of the smaller model is underestimated.

These illustrations support a preference for simple models; see also Sauerbrei (1999). As already stated, the full FP model is not scientifically sensible.

6.9   DISCUSSION

Often, several variables, some continuous, are available as potential predictors of an outcome of interest. Faced with the difficult problem of deciding which selection procedure to use and which variables to include in the model, practitioners frequently pay only limited attention to investigating whether model improvement is possible by choosing a suitable nonlinear effect for a continuous covariate. Three elements are required to develop a successful multivariable model which combines variable selection with the search for a suitable functional relationship: (1) a comprehensible approach, (2) guidance on how to use it, and (3) a suitable computer program. Nonavailability of any of these is a sufficient reason for practitioners either to assume linearity or to categorize a continuous variable and model its effect as a step function. The weaknesses of both approaches are well known and have also been illustrated in our book.

We believe that MFP can improve modelling in many applications. It can be used in a broad spectrum of regression models. It has power to detect nonlinearity and is easy to understand. We give some recommendations for practice in Section 12.2. Our book, together with several published papers, should give sufficient guidance on model building, and programs in Stata, SAS and R are available. Naturally, a study must be large enough for the exacting task of selecting variables and simultaneously determining functional relationships. To be satisfied with the result from MFP, a practitioner should have a philosophy of multivariable modelling similar to our own, but also should be aware of some weaknesses. In Chapter 9, we informally compare MFP with spline approaches. Philosophical differences become apparent.

6.9.1   Philosophy of MFP

There is a consensus that subject-matter knowledge should guide model building. However, such knowledge is usually limited, necessitating data-dependent modelling. The aim of a study should largely determine the values of tuning parameters used in model building. Experience in real applications, simulation studies and investigations of model stability by bootstrap resampling lead to a preference for a simple model unless the data indicate the need for greater complexity. For a dose–response relationship, the simplest function is a straight line. Because general usability and transferability are important aims for the sorts of models we have in mind, an extremely good fit with several local maxima and minima is not desirable. Such features are probably peculiar to the dataset under study and are not likely to be found in data from similar studies.

Typically, many models are found to fit the data about equally well. They have in common some 'strong' factors and differ with respect to weaker factors. For continuous variables, the linearity assumption is often in reasonable agreement with the data, but sometimes the data indicate the need for a nonlinear function. Any model reflecting the principal features of a dataset, i.e. including strong factors and marked nonlinearity where present, and possessing a sensible interpretation, is in our view a candidate for a 'good' model (see also Section 1.7.2). There are always several good models capable of describing the relationships in the data. Discussion of differences and preferences among them requires subject-matter knowledge, partly guided by a preference for parsimony. In the current context, selecting a good model means to incorporate subject-matter knowledge if available, to include all strong factors and to find sensible functions for continuous predictors. Later, we consider additional features of data that require extended modelling, including treatment–covariate interactions (Chapter 7) and time-varying effects in survival data (Section 11.1).

We believe that MFP is well suited to find good models according to the aims expressed above (see also Section 1.7.3; Royston and Sauerbrei (2005); Sauerbrei et al. (2007a)). It is possible to include at least some subject-matter knowledge. Satisfying the aim of a study may be fulfilled by choosing suitable significance levels for selecting variables, functions or confounders. With high probability, BE includes all strong predictors in a model, and the FSP finds nonlinearity in the data if there is sufficient evidence for it. Otherwise, the linear function, our default, is a straightforward and sensible choice. Furthermore, the principles for selecting an MFP model are easily explained and the procedure is applicable without detailed expert knowledge.

6.9.2   Function Complexity, Sample Size and Subject-Matter Knowledge

As with all selection procedures based on P-value criteria, determination of a realistic FP function requires sufficient power to detect a nonlinear relationship. When the aim is to build a model by considering only the inclusion or exclusion of variables, some recommendations as to the required number of EPVs are available (see Section 2.9.1). For a multi-parameter predictor (e.g. an FP2 function), EPV refers to the number of parameters. In the normal-errors model, the number of events equals the sample size, whereas in logistic regression it is the minimum of the number of events or nonevents, and in Cox models it is the number of uncensored observations.

We are not aware of any investigations of EPVs relevant to the MFP procedure. A sufficient sample size is important, because the first step of our FP selection procedure (see Section 6.3) compares an FP2 model with a model excluding the variable. We illustrated the loss of power by applying a test of on FP2 versus null to a variable with a weak linear effect in Section 4.16.1. The same consideration applies within MFP, as illustrated by the diabetes example in Section 6.7.2. Two questions arise.

First, what can be done to ensure that the power loss is small? The power loss is much reduced if only FP1 functions are permitted. Subject-matter knowledge may indicate that some continuous predictors do not require FP2 modelling, e.g. because the function should be monotonic. The MFP algorithm is flexible with respect to the initial complexity of the FP functions allowed for any continuous variable.

Second, is it of concern if a weakly influential variable is excluded? Such an error hardly matters if the aim is to derive a good predictor utilizing several variables (see Sections 2.7.2 and 9.5.2). Reduced power matters more if the aim is explanatory. However, is it reasonable to expect reliably to identify a weak risk or prognostic factor in a multivariable context, ideally with its correct functional form? An increased sample size or reduced power is the price of obtaining better knowledge of the functional form for the effects of several variables.

6.9.3   Improving Robustness by Preliminary Covariate Transformation

In the breast cancer example, Sauerbrei and Royston (1999) introduced a preliminary negative exponential transformation for nodes because the original function was nonmonotonic with a 'hook' at two nodes, a medically implausible functional form. Such an artefact may occasionally arise with FPs in the tails of the covariate distribution. To improve robustness of FP functions at extreme covariate observations, Royston and Sauerbrei (2007a) proposed the initial covariate transformation $g_\delta(x)$ (see Section 5.5). The transformation maps extreme covariate values smoothly to asymptotes, with the bulk of the observations being transformed almost linearly. The first component $g(x)$ of $g_\delta(x)$ reduces the leverage of extreme observations. The second component, a linear transformation, standardizes $g(x)$ to the interval $[\delta, 1]$

and reduces the leverage of covariate values near zero when such values are subjected to logarithmic or negative-power transformations.

Depending on the distribution of x, one or other of the two components may be used on its own. There is a close relationship between $g_\delta(x)$ and the more familiar truncation approach. However, we believe $g_\delta(x)$ is advantageous, since it smooths the fitted function at the extremes and requires no data-dependent decisions on how many observations to truncate. The transformation $g_\delta(x)$ may also provide a convenient mechanism for including in FP models covariates with zero or negative values. Such covariates are not immediately usable in FP models and require, for example, a shift of origin to ensure positive values (see Section 4.7).

6.9.4   Conclusion and Future

The MFP approach has the potential to improve multivariable model-building in many applications. To understand its properties better, further simulation studies are needed, which may also suggest minor modifications and fine-tuning of the procedure. For example, Section 6.5.3 discusses the use of splines to check for local features of a fitted FP curve. This approach could be developed more systematically. Comparison with alternative procedures is needed. In Chapter 9, procedures called the multivariable regression splines (MVRS; see Royston and Sauerbrei (2007b)) and the multivariable smoothing spline (MVSS; see Hastie and Tibshirani (1990)) are introduced. Both are motivated by the principles of MFP, but different types of spline modelling are used. Some comparisons between models arising from MFP, MVRS and MVSS in several datasets are presented in Chapter 9. Although several other procedures using splines have been proposed, our impression is that there is no consensus on how to build a satisfactory multivariable model incorporating these flexible functions for continuous covariates, and little attention seems to have been paid to the question of variable selection.

Additional experience from other users is needed to assess the strengths and weaknesses of MFP modelling. We believe that published papers and our book provide enough guidance for practitioners to use MFP successfully. The simplicity of the basic ideas and the aim of selecting simple models should help the analyst to select a good model, which includes interpretability from a subject-matter point of view.

Because several models usually fit the data about equally well, the chosen variables and 'best' functions are to some extent a matter of chance. Parameter estimates in this scenario are known to suffer from selection bias. The possibility of correcting the bias by applying some kind of shrinkage needs further consideration.

Stability of the selected model is another important issue and is considered in some detail in Chapter 8. Royston and Sauerbrei (2003) used bootstrap resampling to explore variation among the functions selected by MFP in two studies in breast cancer. The analyses suggested some suitable models for the influence of predictors on the outcome. However, much more research on the topic of model stability is required. So far, bootstrap studies indicate that the selected model can be very unstable and that estimates of the variance of the predictor are too small. Application of the 'model uncertainty' concept may provide more realistic estimates of variance in this situation (see Section 11.3.2).

Improving the robustness and stability is an important aim of the recently proposed preliminary covariate transformation $g_\delta(x)$. It has proven its usefulness in examples, but it is an open issue whether it should be applied routinely to all continuous covariates before building MFP models, or whether its role is more as a sensitivity check following a primary analysis. The distribution of the covariate is relevant here. For positive, approximately normally distributed covariates, the transformation is likely to be of little value and seems not to be required. However, for distributions which are positively skew and/or may include zeros or negative values, the transformation can be helpful. Further experience is needed.

CHAPTER 7

Interactions

Summary

1. An interaction is present when the effect of one predictor on the outcome varies according to the value of another.

2. Interactions may be predefined (e.g. in a clinical trial protocol) or emerge as a result of data analysis (hypothesis generation). Caution should be exercised in assessment and interpretation of the latter type. Critical interactions must be confirmed in independent samples (not discussed here).

3. For continuous predictors, categorization is inadvisable because the results depend on the chosen cutpoint(s). Also, power is lost.

4. Extensions of MFP are available for modelling interactions while adjusting for other variables.

5. Two types of interaction with continuous covariates are considered: binary (with obvious extension to categorical) by continuous (MFPI), and continuous-by-continuous (MFPIgen). The treatment effect plot is effective for displaying the former type graphically.

6. Graphical checks of a hypothesized interaction using categorized continuous variables (about four groups) and sensitivity analyses are essential adjuncts to analysis. If these checks confirm the results of the modelling, then the existence of the interaction is more credible.

7. An alternative approach for examining binary-by-continuous interactions, STEPP, is also available. STEPP involves examining interaction effects graphically in overlapping subpopulations of a continuous covariate. Results from STEPP depend on how the subpopulations are defined.

7.1 INTRODUCTION

In this chapter, binary-by-continuous and continuous-by-continuous interactions are considered. Extensions of MFP are described that allow modelling of the two types of interaction, adjusting for other predictors if desired.

Multivariable Model-Building   Patrick Royston, Willi Sauerbrei
© 2008 John Wiley & Sons, Ltd

7.2 BACKGROUND

So far, only main effects of predictors on the outcome have been considered. If a factor z_2 explains (at least partially) the relationship between factor z_1 and the outcome y, then confounding is present. Confounding was found in many multivariable models in the earlier chapters. Another important issue is interaction between two or more predictors in a multivariable model. An interaction between z_1 and z_2 is present if z_2 modifies the relationship between z_1 and the outcome. That means that the effect of z_1 is different in subgroups determined by z_2. For example, an interaction between treatment and stage of disease is present if treatment B prolongs survival time in early stage disease compared with treatment A, whereas at a later stage the two treatments result in similar survival times. In an epidemiological study, the effect of an exposure on the probability of developing a disease may be different among smokers and nonsmokers. In the latter context, the equivalent term 'effect modification' is often used. In some disciplines (e.g. behavioural sciences) the preferred term is 'moderation'. In a clinical trial, the concern is typically whether the effect of a prognostic factor is homogeneous across the treatments (no interaction) or whether heterogeneity (interaction) is present.

The general concept and consideration of assessment and testing for interactions are described in many textbooks for different types of outcome data (e.g. Rothman and Greenland, 1998; Woodward, 1999; Cohen et al., 2003; DeMaris, 2004; Vittinghof et al., 2005) . This basic material is not repeated here. We focus particularly on two-way interactions involving at least one continuous covariate. Higher order interactions, which typically play a role in factorial experiments, are ignored.

7.3 GENERAL CONSIDERATIONS

7.3.1 Effect of Type of Predictor

If z_1 and z_2 are both binary, then no modelling issue arises. When z_1 is binary and z_2 is continuous, a common type of analysis is to categorize z_2 into a number of groups according to one or more cutpoints and to analyse the interaction with z_1 in a model with these main effects and multiplicative interaction terms. A trend test of the effect of z_1 over the ordered categories from z_2 may be performed, and if a trend is present is likely to have more power than the more general unordered test (Becher 2005). All of this raises several issues for the analyst, including: dependence of the statistical significance of the interaction on the number and position of the cutpoints; the interpretation of the results when an unstable model with too many cutpoints is fitted; and, in the case of a trend test, possible loss of power and faulty interpretation if a nonlinear relationship is incorrectly assumed to be linear, because of a poor choice of scores. Another approach is not to apply categorization but to assume linearity in z_2 at both levels of z_1. However, the assumption of linearity may, of course, be incorrect.

If z_1 has more than two (ordered or unordered) categories, then interactions are often handled pairwise or by combining factor levels to produce a binary variable again. If both z_1 and z_2 are continuous, then linearity is usually assumed for both, and the product $z_1 \times z_2$ is tested. However, the assumption could result in an erroneous model if the main effects and/or the interaction are in fact nonlinear.

In real data, a mixture of types of predictor is found. This mixture is reflected in our examples.

7.3.2   Power

Since studies are almost invariably powered to detect main effects of interest, power is usually low to detect even moderately large interactions. For a simple special case in the epidemiological context, Greenland (1993) concluded from considerations of asymptotic power functions and from simulation studies that the power of such tests is low in common situations. Results from efficiency theory (Lagakos, 1988; Farewell et al., 2004) show that the power of tests of interaction is much improved if a factor is continuous rather than binary. It is clear, therefore, that a large sample size is in most cases a prerequisite for a sensible analysis of interactions.

7.3.3   Randomized Trials and Observational Studies

Because of their relevance to treating patients, interactions in a randomized controlled trial are probably the most important case. Because of the randomization, the covariates are (at least in a large study) independent of the treatment variable z_1 by design, and model building involving treatment, therefore, is less of an issue. When z_2 is binary, the analysis is straightforward. With more than two categories, difficulties arise, and no standard approach exists with continuous variables.

In this chapter, we aim to use all information from a continuous covariate while allowing possible nonlinearity in z_2 at all levels of z_1, possibly adjusting for other influential covariates. To do this, an MFP interaction (MFPI) algorithm, an extension of MFP, was suggested for investigating treatment–covariate interactions (Royston and Sauerbrei, 2004a). Adjustment for other covariates enables the methodology to be used generally in observational studies for a binary variable of interest. For variables with more than two categories, see comments in Section 7.3.1. Of course, weaknesses of observational studies also manifest themselves in interactions.

The chapter focuses mainly on using MFPI to analyse treatment–covariate interactions in trials. However, we also consider continuous-by-continuous interactions, a topic that arises naturally in observational studies. Components of MFPI are modified for use in the latter case.

7.3.4   Predefined Hypothesis or Hypothesis Generation

Sometimes, there is literature evidence that two factors may interact, and this hypothesis is to be tested in the study. Ideally, such a hypothesis is prespecified, e.g. in a clinical protocol. Another (more common) situation is a data-driven search for possible interactions to improve model fit or to generate hypotheses for further research. In principle, an analysis of interactions may be done in either setting. The results, however, are interpreted differently. With a prespecified hypothesis, all that is required is a single test of interaction. In the hypothesis generation case, model building is necessary, with attendant issues such as which potential interactions to consider, which adjustment model to use, possible correction for multiple testing, use of selection strategies to determine a model including interactions, etc.

In designed experiments, it is usual to test for all possible two-way interactions, even if main effects are not significant (Bishop et al., 1975). In contrast, in model building in observational studies, variables with no significant main effect are often disregarded when considering possible interactions. However, it is conceivable that a variable without a main effect could interact with another variable. Identifying such an interaction may be of interest and can be done by extending the search for interactions. Of course, the multiplicity issue becomes more severe and interpretation more difficult. The point is discussed and illustrated in examples.

The multiplicity issue is similar to the usual problem of multiple testing. Researchers must interpret P values carefully, and some may prefer to adjust them using procedures such as Bonferroni–Holm (Holm, 1979). However, use of multiplicity adjustment may rob the procedure of most of its (already low) power. It is then likely that even important interactions are missed. The MFPI algorithm was developed to search in a systematic way for possible heterogeneity of treatment effects for a continuous covariate in a randomized trial (Royston and Sauerbrei, 2004a). To generate hypotheses, the MFPI algorithm may also be used for model building in trials and observational studies, interaction terms being added in a forwards stepwise manner. A low nominal P-value, such as 0.01, may be used to control to some extent for overfitting (generating false positive hypotheses).

7.3.5   Interactions Caused by Mismodelling Main Effects

If the main effects of one or both variables of a continuous-by-continuous interaction are incorrectly modelled, e.g. nonlinearity is ignored, then a spurious interaction may be generated. This is a type of residual confounding (Rothman and Greenland, 1998). For example, in the GBSG study, if the main effects of x3 and x5 are erroneously assumed to be linear, then there is an interaction between them significant at the 0.01 level. If the nonlinearity in the effect of x5 is allowed for, then the interaction disappears ($P = 0.2$). Such an effect could also occur if an important adjustment variable was mismodelled. Therefore, attention should be paid to determining an appropriate main-effects model.

7.3.6   The 'Treatment–Effect' Plot

An informative graphical description of the interaction of (binary) z_1 with (continuous) z_2 is a plot of the estimated difference in response between the levels of z_1, as predicted by a model, together with its 95% pointwise confidence band, against z_2. Let $\widehat{f_0}(z_2)$ and $\widehat{f_1}(z_2)$ be the predicted functions at levels 0 and 1 of z_1 respectively. By analogy with a randomized trial, the 'treatment effect' is defined as $\widehat{t}(z_2) = \widehat{f_1}(z_2) - \widehat{f_0}(z_2)$. A significant interaction may be 'qualitative' if $\widehat{t}(z_2)$ crosses zero, because then not only the magnitude but also the *direction* of the treatment effect depends on z_2. In such cases, a treatment may on average be beneficial or harmful depending on the patient's value of z_2. A quantitative interaction means that the treatment effect changes with z_2, but is in the same direction at all relevant values of z_2. See Section 7.5.1 for a detailed example with a plot.

7.3.7   Graphical Checks, Sensitivity and Stability Analyses

To reduce the chance of overfitting and of incorrectly identifying interactions, it is necessary to check the results of complex model-building procedures. Simple graphical checks can be applied to confirm the inferences from the selected model. Sensitivity analyses may involve changing (even removing) the adjustment model or the type of function used to model the interaction. Influential observations may drive an apparent interaction, and this danger should be investigated. In our experience, continuous predictors with a markedly skew distribution are a fertile source of spurious interactions. Application of the robustness transformation to such predictors may help to curb such effects by reducing the influence of extreme observations.

In the case of continuous-by-continuous interactions, further robustification may be achieved by categorizing both variables into, say, three or four groups and graphing the effect estimates against the subgroup means or medians, adjusted for other variables if necessary. One should try to verify the interaction by inspecting the ordering of the results for consistency with that suggested by the model. Disagreement with expectation is an indicator of an erroneous model.

Another possibility is to check the stability of the treatment–effect function by bootstrap resampling. For example, in the kidney cancer example, the interaction between treatment and white cell count was shown to be reproducible when the treatment–effect function was averaged over 1000 bootstrap samples (Sauerbrei and Royston, 2007).

7.3.8   Cautious Interpretation is Essential

Provided that a small number of hypotheses have been prespecified, interpretation of results is clear. However, when model building is used to generate hypotheses or to improve the fit of a model, several types of bias may appear. Currently, the statistical community has no consensus on how to investigate and interpret interactions in clinical trials and observational studies. Furthermore, since power is known to be low, interactions are often not even considered. Many important interactions may be missed.

For studies with a sufficient sample size, we believe that it is worth trying to identify interactions (at least, strong ones). Because of multiplicity, many spurious interactions may be identified by any modelling approach. Therefore, a sceptical attitude to the results of such exercises is essential. 'Reality checks', several of which we describe, must be applied to confirm that interactions are really present in the original data. Interactions should also be checked for consistency with subject-matter knowledge, where available. Finally, confirmation in independent data is always required.

7.4   THE MFPI PROCEDURE

Royston and Sauerbrei (2004a) proposed an extension of the MFP algorithm to investigate interaction between a categorical and a continuous covariate. For simplicity of explanation, z_1 is described as a binary 'treatment' variable, although the methodology applies equally well to any categorical covariate. The context of prognostic factors in a randomized trial is also assumed, though again the restriction is purely linguistic. As above, let z_2 be a continuous covariate and z_1 be a binary treatment variable, coded $\{0, 1\}$. In clinical settings, a variable that is prognostic and that interacts with treatment is often called 'predictive' (of response to treatment). A given variable may be both prognostic and predictive, prognostic only, predictive only, or neither prognostic nor predictive.

Let x be a vector of potential prognostic factors. With a prespecified hypothesis, z_2 is the only covariate to be investigated for interaction with z_1. For hypothesis generation, z_2 is one of several candidate predictive factors; often, z_2 is part of $\mathbf{x}$. The relationship between the outcome and z_2 is modelled by an FP with the same powers but different regression coefficients at each level of z_1. A standard test of interaction is performed on regression coefficients at the final step. To reduce possible confounding, adjustment for $\mathbf{x}$ may be made. Since z_2 cannot belong to the adjustment model, a different such model may be chosen for each z_2 that is investigated. The complete procedure, allowing adjustment for $\mathbf{x}$, is as follows:

1. Apply the MFP algorithm to $\mathbf{x}$ (possibly including z_2) with a P-value threshold of α^* for selecting variables and FP transformations. Let $\mathbf{x}^*$ be the resulting covariate vector, called the adjustment model. $\mathbf{x}^*$ may include (transformed) variables in $\mathbf{x}$ selected by the MFP algorithm. If all variables in $\mathbf{x}$ are uninfluential, then $\mathbf{x}^*$ may even be empty, i.e. no adjustment for members of $\mathbf{x}$ appears to be needed. In some cases, parts or even all of $\mathbf{x}^*$ may be formulated from subject-matter knowledge, reducing or avoiding data-driven searching.
2. Find by maximum likelihood the best-fitting FP2 powers $\mathbf{p} = (p_1, p_2)$ for z_2 with $p_1, p_2 \in S$, always adjusting for z_1 and $\mathbf{x}^*$. Denote the FP2 transformations $z_2^{\mathbf{p}} = (z_2^{p_1}, z_2^{p_2})$.
3. For groups $j = 0, 1$ and powers p_i for $i = 1, 2$, define new predictors $z_{ji} = z_2^{p_i}$ if $z_1 = j$, and $z_{ji} = 0$ otherwise.
4. The test of $z_1 \times z_2$ interaction is a likelihood ratio test between the nested models z_1, z_{01}, z_{02}, z_{11}, z_{12}, $\mathbf{x}^*$ and z_1, $z_2^{p_1}$, $z_2^{p_2}$, $\mathbf{x}^*$. The difference in deviance is compared with χ^2 on two d.f.
5. If an interaction is not found, then z_2 is regarded as a potential prognostic factor only. To investigate if an FP2 function is still needed for z_2, the final model is chosen by repeating step 1, but including z_2 as a potential prognostic factor.

The reason to fit FP2 functions to z_2, rather than simpler functions, is to find the best-fitting specification from a flexible class. In terms of bias/variance trade-off, increased variance incurred through the use of FP2 powers for z_2 may be tolerated as the price of low bias. To avoid excessive overfitting, leading to serious artefacts in the fitted functions, estimation of different powers in each treatment group is not entertained. An FP2 function with the same powers in each treatment group is already a flexible specification.

When z_2 is binary or categorical, the approach reduces to the usual procedure of estimating and testing multiplicative interaction term(s), adjusting for $\mathbf{x}^*$.

7.4.1 Model Simplification

Despite consideration of the bias/variance trade-off, overfitting of interaction terms and the consequent instability resulting from use of the MFPI algorithm with FP2 functions of z_2 may be a real concern. Instead of FP2 functions, the MFPI algorithm just outlined may be implemented using FP1 functions or, previously the 'standard' choice, using linear functions. Owing to their mode of construction, these three interaction models (with FP2, FP1 or linear functions of z_2) are not nested. Nested hypothesis testing, therefore, cannot be used to select a model from among them. Instead, a model may be chosen to optimize a penalized likelihood criterion, such as minimum AIC or BIC. However, since the BIC penalty is quite harsh, the AIC criterion may be preferable to reduce the chance of underfitting and consequent bias.

7.4.2 Check of the Results and Sensitivity Analysis

If a $z_1 \times z_2$ (discrete by continuous) interaction is detected by MFPI, then the treatment–effect plot (see Section 7.3.6) indicates where the differences should lie. This may be checked graphically (e.g. in survival data, by graphs of Kaplan–Meier survival curves) and by estimating the treatment effect in subgroups. This check introduces an element of subjectivity. However, the value of using Kaplan–Meier plots and estimates of treatment effect in a few subgroups has been demonstrated in examples (Royston and Sauerbrei, 2004a; Royston et al., 2004; Sauerbrei et al., 2007d). To show a trend, more than two subgroups must be used. Often, a sensible choice is four subgroups, with cutpoints indicated by the treatment–effect plot and the distribution of the covariate. If an apparent interaction is found following intensive modelling, then it is unlikely that the results in the independent subgroups show a similar pattern. To reduce the chance of artefacts, no subgroup should be too small. This check helps one to identify obvious type I errors. It is also a simple and, it is hoped, convincing way of presenting the main results from MFPI to researchers. When estimating the treatment effect in subgroups, the adjustment model $\mathbf{x}^*$ can also be incorporated. This is less relevant in randomized trials, but important in all types of observational study.

It may be worth additionally performing a sensitivity analysis of interactions to adjustment models of differing complexity. The simplest way to do this is to vary the nominal P-value for selecting a variable. Possible choices are $P = 1$ (full model except z_1 and z_2, perhaps with some additional transformations of continuous covariates), $P = 0.157$ (approximately equivalent to selection by the AIC), $P = 0.05$ (conventional level) and P small, according to the BIC, which depends on the (effective) sample size. At least in a randomized trial, a reliable interaction should survive such modifications to the adjustment model.

7.5   EXAMPLE 1: ADVANCED PROSTATE CANCER

The first example is a well-known trial in patients with advanced prostate cancer, which has been analysed using Cox regression by Byar and Green (1980) and others; the data may be found in Andrews and Herzberg (1985). The main outcome is time to death (overall survival). Of 506 patients randomized to the four treatments under study, only the 475 patients (338 deaths) with complete data on all covariates are considered for this reanalysis. The treatments consisted of a placebo and three dose levels of the synthetic oestrogen drug diethyl stilboestrol. For reasons given by Byar and Green (1980), the placebo and the lowest dose level of diethyl stilboestrol have been combined to give a placebo arm, and the higher doses to give treatment E. This combined treatment variable is called rx. Prior to the main analyses, the implausible value of 6 observed in a single patient for the variable stage was recoded to 4 (assuming it was a typo), making stage binary (3/4), no other stages being represented; and ekg, a categorical covariate with seven levels, was recoded as 0 (normal), 1 (abnormal). The prognostic factors are listed in the first column of Table 7.1.

The analysis is focused on the identification of treatment–covariate interactions. Byar and Green (1980) modelled the seven prognostic factors age, hg, hx, pf, sg, wt and sz (see Table 7.1), and all seven treatment–covariate interactions. All continuous or multicategory factors were categorized into two or three groups. By contrast, in the present analysis, all continuous factors are kept continuous. Some differences between Byar and Green's (1980) results and those from MFPI would, therefore, be expected.

Since no interactions were predefined, the analysis is an example of hypothesis generation. The MFPI algorithm was applied to all binary and continuous factors using $\alpha^* = 1.0, 0.157$ and 0.05 to select variables, and the default 0.05 level was used to select FP functions. Whether or not the main effects entered the adjustment model, all interactions were tested at the $\alpha = 0.05$ significance level in a model including their main effects. The results are given in Table 7.1.

Table 7.1   Advanced prostate cancer data. Prognostic factors and interactions with treatment. The eight factors in the upper portion of the table are continuous; the remainder are binary.[a]

Prognostic factor	Code	α^* (significance level for adjustment model)					
		1.0		0.157		0.05	
		Main	Int.	Main	Int.	Main	Int.
Continuous							
Age at diagnosis	age[b]	✓	0.018	✓	0.022	✓	0.027
Standardized weight	wt[b]	✓		✓		✓	
Systolic blood pressure	sbp	✓					
Diastolic blood pressure	dbp	✓					
Size of primary tumour	sz[b]	✓		✓		✓	
Serum acid phosphatase	ap	✓	0.044		0.030		(0.064)
Haemoglobin (g/100 ml)	hg[b]	✓					
Gleason stage-grade category	sg[b]	✓		✓			
Binary							
Performance status	pf[b]	✓		✓		✓	
History of cardiovasc. disease	hx[b]	✓		✓		✓	
Presence of bone metastases	bm	✓	0.013	✓	0.008		0.003
Stage 4 vs stage 3	stage	✓				✓	
Abnormal electrocardiogram	ekg	✓		✓		✓	

[a] A tick denotes a variable included in the model. Ticked entries under 'Main' denote prognostic variables selected in the adjustment model by the MFP algorithm using a nominal P-value of α^*. Entries under 'Int.' are P-values for treatment–covariate interactions with $P < 0.05$ according to the MFPI algorithm.
[b] Variable selected by Byar and Green (1980).

The adjustment model selected at $\alpha^* = 0.05$ comprises the binary factors pf, hx, stage and ekg, and the continuous factors age, wt and sz (all except age modelled as linear functions). rx × age and rx × bm interactions are found to be significant at the 0.05 level irrespective of the adjustment model. A significant rx × ap interaction is seen for $\alpha^* = 1.0$, 0.157, but not quite with $\alpha^* = 0.05$.

For $\alpha^* = 0.05$, the effect of age was modelled by FP2 functions with powers (3, 3) in each treatment group (results for other α^* were very similar). Figure 7.1 (left panel) shows that for patients older than 75 years the hazard increases in a similar fashion in both groups, whereas younger patients on treatment E appear to have a lower risk. Figure 7.1 (right panel) is a treatment effect plot by age. The benefit of treatment E seems to be substantial for younger patients but may be lost or even reversed for older ones. The confidence intervals outside the range of (60, 80) years are wide.

7.5.1   The Fitted Model

The linear predictor, $\eta(\mathbf{x}^*, z_1, z_2)$, in the Cox model for the age function ($\widehat{\beta}_1 \text{age}^3 + \widehat{\beta}_2 \text{age}^3 \log \text{age}$) in the advanced prostate data, including an interaction with age selected

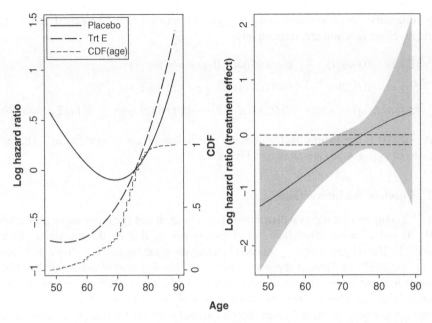

Figure 7.1   Advanced prostate cancer data. rx(treatment) × age interaction. Functions were estimated within a multivariable adjustment model $\mathbf{x}^*$ selected at the $\alpha^* = 0.05$ level and fitted using FP2 functions with powers (3, 3). Left panel: solid line, estimated effect of age in patients on placebo (rx = 0); dashed line, estimated effect of age in patients on treatment E (rx = 1); dotted line and right-hand axis, cumulative distribution function (CDF) of age. Right panel: treatment effect function for age, with 95% pointwise CI. Horizontal dashed lines denote zero and the main effect of rx in a model excluding interaction. (Adapted from Royston and Sauerbrei (2004a) with permission from John Wiley & Sons Ltd.)

at the $\alpha^* = 0.05$ level, is as follows. Let $z_1 = $ rx, $z_2 = $ age/10, $\bar{z}_2 = $ mean(age)/10 $= 7.156$. Then

$$\eta(\mathbf{x}^*, z_1, z_2) = -0.0114(0.0045)\text{wt} + 0.0185(0.0043)\text{sz} + 0.431(0.170)\text{pf}$$
$$+ 0.389(0.136)\text{hx}$$
$$+ 0.239(0.115)\text{stage} + 0.303(0.123)\text{ekg} - 0.180(0.112)\text{rx}$$
$$- 0.0463(0.0288)z_{01} + 0.0204(0.0128)z_{02} - 0.0226(0.0231)z_{11}$$
$$+ 0.0114(0.0100)z_{12}$$

where

$$z_{01} = (z_2^3 - \bar{z}_2^3)I(\text{rx} = 0), \quad z_{02} = (z_2^3 \log z_2 - \bar{z}_2^3 \log \bar{z}_2)I(\text{rx} = 0)$$
$$z_{11} = (z_2^3 - \bar{z}_2^3)I(\text{rx} = 1), \quad z_{12} = (z_2^3 \log z_2 - \bar{z}_2^3 \log \bar{z}_2)I(\text{rx} = 1)$$

and $I(condition) = 1$ if $condition$ is true, 0 otherwise. Standard errors are given in parentheses. Scaling (dividing by 10) and centering has been applied to age (see section 4.11). The subtraction of the constants $\bar{z}_2^3$ and $\bar{z}_2^3 \log \bar{z}_2$ from the FP2 transformations of z_2 centres the FP functions on the mean of age.

The estimated prognostic functions of `age` in treatment (`rx`) groups 0 and 1 and the treatment–effect function are, respectively,

$$\widehat{f_0}(z_2) = -0.0463(z_2^3 - \bar{z}_2^3) + 0.0204(z_2^3 \log z_2 - \bar{z}_2^3 \log \bar{z}_2)$$

$$\widehat{f_1}(z_2) = -0.0226(z_2^3 - \bar{z}_2^3) + 0.0114(z_2^3 \log z_2 - \bar{z}_2^3 \log \bar{z}_2) - 0.180 \qquad (7.1)$$

$$\widehat{t}(z_2) = \widehat{f_1}(z_2) - \widehat{f_0}(z_2) = 0.0237(z_2^3 - \bar{z}_2^3) - 0.0090(z_2^3 \log z_2 - \bar{z}_2^3 \log \bar{z}_2) - 0.180$$

where -0.180 is the main effect regression coefficient of `rx` in $\eta(\mathbf{x}^*, z_1, z_2)$. These three functions of z_2 (i.e. `age`) are plotted in Figure 7.1.

7.5.2   Check of the Interactions

Figure 7.2 compares the survival distributions by treatment and age group using `age` cutpoints of 60, 70 and 75 years. The choice of cutpoints was guided by the treatment effect plot (Figure 7.1). The subgroup `age` $\leq$ 60 years is relatively small but still has 30 events. Figure 7.2 supports the MFPI analysis of the `age` interaction. The log hazard ratios for the effects of treatment in these subgroups, unadjusted and using the $\alpha^* = 0.05$ adjustment model, are shown in Table 7.2. They confirm the trend seen in Figures 7.1 and 7.2.

Since `bm` is binary, plotting Kaplan–Meier curves by `rx` for the two `bm` groups provides a check of the `rx` × `bm` interaction, unadjusted for other factors (see Figure 7.3). There is a large treatment effect (log HR $= -0.89$) in the subset of patients with bone metastases (`bm` $= 1$)

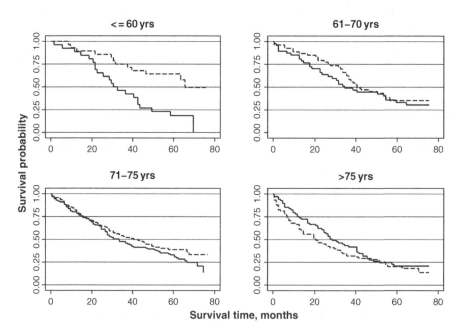

Figure 7.2   Advanced prostate cancer data. Kaplan–Meier survival curves illustrating `rx` × age inter-action with patients divided into four groups by using `age` cutpoints of 60, 70 and 75 years. Solid lines: placebo (`rx` $= 0$); dashed lines: treatment E (`rx` $= 1$). (Adapted from Royston and Sauerbrei (2004a) with permission from John Wiley & Sons Ltd.)

Table 7.2 Advanced prostate cancer data. Treatment effect (log hazard ratio) in age subgroups. 'Adjusted' means adjusting for factors significant at the 0.05 level, as given in Table 7.1.

Subgroup	Age (years)	Patients (%)	Unadjusted		Adjusted	
			$\widehat{\beta}$	SE	$\widehat{\beta}$	SE
1	≤ 60	10	−1.29	0.39	−1.53	0.49
2	61–70	20	−0.15	0.26	−0.34	0.27
3	71–75	43	−0.25	0.17	−0.16	0.18
4	> 75	27	0.27	0.19	0.16	0.21
All patients		100	−0.18	0.11	−0.16	0.11

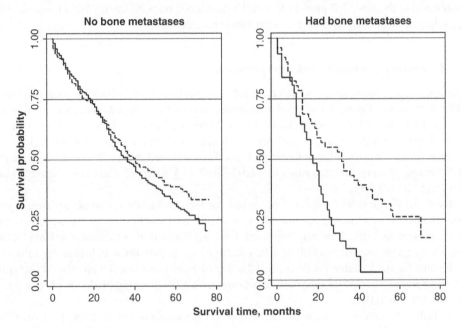

Figure 7.3 Advanced prostate cancer data. Kaplan–Meier curves of overall survival illustrating rx × bm (bone metastases) interaction. Solid lines: placebo (rx = 0); dashed lines: treatment E (rx = 1).

and nearly no treatment effect (HR = −0.16) when bm = 0. The adjusted estimates shown in Table 7.3 and again using $\alpha^* = 0.05$, are similar to the unadjusted.

7.5.3   Final Model

The interactions identified by MFPI appear to be genuine. To determine a final model, a forwards stepwise (FS) method may be used. Here, FS is preferred to BE because BE would begin with a heavily overfitted, unstable model. It is unlikely that several interactions are

Table 7.3 Advanced prostate cancer data. Treatment effect (log hazard ratio) in subgroups by presence or absence of bone metastases. 'Adjusted' means adjusting for factors significant at the 0.05 level, as given in Table 7.1.

Subgroup of bm	Patients (%)	Unadjusted		Adjusted	
		$\widehat{\beta}$	SE	$\widehat{\beta}$	SE
No bone metastases (bm = 0)	84	−0.14	0.12	−0.03	0.12
Had bone metastases (bm = 1)	16	−0.86	0.26	−0.91	0.28

present. Interactions significant at the 0.05 level are added to the main effects model (see Table 7.1). Because it has the smallest P-value (0.003), the rx × bm interaction is added first. After including rx × bm, the rx × age interaction is significant, and both interactions remain significant in the extended model ($P = 0.036$ and 0.005 respectively). With a more rigorous significance level of 0.01, only rx × bm would remain.

7.5.4   Further Comments and Interpretation

The rx × age interaction was also reported as significant at $P = 0.05$ by Byar and Green (1980). However, Figure 7.1 is more informative than analyses based on cutpoints. Byar and Green (1980) defined three age groups ≤ 74, 75–79 and ≥ 80 years and performed conventional tests of interaction. According to the fitted FP2 functions, there is a substantial beneficial effect of treatment E (rx = 1) for younger patients. The estimated treatment effect is −0.72 (SE 0.24) at age 60 years and decreases to −0.03 (SE 0.13) at age 75 years (see Figure 7.1 and $\widehat{t}(z_2)$ in Section 7.5.1).

Byar and Green (1980) reported no results for the continuous covariate ap (serum acid phosphatase), even though this variable was available in the public-domain dataset. According to the present analysis, there may be a weak rx × ap interaction. A treatment–effect plot (not shown) suggests that treatment E may be effective only in patients with higher ap values.

Results for the variable bm (presence/absence of bone metastases) were also not reported by Byar and Green (1980), despite it appearing to be a fairly strong predictor of response to treatment ($P < 0.01$).

Note that all variables were investigated for possible interaction with rx, irrespective of their prognostic significance. The more common practice is to consider only prognostic variables, in which case the interaction with bm would not have been found at the $\alpha^* = 0.05$ level. The same is true for ap at the $\alpha^* = 0.157$ level.

Note also that the significance of the interaction with ap depends on the adjustment model. The interaction may, therefore, be a weak effect and may be disregarded. A final decision would depend on further checking and on clinical reasoning.

None of these apparently predictive factors would be significant following P-value adjustment by the Bonferroni–Holm method.

An alternative approach to analysing the same dataset was presented by Harrell (2001, p. 516). He applied a global test of all interactions between rx and the other variables, and found $X^2 = 12.2$ on 10 d.f. ($P = 0.27$). He states 'so we ignore the interactions'. For clinical application, it is obviously essential to know which, if any, interactions are present. A global test is unlikely to be helpful, and lacks power.

Table 7.4 Advanced prostate cancer data. age × rx interaction model selection according to AIC and BIC, based on partial likelihood from Cox models.

Criterion	Model class		
	FP2	FP1	Linear
AIC	3720.66	3718.26	3718.05
BIC	3747.42	3733.55	3729.52

7.5.5   FP Model Simplification

The dataset was initially analysed using FP2 functions for continuous covariates when estimating interactions with treatment. As discussed in Section 7.4.1, the chance of overfitting the data may be reduced by selecting a simpler FP function, if appropriate. For the interaction between rx and age, Table 7.4 shows the AIC and BIC statistics for FP2, FP1 and linear models, including an adjustment model selected with $\alpha^* = 0.05$. The BIC was calculated using the effective sample size, taken as the number of events (Volinsky and Raftery, 2000). Since both the AIC and the BIC are smallest for the linear model, each criterion would select this model. Also, both criteria suggest that the FP2 model is overfitted. According to the AIC, the FP1 model is a reasonable alternative to a straight line, but the BIC, with its stricter penalty, suggests it is overfitted. The P-value from the likelihood-ratio test for the linear age × rx interaction is 0.0008. Even though the prognostic effects are different, the treatment–effect function is similar to that in Figure 7.1.

7.6   EXAMPLE 2: GBSG BREAST CANCER STUDY

7.6.1   Oestrogen Receptor Positivity as a Predictive Factor

It is well established that oestrogen receptor (ER) status er is a predictive factor for response to hormonal therapy with tamoxifen (tam). The risk of disease recurrence is reduced to a much greater extent by tam in ER-positive patients (Early Breast Cancer Trialists' Collaborative Group, 1998). For present purposes, therefore, the tam × er interaction is an evidence-based hypothesis. However, the choice of cutpoint on er which defines ER-positivity is controversial.

Here, the data are used as an example of applying the MFPI algorithm to investigate a predefined interaction. In Section 7.7, it is used to demonstrate difficulties with categorization when investigating interactions.

7.6.2   A Predefined Hypothesis: Tamoxifen–Oestrogen Receptor Interaction

The adjustment model is considered first. Because of the high correlation between pgr and er, the vector **x** was taken as all the available prognostic factors except for pgr. The adjustment model **x*** selected by MFP comprised an FP2 function for age with powers $(-2, -1)$, $\exp(-0.12 \times nodes)$ and gradd1 (grade 1 versus 2 or 3), at both α^*-levels of 0.157 and 0.05.

For the interaction, MFPI selected an FP2 function for er with powers $(-2, -1)$ or an FP1 function with power -0.5. The AIC values for linear, FP1 and FP2 models were 3459.6, 3448.1 and 3449.1 respectively. The FP1 model may, therefore, be preferred.

The FP1 function of er is shown for the two tam groups in Figure 7.4 (left panel). For clarity, the range of er values in the plot has been restricted to $[0, 100]$. The treatment–effect plot (Figure 7.4, right panel) illustrates the difference clearly. The test for interaction has a P-value of 0.042.

For large er values (say, $> 20 \, \text{fmol}\, \text{l}^{-1}$) the estimated treatment effect is nearly constant (log hazard ratio about -0.5, hazard ratio 0.6), but it changes sharply for small values. To check this result, the tam effect was considered in five subgroups with cutpoints partly determined by the treatment–effect function, partly by the CDF of er, and partly for clinical reasons. Biologically, zero is a special group, and 10 is the cutpoint most often used in the clinical literature. Table 7.5 shows the estimated treatment effect in subgroups of er, unadjusted and adjusted for other factors. The unadjusted estimates summarize the comparison between Kaplan–Meier curves presented in Figure 7.5. For these plots, groups 4 and 5 were amalgamated. Although some of the groups are small, analysis in subgroups confirms the treatment–effect function seen in Figure 7.4. In the small subgroup with $er = 0$ ($n = 82$ with 45 events), the patients treated with tam have a higher hazard, whereas in all other subgroups the hazard is reduced. These checks clearly indicate that the form of dependence of the estimated treatment effect on er is not the result of an artefact generated by the MFPI algorithm.

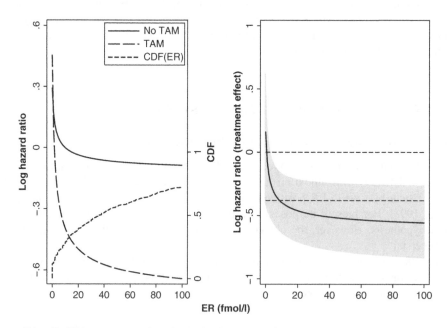

Figure 7.4 GBSG breast cancer data. Analysis of tam × er interaction keeping er continuous, fitted by FP1 functions with power -0.5. Functions were estimated within multivariable models; for details of the adjustment model, see the text. Left panel: solid line, estimated effect of er in patients not treated with tam; dashed line, estimated effect of er in patients treated with tam; dotted line and right-hand axis, CDF of er. Right panel: effect of tam by ER status, with 95% pointwise CI. Horizontal dashed lines denote zero and the main effect of tam in the absence of an interaction.

Table 7.5 GBSG breast cancer data. Effect of tam (log hazard ratio) in er subgroups. 'Adjusted' means adjusted for factors significant at $\alpha^* = 0.05$.

Subgroup	ER (fmol l^{-1})	Patients (%)	Unadjusted		Adjusted	
			$\widehat{\beta}$	SE	$\widehat{\beta}$	SE
1	0	12	0.67	0.31	0.45	0.32
2	1–10	17	−0.61	0.30	−0.41	0.31
3	11–36	21	−0.58	0.30	−0.70	0.32
4	37–115	25	−0.34	0.25	−0.44	0.26
5	>115	25	−0.31	0.26	−0.45	0.27

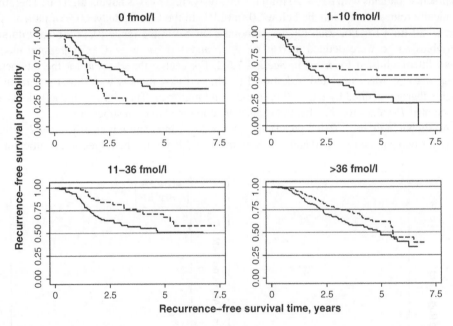

Figure 7.5 GBSG breast cancer data. Kaplan–Meier survival curves illustrating tam × er interaction with patients divided into four groups by using er cutpoints of 0, 10 and 36 fmol l^{-1}. Solid lines: tam; dashed lines: no tam. (Adapted from Royston and Sauerbrei (2004a) with permission from John Wiley & Sons Ltd.)

7.7   CATEGORIZATION

7.7.1   Interaction with Categorized Variables

In many fields, continuous variables are often converted into categorical variables by grouping values into two or more categories. As discussed in Section 3.4, categorization of continuous data is unnecessary for statistical analysis, and is not a natural way of analysing continuous variables. Usually, it is done to make the analysis and interpretation of results simpler. It appears to be the standard approach for handling interactions, at least in clinical research (Assmann et al., 2000).

In this section, the tam × er interaction in the GBSG data is used to illustrate the difficulties caused by categorizing a continuous variable. Often, only one cutpoint is used and it is not specified in advance. It is tempting to search for a suitable cutpoint to answer the two questions 'Does the study show that the effect of the treatment t depend on the continuous variable z?', and 'Which cutpoint on z best separates nonresponders from responders to t?' The first question might be answered by investigating P-values for the $t \times z$ interaction with different cutpoints on z, and the second by exploring possibly different treatment effects associated with each cutpoint.

7.7.2 Example: GBSG Study

In practice, patients with er > 20 fmol l^{-1} are always regarded as having high ER. Therefore, clinically relevant cutpoints lie below 20 fmol l^{-1}. In the GBSG study, 60% of patients had er > 20 fmol l^{-1}. For each integer cutpoint c on er in the range [0, 20], binary dummy variables er_c were defined as 0 if er $\leq c$ and 1 if er $> c$. Cox regression models were fitted, adjusted for $\mathbf{x}^*$ as in Section 7.6.2. For each c the P-value for the interaction tam × er_c was calculated. Additionally, adjusted Cox models were fitted separately in each subgroup defined by er_c (i.e. low er and high er). The regression coefficients and associated P-values for the effect of tam were computed in each subgroup.

As shown in Figure 7.6 (left panel), the tam × er_c interaction is significant at $P < 0.05$ only for two cutpoints: $c = 0$ and $c = 8$. In the high ER group, the regression coefficient for

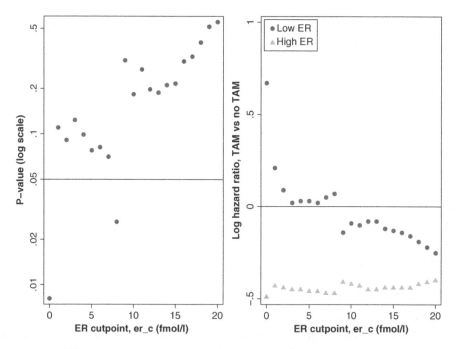

Figure 7.6 GBSG breast cancer data. Cutpoint analysis of interactions in 21 subgroups, adjusting for other prognostic factors. Cutpoints er_c in the range [0, 20] are used to define low and high ER subgroups. Left panel: P-values for tests of tam × er interaction. Right panel: regression coefficients (log hazard ratios) from Cox models for effects of tam in subgroups with ER low (circles) and high (triangles). (Adapted from Royston and Sauerbrei (2004a) with permission from John Wiley & Sons Ltd.)

tam is significantly different from zero for all c, usually with $P < 0.01$ (Figure 7.6, right panel). In the low ER group, the regression coefficient is not significantly different from zero for any $c > 0$. Exceptionally, at $c = 0$ the regression coefficient is significantly positive ($P < 0.05$), suggesting at face value that tam might actually harm patients in this subgroup (see also Figure 7.5).

The predominant impression from Figure 7.6 is of the instability of P-values from cut-point analyses, especially in the important range [0, 10]. Consequently, it is hard to assess how strongly the study supports an influence of er on the effect of tam. The regression coefficients in the ER subgroups show no effect of tam in the low ER group (except possibly at $c = 0$, as just noted), whereas in the high ER group there is a reduction in the log hazard ratio of about 0.5. Although the interaction is of questionable statistical significance, the treatment effect in the high ER group does appear to be real. For higher values of c, a substantial proportion of patients with positive ER values enters the low ER group, causing a small trend towards a tam effect in this subgroup. No allowance has been made in this analysis for the multiple testing implicit in the use of 21 cutpoints, nor for the fact that for extreme cutpoints the subgroups may be small and the estimated regression coefficients imprecise.

7.8   STEPP

A more exploratory approach for the investigation of interactions, called the 'subpopulation treatment effect pattern plot' or STEPP, is based on dividing the observations into overlapping subgroups defined with respect to z_2 and estimating the effect of treatment z_1 separately within each subpopulation (Bonetti and Gelber, 2000). To create the subpopulations, sliding window (SW) and tail-oriented (TO) variants were proposed. STEPP was extended and further illustrated by Bonetti and Gelber (2004). The SW version of STEPP is a type of moving average and is also related to varying-coefficient models (Hastie and Tibshirani, 1993).

To increase the number of patients that contribute to each point estimate, subpopulations are allowed to overlap. This increases the precision of the individual estimates. The subpopulations may be defined in two different ways (SW and TO), as indicated in Figure 7.7 The horizontal axis in Figure 7.7 indexes the various subpopulations for which treatment effects are estimated, and shows the range of covariate values used to define the cohort of patients included in each subpopulation. The TO version has the overall population as the centre group. With increasing distance from the centre, more and more patients with high covariate values (to the left side) or low covariate values (to the right side) are deleted. Subpopulations in the SW version have an overlapping part and a part that differs from neighbouring subpopulations. The number of subpopulations and the percentage of overlapping patients are important parameters of STEPP. To define the size of the subpopulations, the SW version has two parameters: n_1 and n_2. A subpopulation must have at least n_2 patients, of which at least $(n_2 - n_1)$ patients are required to be different between neighbouring subpopulations. The degree of discreteness of a continuous variable determines the size of a subpopulation. The TO version has a parameter g giving $(g - 1)$ subpopulations, where patients with larger values are eliminated and $(g - 1)$ subpopulations excluding patients with smaller values. For further details of how the subpopulations are created, see Bonetti and Gelber (2000).

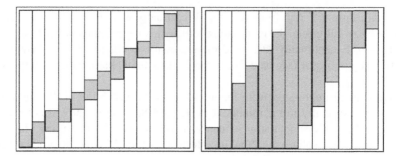

Figure 7.7   Schematic depiction of the two sets of subgroups used in STEPP. SW (left) and TO (right). The horizontal axis indexes the various subpopulations for which treatment effects are estimated, and shows the range of covariate values (vertical axis) used to define the cohort of patients included in each subpopulation. The TO version has the overall population as the centre group. (Adapted from Sauerbrei et al. (2007d) with permission from Elsevier Limited.)

The estimated treatment effects in the subpopulations defined by z_2 should be similar to the treatment effect in the overall population if z_2 has no influence on the treatment effect, i.e. no interaction exists between z_1 and z_2. Plots showing the estimated treatment effect with corresponding CIs in the subpopulations and tests based on the deviation of treatment effects in the subpopulations from the corresponding estimate in the overall population may be used to investigate an interaction between z_1 and z_2. For z_1, each subpopulation is represented by its mean. For more details see Bonetti and Gelber (2004).

7.9   EXAMPLE 3: COMPARISON OF STEPP WITH MFPI

7.9.1   Interaction in the Kidney Cancer Data

Royston et al. (2004) used the kidney cancer data from the MRC RE01 trial (see Appendix A.2.9) with the MFPI procedure and found a significant interaction between treatment `trt` and white cell count `wcc`. Altogether, they considered 10 potential predictive factors, of which six were continuous. Interest centres on the `trt` × `wcc` interaction.

Figure 7.8 displays the results of several STEPP analyses of the interaction between treatment `trt` and white cell count `wcc`. It is clear that use of small subpopulations ($n_1 = 25$, $n_2 = 40$; upper left-hand plot) with the SW method results in considerable variation caused by overfitting the data. Increasing the sample size in each subpopulation reduces the variation and leads to treatment estimates which show a similar dependence on `wcc`. For example, for $n_1 = 50, n_2 = 80$, there is only one additional 'blip' for `wcc` around 7. The lower panel clearly indicates that results from the TO version are less noisy and, hence, easier to interpret. The graph with $g = 4$ (lower right-hand plot) clearly indicates an interaction, and can be regarded as a rough approximation to the treatment effect function from MFPI (see Figure 7.9).

7.9.2   Stability Investigation

(In)stability, loosely defined as a vulnerability of modelling results to small changes in the data, is a critical issue when working with flexible models (Breiman, 1996b). See Chapter 8

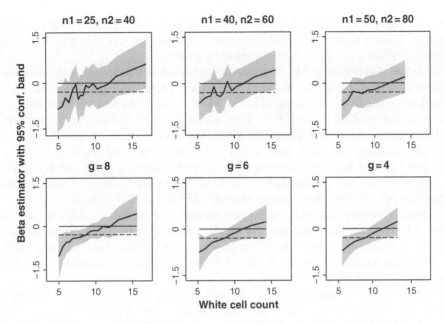

Figure 7.8 Kidney cancer data. STEPP plots for the interaction between `trt` and `wcc`, constructed with several choices of parameter values. Upper panel: SW; lower panel: TO. The plotted points represent the estimated treatment effects in each subgroup, with 95% CIs (faint dashed lines). The solid horizontal lines at zero represent no treatment effect and the dashed lines the estimated overall treatment effect. (Adapted from Sauerbrei et al. (2007d) with permission from Elsevier Limited.)

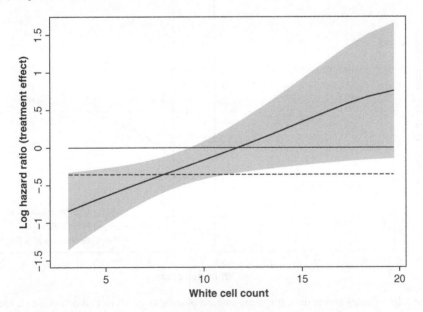

Figure 7.9 Kidney cancer data. Treatment effect plot for `wcc` (white cell count) from an MFPI analysis. Estimated treatment effect with pointwise 95% CI. The dashed line denotes the overall effect of treatment. (Adapted from Sauerbrei et al. (2007d) with permission from Elsevier Limited.)

for more details and examples. When estimating a treatment effect function for a continuous covariate, a small number of influential points may drive the function, thus indicating an interaction which is mainly a result of overfitting the data.

To explore the stability of MFPI, the treatment effect function was estimated in bootstrap replications in the kidney cancer example. Included as potential confounders were all variables other than the one under investigation (wcc). MFP was used to select the confounder model with a nominal significance level of $\alpha = 0.05$. wcc is the only variable identified by MFPI as a predictive factor. The stability of the function chosen was investigated, adjusting for the other variables. As in Royston and Sauerbrei (2003), a mean function and empirical bootstrap CIs were determined. Bootstrap resampling was also used to investigate the stability of STEPP, but without adjusting for covariates. For each STEPP group, the bootstrap mean and empirical 95% CI for the treatment effect were computed, using SW and TO variants.

The stability of the treatment effect function and of the corresponding STEPP functions were assessed in 1000 bootstrap samples. For the MFPI analyses, the multivariable adjustment model was selected, and then the FP2 treatment effect function for wcc (with preliminary robustness transformation; see Section 5.5). Selecting a new adjustment model in each bootstrap replication increases instability.

As an illustration, 20 randomly selected curves from MFPI analyses are shown in the left-hand panel of Figure 7.10. Most of the individual curves are similar to the curve from the original analysis presented in Figure 7.9. The mean of the 1000 bootstrap replications gives a nearly identical curve, with small differences appearing for more extreme values (wcc < 5

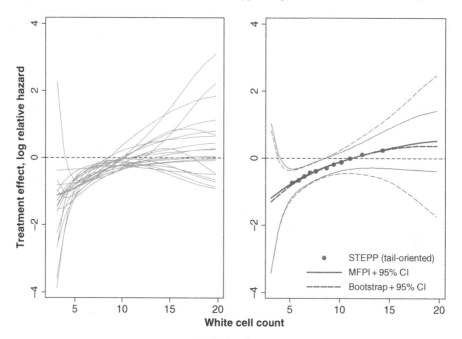

Figure 7.10 Kidney cancer data. Bootstrap analysis of stability of interaction between trt and wcc. Left panel: treatment effects plots in 20 bootstrap replications, using MFPI with data-driven adjustment model. Right panel: mean and 95% CI of treatment effect function from 1000 bootstrap replications, also showing the treatment effect function on the original data with 95% CI, and TO variant of STEPP. (Adapted from Sauerbrei et al. (2007d) with permission from Elsevier Limited.)

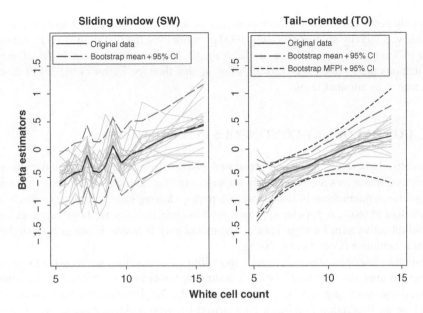

Figure 7.11   Kidney cancer data. Results from 1000 bootstrap replications of STEPP analysis of wcc. Left panel: SW ($m = 40$, $n = 60$); right panel, TO ($g = 6$). Thick lines represent the original data, bootstrap mean and 95% CI. Note that the bootstrap mean and original estimates are indistinguishable. Thin lines are results from 20 bootstrap replications selected at random. (Adapted from Sauerbrei et al. (2007d) with permission from Elsevier Limited.)

or > 15). The estimated effects from 11 subpopulations using the TO version of STEPP with $g = 6$ agree very closely with the functions from MFPI. For the bulk of the distribution of wcc values, the 95% pointwise CI derived from the 1000 bootstrap replications is a little wider than the interval from the original analysis. For larger values (say, > 12), the data become sparse and the bootstrap intervals become much wider, reflecting greater uncertainty in the FP2 functions selected by MFPI.

Figure 7.11 shows a random sample of 20 curves from 1000 bootstrap replications using STEPP with $n_1 = 40$, $n_2 = 60$ (SW) or $g = 6$ (TO). Major instability is apparent for the SW version. Functions for the TO version are more variable than the functions from MFPI, but the trend representing the result from the original analysis stands out clearly. For TO, the bootstrap interval is narrower than the corresponding interval from MFPI, where selection and estimation of the treatment effect function was repeated in each bootstrap replication. The wider intervals are partly due to the additional variation introduced in the MFPI analyses by selecting the adjustment model. Note that with STEPP the range of the covariate is restricted in the tails by the grouping process. No information about the treatment effect function is available beyond the range of the mean covariate values in the extreme groups.

7.10   COMMENT ON TYPE I ERROR OF MFPI

Little is known about the type I and type II errors of MFPI. In one example (Sauerbrei et al., 2007d), the values of the continuous variable haemoglobin in the kidney cancer dataset were repeatedly permuted at random. Independence of the effects of treatment and the continuous variable was thereby simulated. Using 1000 permutations, the distribution of P-values from a test of interaction was close to uniform on $(0, 1)$ (data not shown). In 54 of the 1000 permutations the P-value was < 0.05, showing that the type I error of the MFPI procedure was close to its nominal level.

7.11   CONTINUOUS-BY-CONTINUOUS INTERACTIONS

As has already been discussed and illustrated, discrete-by-continuous interactions involving randomized treatments are particularly important in clinical trials. The topic of continuous-by-continuous interactions is also of interest, perhaps having more relevance to observational studies than to trials. A popular approach is to assume linearity for both variables and test the multiplicative term for significance. The model may fit poorly if one or both of the main effects is nonlinear (Cohen et al., 2003).

Despite its importance in many areas, modelling of interactions seems often to be ignored in practical analyses (Ganzach, 1998). Uncertainty about how to proceed in the context of multivariable modelling may be a reason. Even in the simplest case, the decision on whether to include an interaction requires a comparison between additive models $\beta_1 z_1 + \beta_2 z_2$ and interaction models $\beta_1 z_1 + \beta_2 z_2 + \beta_3 z_1 z_2$. Three questions immediately arise:

1. Is the sample size sufficient to allow detection of a 'nonnegligible' interaction?
2. Is the assumption of linear effects of z_1 and z_2 justifiable?
3. Is the increased complexity of the model resulting from including the interaction worthwhile?

Certainly, the presence of an interaction complicates the interpretation and presentation of the model.

Because of lack of power, a small sample size may result in erroneous rejection of the interaction (Greenland, 1983). Incorrectly assuming linear effects for z_1 and z_2 may lead to a wrong decision to include the interaction. The severity of the problem and the probability of selecting the interaction model instead of a model with nonlinear terms for z_1 and/or z_2 depend on several factors. They include the correlation between z_1 and z_2, the magnitude of β_1, β_2 and β_3, the sample size, and possible measurement errors in z_1 and z_2. MacCallum and Mar (1995) reported a large simulation study looking at factors influencing the chance of selecting the correct model. Whereas the interpretation of estimates from an additive model is straightforward, e.g. increasing z_1 from a to b leads to an increase in $E(y)$ of $\beta_1(b - a)$, it is much harder to interpret the increase in an interaction model where it depends on z_2, e.g. $(\beta_1 + \beta_3 z_2)(b - a)$. The meaning of β_1 changes when the product term is included (Greenland, 1989).

In real analyses, more than two predictors must be considered, increasing the difficulties. If one variable, z_1 say, is of particular interest, e.g. an exposure variable in an epidemiological study, then the analysis strategy is more straightforward. The main interest resides in the estimation of a satisfactory functional form for z_1, adjusted for possible confounders. The potential addition of interactions between z_1 and confounders can be done in a second step. If a product term is to be included, then so also should be the relevant main effects (the hierarchy principle: see Bishop et al. (1975)). Including several interactions produces a complex and perhaps uninterpretable model. Subject-matter knowledge should be taken into account when considering such model extensions, and could result in adding nonsignificant interaction terms (Pearce and Greenland, 2005, p. 388).

For general model building without any particular variable of interest, all two-way interactions may be considered. Since the main-effects model may include nonlinear functions, products of these functions should be included as candidates. To handle this, we propose a procedure which is a natural extension of MFPI. More model building and data-dependent decisions are involved, leading to possible overfitting. Several 'reality checks' are needed to assess whether or not significant interactions are artefacts of mismodelling. The aim should be to select models which are 'as simple as possible' but which nevertheless show the important dependencies in the data. Most of the 'significant' interactions are not expected to reveal any crucial aspect of the data and, hence, are not required in a final model. However, a detailed search for interactions may improve model fit and may display interesting features of the data. Whether these features arise by chance or are reproducible must be assessed in external validation studies.

7.11.1   Mismodelling May Induce Interaction

Mismodelling a curved regression relationship, say as linear, may induce a spurious interaction with another variable (Lubinski and Humphreys, 1990; MacCallum and Mar, 1995). We illustrate this effect by considering predictors of 10-year all-cause mortality all10 in the Whitehall I dataset in a logistic regression analysis. In an MFP model for all10, both age and weight are significant predictors of outcome. We consider only these two predictors and their possible interaction. MFP selects a model with linear age and an FP2 transformation of weight with powers $(-1, 3)$. The interaction between age and weight is computed by including the multiplicative terms age $\times$ weight^{-1} and age $\times$ weight^3 in the logistic model. The likelihood ratio test (two d.f.) for this interaction has $\chi^2 = 5.27$ ($P = 0.07$), i.e. is not significant at the 0.05 level. However, if weight is erroneously modelled as linear, then age $\times$ weight is highly significant ($\chi^2 = 8.74$, $P = 0.003$), suggesting that mismodelling of the main effect of weight has induced a spurious interaction.

We first show that there is no strong interaction between age and weight when weight is modelled as a nonlinear function. To check the model and illustrate the effect, we divide age into four equal (quartile) groups and compute a running line smooth of the binary outcome all10 on weight in each age group. Because the definitive analysis involves a logistic regression model, the smoothed probabilities are transformed to the logit scale and the results plotted against weight. With no interaction between age and weight, we would expect the curves to be roughly parallel. The results (see Figure 7.12) show that the logits of the smoothed probabilities are indeed approximately parallel across age groups, suggesting no (strong) interaction. Also shown in Figure 7.12 are the estimated slopes from logistic regressions on weight in each age group, erroneously assuming linearity. The lines are clearly not parallel. The sign of the slope even changes across age groups, explaining why a significant interaction is found when the effect of weight is assumed linear.

Other types of mismodelling, such as omission of correlated influential variable(s), may also introduce spurious interactions.

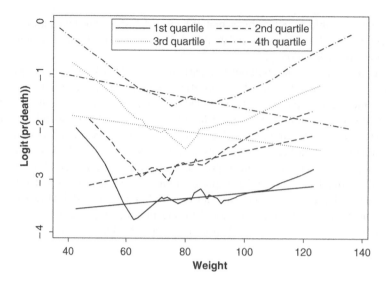

Figure 7.12   Whitehall I data. Graphical exploration of possible interaction between age and weight. The four pairs of lines show running line smooths (jagged lines) and linear fits (straight lines) in each of the four quartile groups by age. The changing slopes of the linear fits erroneously indicate an interaction, which disappears when a nonlinear function of weight is fitted.

7.11.2 MFPIgen: An FP Procedure to Investigate Interactions

With MFPI, a model for discrete z_1 and continuous z_2 is determined by finding the best FP transformation of z_2 and forming multiplicative interaction terms with the dummy variable(s) for z_1. A joint test of interaction involving the FP terms for z_2 and all dummy variables for z_1 is performed.

For general z_1 and z_2 (both possibly continuous) and confounders **x**, a new procedure called MFPIgen, in the same spirit as MFPI, is as follows:

1. Apply MFP to **x**, z_1, z_2 with significance level α^* for selecting members of **x** and FP functions of continuous variables. Force z_1 and z_2 into the model and apply the FSP to them. In the notation of Chapter 6 (see Section 6.1), MFP(α^*, α^*) is applied to **x**, while simultaneously MFP(α_1, α_2) with $\alpha_1 = 1$ and chosen α_2 is applied to z_1, z_2. This step requires a single run of MFP.
2. Multiplicative interaction terms are calculated between the FP transformations selected for z_1 and z_2, or between untransformed z_1 and z_2 if no FP transformation is needed. For example, if both variables need FP2 transformation, then four interaction terms are created.
3. The model selected on **x**, z_1, z_2 is refitted with the interaction terms included. The latter are tested in the usual way using a likelihood ratio test. If k interaction terms are added to the model, then the interaction χ^2 test has k d.f. For example, if FP2 functions were selected for both z_1 and z_2, then $k = 2 \times 2 = 4$.
4. All pairs of predictors are considered for possible interaction, irrespective of the statistical significance of the main effects in the MFP model. If z_1 and/or z_2 is binary or forced to be linear, then the procedure simplifies to the usual situation. If z_1 and/or z_2 are categorical,

then joint tests on all dummy variables are performed. An option is to treat the dummy variables as separate predictors.

5. All interactions should be checked for artefacts (see Section 7.4.2) and ignored if they fail the check.

6. If more than one interaction is detected, then a forward stepwise procedure can be used to extend the main-effects model.

There is one difference between this algorithm, MFPIgen, and MFPI. In MFPI, the confounder model $\mathbf{x}^*$ is selected independently of z_1 and z_2, whereas a joint model is selected in MFPIgen. The reason for the difference is that MFPI is principally intended for use with data from a randomized trial in which the effect of the treatment covariate z_1 is by design independent of other covariate effects. Therefore, adjustment by $\mathbf{x}^*$ is less important (see, for example, Table 7.1). In observational studies, however, it may be necessary fully to adjust the effects of z_1 and z_2 for confounders before investigating their interaction.

Since MFPIgen addresses dozens of potential interactions, multiple testing is a major issue. Results must be checked in detail and interpreted cautiously as hypotheses only. See Section 7.3 for further comments.

7.11.3   Examples of MFPIgen

Simplest Case
We consider the simplest case of a continuous-by-continuous interaction, i.e. that of a continuous outcome and two continuous covariates. Cohen et al. (2003) describe a study in 250 individuals of the intention to quit smoking y as a function of a measure of the fear of health ill-effects of smoking x and of self-efficacy for quitting smoking z. The correlation between x and z is 0.3. On applying MFPIgen, linear functions are selected for x and z and the x $\times$ z interaction is significant ($P = 0.008$).

Figure 7.13 shows the relationship between y and z in the four quartile groups of x. The relationship between y and z appears linear in each group. The regression slopes on z are nearly the same in the first and second quartile groups but subsequently increase with x. The largest slope is seen in the highest quartile group.

The MFPIgen model selected for this dataset is $E(y) = \beta_0 + \beta_1 x + \beta_2 z + \beta_3 xz$, whereas Cohen et al. (2003) included additional quadratic terms x^2 and $x^2 z$. Although x^2 and $x^2 z$ are jointly significant at the 0.05 level, they complicate the interpretation of the model and add little to R^2, which increases from 0.52 to 0.54.

More Complex Cases (1): Prostate Cancer Dataset
Table 7.6 shows the results of applying MFPIgen to all 21 pairs of variables in the prostate cancer dataset. In principle, each of the 21 analyses could involve a different main-effects model, to which is added one interaction term. Consider, for example, the interaction between age and pgg45. MFP is applied to all variables, with age and pgg45 forced into the model and the other variables selected at the 0.05 significance level. FP functions for all variables are selected in the usual way at the 0.05 level. Finally, the interaction between the selection functions for age and pgg45 is tested.

Linear terms were selected for all variables considered in interactions except for cavol, for which a log transformation was always selected. Six and two of the 21 interactions are significant at the 0.05 and 0.01 levels respectively, the strongest being pgg45 $\times$ cp ($P = 0.003$).

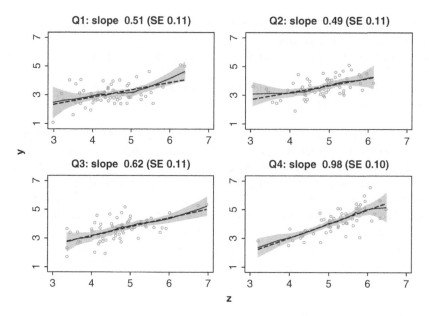

Figure 7.13   Quit-smoking study. Investigation of x by z interaction in four quartile groups of x (Q1–Q4). Solid lines, running line smooth with pointwise 95% CI; dashed line, fit from linear regression.

Table 7.6   Prostate data. Analysis of two-way interactions. Entries are P-values from tests of interaction using the MFPIgen procedure.

Variable	age	pgg45	cavol[a]	bph	cp	weight	svi
age	–						
pgg45	0.3	–					
cavol[a]	0.2	0.2	–				
bph	0.8	0.2	0.1	–			
cp	0.04	0.003	0.2	0.006	–		
weight	0.9	0.06	0.03	0.1	0.09	–	
svi	0.3	0.06	0.9	0.02	0.6	0.03	–

[a] cavol log transformed, all other variables untransformed.

Checking the Model   The possible effect of influential observations on the six interactions age×cp, pgg45×cp, bph×cp, log cavol×weight, svi×bph and svi×weight was assessed. DFBETA (Belsley et al., 1980) is a measure that quantifies the effect of removing an observation on the regression coefficient of a predictor. Leverage is a measure of the influence of an observation on the predicted values from the model. Points that appear to be outliers in a scatter plot of absolute values of DFBETA against leverage may indicate undue influence of the relevant observations. Figure 7.14 shows such plots for the multiplicative interaction terms in the six models under consideration. Observation 47 appears influential for interactions (a), (b), (c) and (e). Observation 89 appears influential for (d) and (f). Observations 47 and 89 were omitted and all six interactions were retested. The resulting P-values were 0.2, 0.08, 0.1, 0.1, 0.3 and 0.2 respectively. From this it may be concluded that there are no important interactions in the dataset.

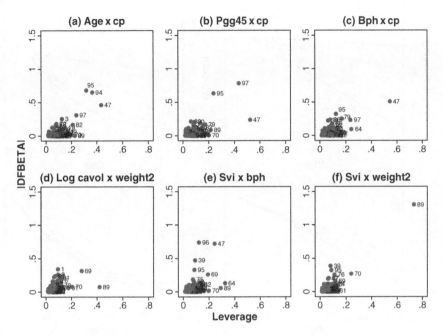

Figure 7.14   Prostate data. Influence plots for six interactions significant at the 0.05 level according to MFPIgen. Observation numbers are used to label points.

More Complex Cases (2): Whitehall I

An analysis of the CHD outcome in the Whitehall I study was described in Section 6.7.3. Here, we focus on analysing 10-year all-cause mortality all10, a binary outcome, using logistic regression.

An MFP(0.05) analysis of the covariates cigs, sysbp, age, height, weight, chol and grade (four-level categorical variable, three dummy variables tested jointly) gives FP powers 0.5 for cigs, $(-2, -2)$ for sysbp, $(-2, 3)$ for weight and linear terms for the other continuous variables. Main effects for all variables are required; therefore, the adjustment model used in assessing interactions in turn is always the same. Pairwise interactions among these variables are now considered (see Table 7.7). The grade × weight interaction is

Table 7.7   Whitehall I data. Analysis of two-way interactions. Entries are P-values from tests of interaction using the MFPIgen procedure.

Variable	cigs[a]	sysbp[a]	age	height	weight[a]	chol
cigs[a]	–					
sysbp[a]	0.7	–				
age	0.9	0.2	–			
height	0.1	0.5	1.0	–		
weight[a]	0.9	0.5	0.1	0.4	–	
chol	0.2	0.07	0.001	0.8	0.2	–
grade	0.2	0.2	0.2	0.2	0.04	0.4

[a] With FP transformation (see text), all other variables untransformed.

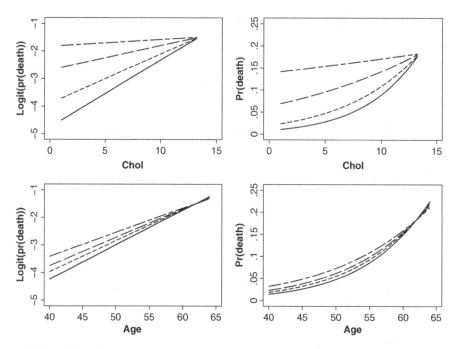

Figure 7.15 Whitehall I data. Graphical presentation age × chol interaction. Left-hand panels show the logistic scale, right-hand panels the probability scale. Upper panels show results for the 10th, 35th, 65th and 90th centiles of the distribution of age, lower panels for the same centiles of chol. Adjusted for other covariates.

significant at the 0.04 level but is not considered further. One interaction stands out: age × chol. The relationship between mortality predicted by the fitted interaction model and chol at the 10th, 35th, 65th and 90th centiles of the distribution of age is presented in Figure 7.15. The age by chol relationship is adjusted for other covariates centred on their means (see Section 4.11). The results show that the risk gradient on chol is much steeper in younger men than in older ones. The lines converge at a chol of 12 mmol l^{-1}. A very high cholesterol level (12 mmol l^{-1}) is associated with a probability of 10-year mortality of about 18%, irrespective of age. The lower right panel shows that the risk gradient on cholesterol disappears at about age 62 years.

The validity of this putative interaction was checked in four equal-sized age groups as follows. The slopes from logistic regression on chol were computed, adjusted for other factors in the MFP model (see above). The results are shown in Figure 7.16. The linearity assumption in each age group was checked by cubic regression splines (four d.f.) and confirmed to be well supported by the data. However, the slopes on chol are not monotonically ordered across the age groups, suggesting a possible lack of fit of the linear-by-linear interaction model.

In the MFP model, an FP1 term for age with power -1 is close to significant at the 0.05 level. To try to improve the fit, a model including age^{-1}, chol and their product was fitted. The deviance was reduced by 7.6 compared with that for the model with age, chol and their product. A common approach is to consider quadratic terms. A similar deviance reduction was obtained by adding quadratic interaction terms (age^2, age^2 × chol) to the linear by linear interaction model. The quadratic terms were highly significant ($\chi^2 = 9.67$, two d.f.,

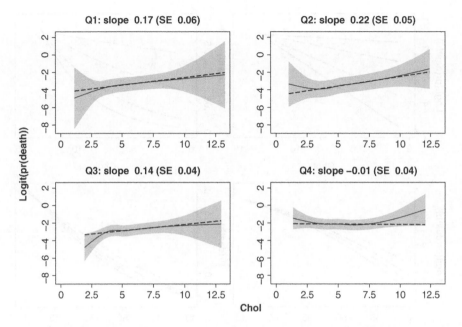

Figure 7.16 Whitehall I data. Interaction between age and chol. The relationship between all10 and chol is explored in four equal-sized age groups. Solid lines: fit and 95% pointwise CIs from cubic spline regression with four d.f. Dashed lines: fitted line on chol from linear logistic regression model.

$P = 0.008$). Since the FP1 interaction model is more parsimonious and its fit is similar to the quadratic model, it is preferable.

Figure 7.17 repeats Figure 7.15 but includes the interaction between age^{-1} and chol. The slight curvature on age can be seen, but the overall message from the revised model is not very different. It requires a large sample size to be able to distinguish between such subtly different models.

The example is instructive in that MFPIgen found a highly significant interaction between age and chol, which subsequent checking showed to be susceptible to some improvement. In a complete analysis of the dataset, the researcher might go on to consider other extensions of the age function and its interaction. The lesson to learn is that simple approaches to interaction modelling may not always be sufficient. Care is needed in checking the fit of the model and improving it if necessary.

7.11.4   Graphical Presentation of Continuous-by-Continuous Interactions

Since a two-way interaction involves three variables, a three-dimensional surface plot might be considered the preferred way to present it. However, such a plot may be hard to interpret, since its appearance depends critically on the orientation of the axes. Also, not all statistical software can produce three-dimensional graphics.

In our view, a better option is a 'sliced' plot, as in Figures 7.15 and 7.17. The fitted function at selected centiles of the distribution of the first variable is plotted against the second variable. The plot is repeated, reversing the roles of the two variables. In the Whitehall I example,

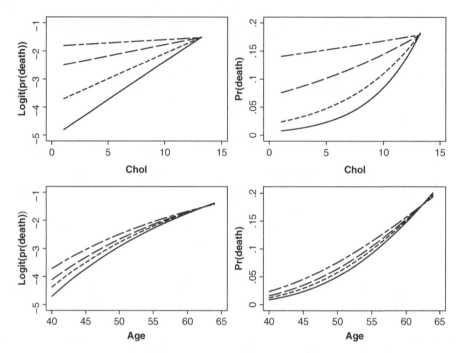

Figure 7.17   Whitehall I data. Graphical presentation of $\texttt{age}^{-1} \times \texttt{chol}$ interaction. Details as for Figure 7.15.

we chose to plot the fit at the 10th, 35th, 65th and 90th centiles. Probably four centiles is enough in most cases. Since distributions of observations are usually more 'bunched' around the centre than at the extremes, we selected two values near the extremes and two surrounding the median.

7.11.5   Summary

We have suggested MFPIgen, an extension to MFPI, for modelling continuous-by-continuous interactions in a multivariable context. We limit our analyses to first-order interactions because higher order interactions are unstable and hard to interpret. In the simplest case, all variables have a linear main effect. An interaction between two variables, z_1 and z_2, can be assessed by testing the estimated regression coefficient of the product $z_1 z_2$ for significance. Such product terms may introduce collinearity, resulting in difficulties in computation and interpretation. Centring continuous variables (e.g. on their means) improves matters. A more detailed discussion is given in standard textbooks, such as Vittinghof et al. (2005), DeMaris (2004), or Cohen et al. (2003).

As elsewhere in our book, we examine the data systematically for convincing evidence of nonlinearity of the main effects of continuous variables. By considering just two continuous variables in the Whitehall I study, we demonstrated that mismodelling a main effect by erroneously assuming linearity can introduce a spurious interaction. We propose graphical checks of whether an interaction is genuine or may have arisen from mismodelling. Such checks are essential in a multivariable context where several potential interactions are considered.

For example, the interaction between `age` and `weight` in the Whitehall I example was caused by mismodelling the main-effect function for `weight` as linear when it should have been an FP2. The significance of several interaction terms in the prostate cancer data depended critically on two influential observations among the 97 patients. The analysis of 10-year all-cause mortality in Whitehall I revealed a linear-by-linear interaction for `age` × `chol` undetected in earlier analyses (including our own). Because of the large sample size (1670 events), the interaction is highly significant. Checks suggest that it is real, but also hint at a nonlinear effect of `age`. Even with its substantial power, the Whitehall study does not provide assurance as to whether the effect of `age` is linear or nonlinear.

In general, we hypothesize that many interactions remain undetected because nonlinearity is not considered or because the sample size is too small. Simulation studies are necessary to gain more insight into the characteristics (e.g. the power) of MFPI and MFPIgen.

7.12   MULTI-CATEGORY VARIABLES

The natural extension for an unordered multi-category variable z_1 with $k > 2$ levels is to estimate a function for each level, using the FP2 power terms from the main-effects model for z_2, adjusting for $\mathbf{x}^*$ (see Section 7.4). As before, the test of interaction compares deviances between the main-effects model and the extended model with parameters estimated for each category. The d.f. for the test are $2(k - 1)$. Comparisons between functions at different levels depend on the reference category. Treatment-effect plots are constructed as for the binary case (see Section 7.3.6), one for each of the $k - 1$ comparisons.

For an ordinal covariate, several options are available. The first option (which has serious drawbacks) is to ignore the ordering and proceed as above. The second option is to define scores (coding schemes) z_1^* for z_1 (see Section 3.3), analyse z_1^* as though it were continuous, and then use MFPIgen. If MFPIgen finds that an FP transformation of z_1^* improves the fit, then in effect the scores are redefined. Finally, the theory of trend tests and their associated contrasts may be used to define suitable tests based on the functions estimated for each level of z_1, as described above for the unordered case.

7.13   DISCUSSION

In this chapter, we have considered methods to identify interactions with continuous variables. Regarding treatment effects in randomized trials, current practice still seems to be either to assume linearity or to categorize the variable into a number of groups according to one or more cutpoints. The linearity assumption may be incorrect, and categorization reduces power and raises questions of the number and position of the cutpoints. MFPI can in principle be used for any binary variable (not just treatment).

In observational studies, modelling of confounder variables is important and a careful interpretation of the 'treatment effect' function is required. The confounder issue is not specific to our modelling approach, but relates to the fact that, in randomized trials, treatment is by design independent of all other covariates, whereas treatment and other such variables are almost always confounded by other factors in observational studies.

We emphasize the importance of several model checks to ensure that a postulated interaction is supported by the data and not a result of mismodelling or driven by a few influential observations. We also draw attention to the important distinction between a prespecified interaction and one that is identified in a data-dependent fashion. The strength of evidence for an interaction differs considerably between these two cases.

STEPP has been used several times in the literature with data from randomized trials to demonstrate interaction between treatment and a continuous covariate. The approach compares treatment effects in overlapping sub-intervals of the covariate. It may be considered as midway between the still-popular dichotomization with the comparison of treatment effects in two subgroups and our method, which estimates a treatment effect function. In one example using the TO version of STEPP with a smaller number of subgroups, the results from MFPI and STEPP agreed remarkably closely.

The principles of MFPI have been extended to investigate continuous-by-continuous interactions. We have shown in examples the importance of considering possibly nonlinear main effect functions. If nonlinearity is ignored by using the common approach of testing linear-by-linear product terms, then spurious interactions can be introduced into a model. Models derived with MFPIgen may not only fit the data better, but if an interaction term is not required, they may even give results which are easier to interpret. However, the sample size may not be large enough to discriminate well between several models with similar fits. Subject-matter knowledge plays an important role here.

For detailed discussion of various aspects of interactions between different types of variables, see for example DeMaris (2004, chapters 4 and 5) and Cohen et al. (2003, chapters 7 and 9). A fuller discussion of interaction among continuous variables, including higher order interactions, is given by Aiken and West (1991). A more flexible approach using semi-parametric modelling for investigating interactions with continuous predictors is described by Ruppert et al. (2003, chapter 12). None of these books considers the FP framework.

CHAPTER 8

Model Stability

Summary

1. Model instability occurs when selected variables and functions are sensitive to a small change in the data.
2. An assessment of the stability of a multivariable model is important, but rarely done in practice. The chapter provides some practical guidance.
3. Bootstrap resampling may be used to study stability of MFP models. Large numbers of bootstrap replications are required.
4. Typically, many different models are selected by MFP across bootstrap samples. The bootstrap inclusion fractions (BIFs) provide information on the relative importance of each predictor. BIFs for pairs of variables should be considered. For correlated variables, BIFs are related.
5. The BIFs for types of simplified functions (excluded/monotonic/nonmonotonic; excluded/linear/nonlinear) and plots of functions estimated in bootstrap samples help one to assess the functional form required for a continuous predictor.
6. With an insufficient sample size, the default linear function is often selected even if the underlying function is nonlinear.
7. Quantitative measures to assess stability of functions are described.
8. Dependencies among BIFs may be further studied by log-linear modelling.

8.1 INTRODUCTION

In this chapter, we discuss the model instability that occurs when selected variables and FP functions are sensitive to a small change in the data. We show how instability can be assessed by using bootstrap resampling, and consider the complications that arise when MFP is used to model continuous predictors.

Multivariable Model-Building   Patrick Royston, Willi Sauerbrei

8.2 BACKGROUND

Whether linearity is assumed (Chapter 2) or not (Chapter 6), the result of applying a model selection procedure to a set of candidate variables is a single model. A very low P-value for including a predictor indicates that that predictor is 'stable' in the sense of being selected with high probability in similar datasets. For less significant predictors, selection may be more a matter of chance. In reality, several competing models are likely fit the data about equally well, but these potential competitors are usually ignored. The particular model chosen may depend on the characteristics of a small number of observations in one or more dimensions. If the data are slightly altered, then a different model may be selected. This is an issue for all selection procedures.

When the same selection procedure is used with the original data and in an 'ideal' validation study, e.g. on a population defined by the same inclusion and exclusion criteria, then (nearly) the same variables should be selected. This is sometimes called 'replication stability' (Gifi, 1990). Differences between variables selected in two different samples from similar populations may increase when procedures which tend to produce more complex models are used. Variables with a 'strong' effect are usually included, whereas those with a 'weak' or 'borderline significant' effect may enter to a greater extent at random (Sauerbrei and Schumacher, 1992).

Although model stability, or rather lack of it, is a long-standing issue (Harrell et al., 1984) most data analysts continue to ignore it as a worthwhile criterion of the quality of a selected model and as a useful technique for model criticism. Often, the result of a complex model-building process is presented without mentioning that minor changes to the data may result in major changes to the model, or at least to parts of it.

The seriousness of the issue was first demonstrated more than two decades ago (Gong, 1982; Harrell et al., 1984; Chen and George, 1985). Harrell et al. (1984) used data splitting to investigate the stability of Cox models for survival data chosen by stepwise selection. They found 'great variation in the factors chosen' in three randomly selected subsamples. Gong (1982) demonstrated the instability of a selected logistic regression model by bootstrap techniques. In an analysis of survival data, Chen and George (1985) were able to reproduce the original model in only two of 100 bootstrap samples. Five of the six variables in the original model were included in the model in 64% to 82% of bootstrap replications. The bootstrap inclusion frequencies of variables not entering the original model were much smaller, suggesting that the most important variables were in fact selected. Others have reported similar experiences (Altman and Andersen, 1989; Sauerbrei and Schumacher, 1992; Sauerbrei, 1999).

Recently, the use of resampling techniques for stability analysis seems to have become more popular. Qiu et al. (2006) use resampling as a tool for reducing the set of initially selected genes in microarray data and to assess the variability of different performance indicators. Similar to the approach by Sauerbrei and Schumacher (1992), Austin and Tu (2004a) use bootstrap resampling in conjunction with automated methods for variable selection to develop prediction models. They concluded that their approach could identify a parsimonious model with excellent predictive performance. Although this result is consistent with our own experience (Sauerbrei, 1999), we criticized Austin and Tu for inappropriate simplification of the model selection procedure (Sauerbrei et al., 2005).

Several workers have used the bootstrap to attempt to incorporate model selection uncertainty in the model-building process (Buckland et al., 1997; Augustin et al., 2005; Faes et al., 2007). The latter considered model averaging for functions (see Section 11.3.2).

Here, the assessment of model stability is explored using the bootstrap. The stability of models with linear covariate effects is considered in Section 8.3, and the extensions necessary when allowing selection of functions of continuous covariates in Section 8.7. The issues are illustrated in examples.

8.3   USING THE BOOTSTRAP TO EXPLORE MODEL STABILITY

8.3.1   Selection of Variables within a Bootstrap Sample

An appropriate method for studying model stability is nonparametric bootstrap sampling. A random sample with replacement is taken from the numbers $1, \ldots, n$ which index the observations. The complete observations (i.e. the response, covariates and, if relevant, censoring indicator) associated with the selected indices form the bootstrap sample. Some of the indexes (hence, some of the original observations) are omitted from the bootstrap sample and others are included once or several times. An example of a bootstrap dataset for $n = 10$ is the cases 8, 6, 3, 4, 10, 6, 3, 4, 9, 4. Cases 1, 2, 5 and 7 do not enter this particular bootstrap sample, whereas case 4 is represented three times and cases 3 and 6 twice. The expected proportion of original observations entering a bootstrap sample at least once is $1 - e^{-1}$ or about 63% (Efron and Tibshirani, 1993).

A bootstrap sample or dataset is a random sample from the CDF of the observations. Characteristics of the original sample may affect the method of resampling, e.g. separate random sampling in the treatment groups of a randomized trial may be preferable (Davison and Hinkley, 1997) – see, for example, Section 8.4. A collection of bootstrap samples is a suitable vehicle for exploring variations among possible models for the original dataset. Suppose multivariable model-building with k covariates is envisaged. In the simplest case of the linear model, for which the inclusion or exclusion of variables is the only model selection issue, the application of the bootstrap to study instability goes as follows:

1. Draw a bootstrap sample of size n.
2. Apply the model selection procedure within this bootstrap sample.
3. For each covariate x_j, $(j = 1, \ldots, k)$, record whether x_j is selected in the model or not.
4. Repeat steps 1–3 a large number B of times.
5. Summarize the results.

The results can be presented as a matrix with B rows and k columns. The ith row and jth column element is an indicator variable $I(x_j)$ taking the value one if x_j was selected in the ith bootstrap sample and zero if not. The bootstrap inclusion (relative) frequency or BIF for x_j is the sum of $I(x_j)$ over the B bootstrap samples, divided by B.

In a particular example, Sauerbrei and Schumacher (1992) studied the effect of the number of bootstrap replications by varying B between 50 and 1000. They concluded that even $B = 100$ can give reasonable results, although in practice they suggested working with several hundred.

Since computer power has greatly increased since their investigation, $B = 1000$ or even more is practicable nowadays.

8.3.2   The Bootstrap Inclusion Frequency and the Importance of a Variable

Generally, important variables should be included in most of the bootstrap replications, since it is assumed that each replication, being a random sample from the observations in the study, reflects the underlying structure of the data. The BIF may be used as a criterion for the importance of a variable. A variable which is approximately uncorrelated with others, and which is just significant at the chosen nominal level α in the full model, is selected in about 50% of bootstrap samples. If its P-value in the full model is $< \alpha$, then the BIF should be $>50\%$; and if the P-value is very small, then the BIF tends towards 100%. For the special case of one candidate variable ($k = 1$), further aspects of the relationship between the significance level and the BIF are discussed by Sauerbrei and Schumacher (1992). In the normal-errors model, these considerations are also valid for uncorrelated variables.

A difficulty with the BIF as a criterion of the importance of a variable is how to cope with correlated variables. Often, only one variable of a correlated set is selected in a particular bootstrap replication. Sauerbrei and Schumacher (1992) considered the inclusion frequencies of all possible pairs of variables. In an extreme example, one of two highly correlated variables may always be selected, but the BIF of each variable may be only about 50%. Failure to recognize the correlation between the inclusion frequencies of two variables, together with the aim of building simpler models, may result in inappropriately eliminating both variables from the model. Furthermore, higher dimensional dependencies among the inclusion frequencies often occur. Sauerbrei and Schumacher (1992) considered only two-dimensional inclusion frequencies in 100 bootstrap replications, whereas for a detailed investigation of higher dimensional relationships, more replications are required – see, for example, Royston and Sauerbrei (2003). The issue is discussed in more detail in Section 8.7.

With the bootstrap approach, one is in effect attempting to estimate and interpret the multivariate distribution of inclusion frequencies of the candidate variables. Such an approach lacks theoretical underpinning, and several variants are possible. The distribution of inclusion frequencies depends critically on the sample size (or number of events) and the significance level of the selection procedure. Nevertheless, some understanding of the distribution of inclusion frequencies of variables and their relationships is a useful complement of the model-building process.

8.4   EXAMPLE 1: GLIOMA DATA

The main results of several strategies for variable selection in the glioma data are briefly summarized in Section 2.7.3. More details are given by Sauerbrei and Schumacher (1992). Table 8.1 gives the standardized regression parameter estimates from the full model and from the model selected by BE (0.05) in the original data. BE(0.05) selected five of the 15 variables. Some of the absolute standardized regression estimates increase in comparison with the full model, which is mainly a result of excluding correlated variables.

Table 8.1   Glioma data. Columns 2 and 3: Standardized regression coefficients in the full model and in the model selected by BE(0.05). Columns 4–6: variables selected by BE(0.05) in the first three bootstrap replications. Final column: BIF (percentage of 100 replications).

Variable	$\widehat{\beta}/SE$		Bootstrap reps			BIF (%)
	Full model	BE(0.05)	1	2	3	
x_1	−1.36			✓		30
x_2	−0.91					20[a]
x_3	3.18	4.99	✓	✓	✓	92
x_4	1.35		✓			33
x_5	5.75	5.91	✓	✓	✓	100
x_6	−2.27	−2.81			✓	63
x_7	−0.23		✓			5[a]
x_8	−4.90	−5.70	✓	✓	✓	100
x_9	−1.55				✓	34
x_{10}	0.69					11
x_{11}	1.89		✓		✓	59
x_{12}	−1.80	−2.09	✓	✓	✓	52
x_{13}	0.49		✓	✓		20[a]
x_{14}	1.53				✓	49
x_{15}	−1.08				✓	22

[a] Sign of the regression coefficient after selection of the variable varies across bootstrap replications.
(Adapted from Sauerbrei (1999) with permission from John Wiley & Sons Ltd.)

The results of the first three bootstrap replications and the inclusion frequencies of 100 bootstrap samples are also given in Table 8.1. In the bootstrap analysis, separate samples were taken from the two treatment groups of the randomized trial to account for the design of the study, an issue not considered in an earlier bootstrap analysis by Sauerbrei and Schumacher, (1992). Major differences occur between the models selected in the different bootstrap datasets. Only three variables are selected in nearly all replications. Several variables are selected in about half of the replications and some are selected only occasionally. For some of the latter, the signs of the regression coefficients in the selected models are positive in some replications and negative in others. Such variables should clearly be excluded from a sensible model. Apart from the three 'dominating' variables x_3, x_5 and x_8, no other variable seems to be necessary in a model including only strong factors. However, investigation of two-dimensional inclusion frequencies shows that the inclusion of x_6 depends on that of x_{12}. Neither of these two variables is included in more than two-thirds of the replications, but one or both of them are included in 86%. Even if the aim is to select a simple model with strong factors only, at least one of these two variables should enter. The empirical results presented in Table 8.1 point to the inclusion of x_6, but subject-matter knowledge may lead one to prefer x_{12}. Dependencies between the inclusion of correlated variables and the choice of variables for a final model are discussed in more detail by Sauerbrei and Schumacher (1992). They suggested two strategies, one to include both weak and strong factors, and the other to select strong factors only. The issue is not considered further here.

All variables from the BE(0.05) model have BIF > 50%. Among the excluded variables, only x_{11} (59%) and x_{14} (49%) have relatively large BIFs. The final decision on whether to include them or not may depend on subject-matter considerations. Altogether, the stability investigations support the BE(0.05) model and point to possible extensions of the model if weaker factors should also be included.

8.5   EXAMPLE 2: EDUCATIONAL BODY-FAT DATA

The MFP procedure (chapter 6) was applied to the educational body fat data to build models at the 5% significance level in each of 1000 bootstrap replications. The outlying case 39 was excluded. A linear effect was assumed for all continuous covariates. Since MFP combines backward elimination with a search for non-linear effects, the restriction to linearity reduces MFP to a BE procedure. Because MFP fixes the order in which candidate variables are considered for inclusion/exclusion, it is somewhat less flexible than BE procedures found in most software packages. This use of MFP is called by Stata 'hierarchical' BE, and the abbreviation 'BE-hier' is used for it.

BE-hier selects the variables x_2, x_6 and x_{13}, while BE selects x_1, x_3, x_6 and x_{13}. The explained variation of the two models is similar ($R^2 = 0.736$ and 0.741 respectively), and only slightly lower than that of the full model with 13 variables ($R^2 = 0.753$).

Results from the first five bootstrap replications, given in Table 8.2, exhibit typical variation in the model selected. Variables x_1, x_6 and x_{13} are selected in all five replications, otherwise the models are all different. Six of the other 10 variables are included in one, two or three of the five models. The eighth column of Table 8.2 gives the BIF for BE-hier in the 1000 bootstrap replications. The variable x_6 is selected in 100% and x_{13} in 91% of replications. These two dominating variables are also included in the original analysis using BE or BE-hier. None of the other variables appears to be essential, since their BIFs are all $< 60\%$.

Table 8.2   Educational body-fat data, case 39 excluded. Results of the first five bootstrap replications and summaries by BIF for 1000 replications of the model selection procedure BE-hier. Columns 2–8 relate to all variables, whereas columns 9 and 10 are for models excluding x_6. Standardized regression coefficients are for the original data (see Table 2.3).

Variable	Bootstrap rep.					All variables		Excluding x_6	
	1	2	3	4	5	$\widehat{\beta}/\text{SE}$	BIF (%)	$\widehat{\beta}/\text{SE}$	BIF (%)
x_1	✓	✓	✓	✓	✓	2.31	57	6.20	100
x_2	✓					−0.28	36	3.08	93
x_3			✓	✓	✓	−1.30	46	−4.32	98
x_4			✓	✓		−1.68	34	−1.36	31
x_5						−1.10	21	1.21	19
x_6	✓	✓	✓	✓	✓	9.90	100	—	—
x_7						−1.02	16	0.98	31
x_8				✓		1.22	29	1.33	29
x_9						−0.17	6	−0.33	10
x_{10}						0.85	18	−0.21	4
x_{11}	✓	✓				1.04	26	−0.33	6
x_{12}			✓		✓	1.34	25	0.24	4
x_{13}	✓	✓	✓	✓	✓	−3.46	91	−4.34	99

However, investigation of two-dimensional inclusion frequencies reveals that one or both of x_2 and x_3 are included in 80% of the replications, so that one of them seems to be needed. In the original data, BE-hier selects x_2 whereas BE selects x_3. In the bootstrap replications, the inclusions of x_1 and x_3 are nearly independent, whereas those of x_1 and x_2 are correlated. Since BE-hier includes x_2, the need to include x_1 as well is reduced. x_1 enters the model in only 40% of the replications in which x_2 is included, suggesting why BE-hier does not select x_1. In contrast, x_1 is included in 55% of replications in which x_3 is selected, perhaps explaining why BE also selects x_1. Although further strong dependencies between the inclusion of pairs of variables are present in the data, no further variables seem to be required in the model. In the rare cases (9%) in which x_{13} is excluded, x_4 is included as a substitute for it in 79% of these replications.

As in Section 2.7.2, we repeated the bootstrap investigation without the dominating variable x_6. The results for the other variables change substantially. Now x_1, x_2, x_3 and x_{13} are included in nearly all replications and are obviously important, but none of the other variables seems to be required. The only other interesting feature is the high correlation between the inclusion frequencies of x_2 and x_3. As expected from Table 8.2, BE and BE-hier both select the model x_1, x_2, x_3, x_{13} on the original data.

8.5.1 Effect of Influential Observations on Model Selection

Working again with 1000 bootstrap replications, the influence on the BIFs of case 39, which has an outlying value of the important predictor x_6, was examined. The number of occurrences of case 39 in each bootstrap sample was noted. Table 8.3 summarizes the BIFs for all the variables according to the relative frequency of case 39. It is clear that the number of times

Table 8.3   Educational body-fat data, case 39 included. Results of 1000 bootstrap replications of the model selection procedure BE-hier. BIFs are given overall and according to the frequency of occurrence of case 39 (0, 1, 2, 3 or more) in the bootstrap samples. Spearman r_S is the rank correlation between the inclusion/exclusion of a given variable in the model selected in bootstrap samples and the frequency of case 39.

Variable	Frequency of case 39 (% of bootstrap reps)					Spearman r_S
	Any (100)	0 (36)	1 (37)	2 (18)	≥ 3 (9)	
x_1	50	57	51	40	36	−0.15
x_2	56	30	64	74	88	0.40
x_3	28	47	21	13	16	−0.29
x_4	46	32	53	55	56	0.20
x_5	14	25	7	8	5	−0.23
x_6	100	100	100	100	100	0.00
x_7	28	18	34	37	29	0.14
x_8	35	23	40	45	46	0.19
x_9	7	6	7	8	10	0.03
x_{10}	14	14	12	19	15	0.03
x_{11}	24	27	22	24	23	−0.03
x_{12}	60	22	74	88	97	0.57
x_{13}	86	90	87	81	79	−0.11

that case 39 enters the bootstrap sample has a major influence on the model that is selected, strikingly so for x_2 and x_{12}. Neither variable is important with case 39 excluded; the more often case 39 is included, the higher the BIF for these two variables. On the other hand, x_3 is less often selected if case 39 is included. Such bootstrap investigations may offer an alternative approach to identifying influential observations.

In the original data, BE-hier selects x_1, x_3, x_4, x_6, x_{12}, x_{13} if case 39 is included and x_2, x_6, x_{13} if it is excluded. Thus, selection of variables in this example depends critically on the presence or absence of a single influential observation among 252.

8.6   EXAMPLE 3: BREAST CANCER DIAGNOSIS

Sauerbrei and Royston (1999) used MFP to develop a logistic regression model (called model III) for the probability of a patient having breast cancer in the setting of a multifactorial diagnostic test. The estimated linear predictor was

$$-11.71 + 0.0684\text{age} + 5.323\log(\text{art} + 1) - 0.685\text{artcl}$$

where age is age, art is the number of tumour arteries and artcl is the number of contralateral arteries. None of three blood velocity measures (avv, sumv, maxv) was included in the model. The BIFs for age, art, avv, sumv, maxv and artcl in 5000 bootstrap replications were 98.3%, 97.5%, 19.4%, 18.7%, 14.2% and 90.4% respectively, indicating a clear distinction between important and unimportant variables. A model including only age, art and artcl was chosen in 2906 (58%) of replications. Using a log-linear analysis of the six-dimensional variable inclusion/exclusion matrix, a strong three-way interaction between the inclusion fractions for avv, sumv and maxv was identified. One or more of these three correlated variables was selected in only about one-third of replications, supporting the conclusion that none of them was essential to the final model.

Further analysis involved distinguishing between linear and nonlinear functions for age, art and artcl in the 2906 replications with no additional variables selected. A linear function was chosen for age in about 90% of this subset. Linear functions of art and artcl were chosen in about 40%. Log-linear modelling of the result matrix for linear/nonlinear functions revealed a strong negative association between the types of function (linear, nonlinear) chosen for art and artcl. That is, linear for one usually coincided with nonlinear for the other. The variables were simultaneously linear in only 11% of the 2906 replications, indicating a need for one of them to be modelled as a nonlinear function.

The bootstrap investigation lent support to Sauerbrei and Royston's (1999) models III and IV, both of which were derived using only the original data. Model IV comprised age linear, a negative exponential transformation of art and an FP1 transformation of artcl. In both models age was modelled linearly and art nonlinearly, whereas artcl was linear in model III and nonlinear in model IV. The combinations of age linear and art nonlinear with artcl either linear or nonlinear were chosen in 20.2% and 22.5% of the 5000 replications respectively.

In addition to the models derived from the original data, the bootstrap analysis suggested a new, third model (age and art linear, and artcl nonlinear) with a slightly higher inclusion fraction of 22.7%. This model was not identified in the original analysis (Sauerbrei et al.,

1998). Preference for a particular model may rest on external criteria, such as simplicity in practical use.

8.7   MODEL STABILITY FOR FUNCTIONS

Selecting a functional form for a continuous covariate in bootstrap replications greatly increases the number and types of candidate models available. A further issue which arises with functions is how to define similarities or differences between them and how to summarize them across bootstrap replications. Using BIFs alone is not adequate for assessing the effect of a covariate. In some replications a variable may be excluded, and in others it may be included as linear or as a nonlinear, monotonic or nonmonotonic FP function. Investigating interdependencies of inclusion and function selection for pairs of variables is no longer possible with 2×2 tables. Methods for summarizing variation between curves and a measure of curve instability are required. A measure recently proposed by Royston and Sauerbrei (2003) is presented. The key components of a model stability analysis for functions are then discussed and illustrated in a simple example. Finally, the issues in such an analysis are addressed in the GBSG breast cancer study, a realistic example.

8.7.1   Summarizing Variation between Curves

Each bootstrap replication in which a continuous covariate x enters the model generates a straight line or curve representing one estimate of the function. In the Cox model there is no intercept term (risks or hazards are estimated relative to an unspecified baseline hazard function); therefore, each estimated function must first be standardized. In normal error models and logistic regression models, variation in the intercept term across replications is of no interest and may also appropriately be removed by standardization. Royston and Sauerbrei (2003) standardized each log relative risk (or log relative odds) function to have mean zero when averaged over the empirical distribution of x for observations in the original data. In this case, each estimated risk in a given bootstrap replication is a multiple of the geometric mean risk (or OR) according to the particular function chosen. If $\widehat{f}_b(x)$ is the estimated function in the bth bootstrap replication, then the standardized function $\widetilde{f}_b(x)$ is given by

$$\widetilde{f}_b(x) = \widehat{f}_b(x) - \frac{1}{n} \sum_{i=1}^{n} \widehat{f}_b(x_i) \tag{8.1}$$

where x_i is the value of x for the ith observation in the original dataset.

To give a graphical summary of the variation among bootstrapped functions, Royston and Sauerbrei, (2003) calculated the cross-sectional mean (i.e. $q^{-1} f_{\text{bag}}(x)$; see Section 8.7.2) and an uncertainty band comprising the 5th and 95th centiles of $\widetilde{f}_b(x)$ across the bootstrap replications. The mean and uncertainty band were plotted against x at sample values x_i. Note that a replication in which $\widetilde{f}_b(x_i)$ is extreme at x_i relative to the selected functions in other replications propagates uncertainty across the whole of $\widetilde{f}_b(x)$ and, hence, tends to increase the width of the uncertainty band at x_i and possibly also at other values of x.

8.7.2 Measures of Curve Instability

Royston and Sauerbrei (2003) assessed the instability of a multivariable model of interest (the 'reference model') by considering variability among the bootstrap-generated curves $\tilde{f}_b(x)$. Some of the variability is uncertainty associated with estimating the functional form (many different forms are available within a flexible class such as FPs), some reflects the influence of other predictors which may or may not happen to be selected, and some is just random variation around the curve. Instead of deriving global instability measures across all covariates, Royston and Sauerbrei (2003) defined the following measure univariately for a continuous x of interest.

Let $f_{\text{ref}}(x)$ be the estimated function of x in the reference model, standardized as in Equation (8.1). Define $f_{\text{ref}}(x) = 0$ if x does not enter the reference model, and $\tilde{f}_b(x) = 0$ if x does not enter the model obtained in replication b. For a sample R of replications, a bootstrap summary of the function is obtained by using Breiman's (1996a) average or 'bagged' (bootstrap-aggregated) estimator

$$f_{\text{bag}}(x) = \frac{1}{|R|} \sum_{b \in R} \tilde{f}_b(x)$$

where $|R|$ denotes the number of members of R. By considering the decomposition

$$\frac{1}{|R|} \sum_{b \in R} [\tilde{f}_b(x) - f_{\text{ref}}(x)]^2 = \frac{1}{|R|} \sum_{b \in R} [\tilde{f}_b(x) - f_{\text{bag}}(x)]^2 + [f_{\text{bag}}(x) - f_{\text{ref}}(x)]^2$$

which may be written as

$$T(x) = V(x) + D^2(x)$$

the total variation $T(x)$ of bootstrapped functions around the reference function may be expressed as the sum of the within-subset variance $V(x)$ and the squared deviation $D^2(x)$ between the bagged and reference curves. Large values of $D^2(x)$ may indicate that the shapes of the reference curve and the bagged function differ. The function $V(x)$ quantifies the random variation and other contributions to variability of the individual curves $\tilde{f}_b(x)$ around the bagged function.

An additional stability measure is the conditional variance $V_{\text{cond}}(x)$ of the bootstrap functions in the subset R_x in which x enters the model:

$$V_{\text{cond}}(x) = \frac{1}{|R_x|} \sum_{b \in R_x} \left[\tilde{f}_b(x) - \frac{1}{q} f_{\text{bag}}(x) \right]^2$$

where $q = |R_x|/|R|$ is the bootstrap inclusion fraction for x with respect to R. Division by $q \leq 1$ inflates $f_{\text{bag}}(x)$ appropriately to allow for exclusion of the replicates in which $\tilde{f}_b(x) = 0$. If $q = 1$, then $V(x)$ and $V_{\text{cond}}(x)$ are identical, whereas for uninfluential variables (low q), $V(x)$ and $V_{\text{cond}}(x)$ may differ considerably.

Each stability measure as defined above is in fact a function of x. For practical application, summary measures are defined by averaging over the empirical distribution of x in the original sample:

$$V = \frac{1}{n} \sum_{i=1}^{n} V(x_i), \quad D^2 = \frac{1}{n} \sum_{i=1}^{n} D^2(x_i), \quad V_{\text{cond}} = \frac{1}{n} \sum_{i=1}^{n} V_{\text{cond}}(x_i)$$

8.8   EXAMPLE 4: GBSG BREAST CANCER DATA

Royston and Sauerbrei (2003) applied the MFP algorithm to the GBSG breast cancer data in 5000 bootstrap replications. Both the selection and the transformation of a variable of interest may affect the inclusion of other variables and the functional form selected for them. To make the problem tractable, Royston and Sauerbrei (2003) explored interdependencies among the inclusion or exclusion of binary variables and four simplified types of function for continuous variables: exclusion, linear functions, or FP1/FP2 functions. Further simplification was achieved by dimension reduction: variables which almost always or hardly ever entered the model were not required for the analysis of interdependencies. For example, the exponentially transformed variable enodes = exp($-0.12 \times$ nodes) was included in 100% and pgr in 98.8% of the replications, whereas gradd2 was selected only in 9.0%. It is clear that such variables must either be in a preferred model or out of it. Interdependencies are probably best investigated in subsets excluding replications with 'unusual' results (e.g. in this case, with pgr excluded or gradd2 included). This can help to reduce the number of variables and, therefore, the dimensionality of the analysis.

8.8.1   Interdependencies among Selected Variables and Functions in Subsets

For the remaining variables and simplified functions, Royston and Sauerbrei (2003) used hierarchical log-linear modelling to identify interrelationships within subsets of bootstrap replications. Several steps were required. For example, age was included in 4606 (92%) of the bootstrap replications, so the result matrix for the 4606 replications with age included was analysed. 'Unusual' results for age (e.g. age excluded) were thereby eliminated, further simplifying the remaining types of age function. With other variables also considered, a subset of 3005 replications was found in which the type of function (FP1 or FP2) chosen for age was independent of the inclusion of other variables. The predictors age, enodes and pgr were always included in this subset, and meno and er were never included. Royston and Sauerbrei (2003) called it a 'stable subset' of replications.

Only 24 types of simplified model seemed to fit the data in the stable subset best: enodes included as linear; age as FP1 or FP2; pgr as linear or nonlinear; gradd1 in or out; and three categories (exclusion, linear function, FP1/FP2 function) for size. Tabulation of the 24 frequencies showed that just three combinations of factor levels represented 58% of the 3005 replications, having in common an FP2 function for age and exclusion of size.

8.8.2   Plots of Functions

Figure 8.1 shows random samples of fitted curves for age, size, nodes and pgr from the stable subset of 3005 replications. All selected models for nodes were FP1 for enodes with power 1 (i.e. linear) and pass through a single point. The ranges have been restricted for plotting purposes. The curves for age and nodes are reasonably consistent across the replications, whereas those for size and pgr exhibit considerable variability. Bundles of curves from different FP models (especially linear and FP1 with various powers) may be discerned, particularly for pgr, where the standardization leads to a rather inhomogeneous appearance.

Figure 8.2 shows curve summaries (means, 5th and 95th centiles) for the effects of age, size, nodes and pgr over the entire stable subset. The graphs for age and nodes show

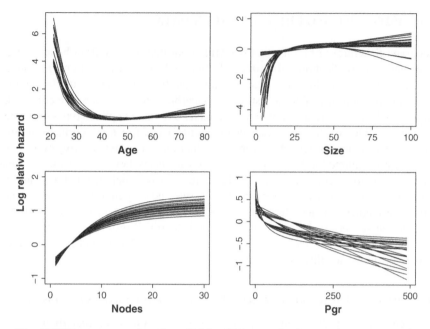

Figure 8.1   GBSG breast cancer data. Sets of 20 fitted FP curves for four continuous prognostic factors. Random samples from the stable subset of 3005 bootstrap replications are shown. (Adapted from Royston and Sauerbrei (2003) with permission from John Wiley & Sons Ltd.)

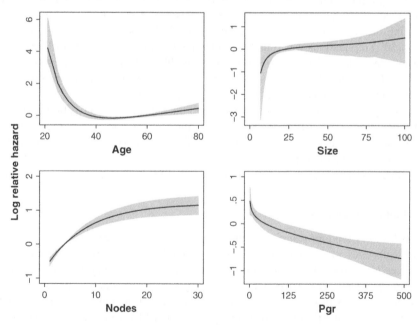

Figure 8.2   GBSG breast cancer data. Curve summaries of four continuous prognostic factors based on models fitted in the stable subset of 3005 bootstrap replications. Solid lines, means; shaded areas, 5th to 95th centiles of fitted curves. (Adapted from Royston and Sauerbrei (2003) with permission from John Wiley & Sons Ltd.)

well-defined effects with relatively narrow uncertainty bands. pgr is included in nearly every bootstrap replication. Its effect appears to be fairly precisely estimated at low values but very uncertain at high values.

The individual curves show that size is either included as a linear function (always with a small positive effect estimate) or as a nonlinear function with a steep increase for small tumours and a flattening off from about 20 mm. The latter type of function agrees well with clinical thinking and is confirmed by other analyses (e.g. Vorschraegen et al., 2005). Several studies indicate that this functional form plausibly describes the biological behaviour. In the GBSG data the power is insufficient to select such a nonlinear function for size in a multivariable context, particularly when adjusting for the dominating variable nodes as a continuous function. In 60% of the replications size did not enter the model at all. When it did enter, power was often too low for a nonlinear effect to be significant. Therefore, the default, a linear function, was selected. However, when a nonlinear function was selected (which occurred in about 1000 replications), the functional form was always very similar. The bootstrap analysis indicates that Sauerbrei and Royston's (1999) original model, in which size was excluded, may reasonably be extended with a nonlinear function of size. Without the bootstrap investigation, such an extension would presumably not have been considered. A decision on the final model needs discussion with clinical experts and consideration of similar studies in the literature. Here, the specific question of whether to include size is discussed, but such a situation is likely to be commonplace in many studies with insufficient power to provide convincing evidence of a nonlinear function.

8.8.3   Instability Measures

In Chapter 2, the role of the full model was discussed in detail. Those arguing for it on the grounds of claimed lower bias do not usually consider transformation of covariates. Accepting the importance of looking for transformations, estimating the best FP2 function for each continuous variable is a natural extension of the full model approach. Even for the 'usual' (linear) full model, there is already a cost in terms of the complexity and hence plausibility and practical applicability of the model. This issue is even more important in the full FP model, comprising all binary predictors and FP2 functions for all continuous predictors (Royston and Sauerbrei, 2003). In the GBSG breast cancer example, nodes is an exception, since it was restricted to FP1 with positive powers of the pre-transformed variable enodes, as described by Sauerbrei and Royston (1999). In Table 8.4, the instability measures V, D^2 and V_{cond} are reported for the continuous predictors age, size , enodes, pgr and er.

In the full model there is no selection of variables; therefore, V_{cond} is identical to V here. Only age, enodes and pgr entered the reduced model selected by MFP(0.05, 0.05).

The instability of $f(\text{size})$, $f(\text{enodes})$, $f(\text{pgr})$ and $f(\text{er})$ in the full FP model is very large compared with the values for the reduced model in the complete set of 5000 replications. Only for age is the instability of the full model less than that of the reduced model. This finding is not surprising, since the effect of age is best estimated by an FP2 function. Such a function is fitted in all the full FP models, but in reduced models age is excluded in some replications, resulting in a zero function, and is modelled as an FP1 in others. Furthermore, the selection or omission of other variables, especially meno, may destabilize the function for age.

All the instability measures are smallest for enodes in the reduced model, reflecting the stability of the functions chosen in the replications and the close agreement with the

Table 8.4 GBSG breast cancer data. Instability measures ($\times 10^4$) for estimated functions of each continuous predictor.

Variable x	Full FP model[a]		Reduced model[b]					
	V	D^2	Subset R	V	D^2	V_{cond}	$|R|$	$|R_x|$
age	145	4	All	245	28	219	5000	4606
			Stable	72	5	72	3005	3005
enodes	75	11	All	46	1	46	5000	5000
			Stable	45	1	45	3005	3005
pgr	5808	157	All	1013	37	1005	5000	4941
			Stable	874	32	874	3005	3005
size	2813	159	All	1207	146	2464	5000	2008
			Stable	1329	138	2910	3005	1140
er	283	7	All	83	4	352	5000	955
			Stable	–	–	–	3005	0

[a] Full FP model (based on 1000 bootstrap replications) was obtained with selection of best FP2 function.
[b] Reduced model (based on 5000 bootstrap replications) was obtained by the MFP procedure with nominal
  P-value of 0.05 for variable and function selection. 'Stable' refers to the stable subset of 3005 replications
  in which meno and er did not enter the model but age, nodes and pgr were included. $|R|$ and $|R_x|$
  denote respectively the number of replications in the subset R and the number in which x entered the
  model.
(Adapted from Royston and Sauerbrei (2003) with permission from John Wiley & Sons Ltd.)

reference function. This may be the result of constraining the functions to be monotone. In addition, enodes was selected in all 5000 replications. All instability measures for age are considerably reduced in the stable subset and the variance V is also much smaller than in the full FP model. The variance measure V varies widely between predictors, being highest for the weak predictor size and nearly as high for pgr. The surprising finding that all measures for er (which did not enter the reference model) are much lower than for pgr (which did) is explained by the very stable reference function chosen for er (identically zero) and the generally small marginal effect of er in multivariable models. All measures for age and pgr are reduced in the stable subset, an effect not seen with size. The conditional variance measure V_{cond} indicates high instability of functions chosen for size in the 40% of replications in which it entered the model (see also Figures 8.1 and 8.2), comparable to V for the full FP model.

8.8.4 Stability of Functions Depending on Other Variables Included

Table 8.5 summarizes the effect of inclusion or exclusion of meno and er on the stability of the age function, $f(\text{age})$, in the reduced model. Results for 'Stable' are for the subset of 3658 replicates in which meno did not enter the model. Since age and meno are highly correlated, the inclusion or exclusion of meno has a large effect on the shape of the function for age, reflected by the large value of D^2 (see rows 'meno in' and 'meno out'). The values of V and V_{cond} are about twice as large when meno is selected than otherwise. In the subset with meno excluded, the effects on $f(\text{age})$ of inclusion or exclusion of er is much smaller. Inclusion or exclusion of size or gradd1 or gradd2 had no effect (data not shown).

Table 8.5   GBSG breast cancer data. Instability measures ($\times 10^4$) for estimated functions of age. q is the bootstrap inclusion fraction for the relevant predictor with respect to R. Other details as for Table 8.4.

Subset R	V	D^2	V_{cond}	$|R|$	$|R_x|$	q (%)
All	245	28	219	5000	4606	92.1
Stable	72	5	72	3005	3005	100
meno in	212	678	184	985	948	96.2
meno out	122	0	76	4015	3658	91.1
meno out, er in	144	1	90	728	647	88.9
meno out, er out	117	0	73	3287	3011	91.6

(Adapted from Royston and Sauerbrei (2003) with permission from John Wiley & Sons Ltd.)

The value of q and the type of reference function must always be borne in mind when interpreting the unconditional instability measures. Low values of q *per se* represent stability in the sense that the variable was rarely selected and its reference function is probably close to zero. Therefore, the judgement that $f(\mathrm{er})$ is fairly stable with respect to its reference function of zero is reasonable, even though the value of V_{cond} is not particularly small. By contrast pgr, although a strong and highly statistically significant predictor, has a rather unstable estimated function. The results for the full FP model (see Table 8.4), in which all the reference functions are nonzero, support this conclusion.

8.9   DISCUSSION

The emphasis in this chapter on assessing the instability of each individual predictor reflects a medical or epidemiological concern that focuses on the role (e.g. the functional form) of such factors in influencing the prognosis or the risk of an event. By contrast, in the context of prediction, assessing the instability of the index $\widehat{\eta}$ or a transformation of it to another scale (e.g. probability) is more relevant. The techniques outlined here may also be used with the index.

It has been known for many years that variable selection procedures yield unstable results. This has been shown by data splitting (Harrell et al., 1984) and bootstrap resampling (Chen and George, 1985; Altman and Andersen, 1989; Sauerbrei and Schumacher, 1992). These investigations have shown that variables with a strong (meaning, in this context, statistically highly significant) influence on the outcome enter in nearly all replications, but variables with a weak effect enter in a much smaller fraction, say <40%. Since there are usually several such variables, different models are obtained in nearly every replication.

Austin and Tu (2004b) used techniques similar to those of Sauerbrei and Schumacher (1992) to develop models to predict 30-day mortality of patients with acute myocardial infarction. They used BE within each of a large number of bootstrap samples and determined the BIF for each variable. They concluded that bootstrap sampling in conjunction with automated model selection methods could identify a parsimonious model with excellent predictive performance. Their conclusion confirms that simple models including all 'strong' predictors have discriminative ability similar to that of more complicated models using more variables (Sauerbrei, 1999; Hand, 2006).

8.9.1   Relationship between Inclusion Fractions

When selecting the functional form for a continuous covariate, instability is evident even when considering just one predictor. For example, the FSP (see Section 4.10) chooses among eight FP1 and 36 FP2 models. When several continuous covariates are available for transformation, application of the MFP procedure selects within a space of many hundreds or thousands of types of model (5760 'simplified' functions in the GBSG study). Selecting from a large class of permissible functions may result in a predictor which overfits the data, whereas use of a small class raises the danger of underfitting. Statisticians need to find a compromise, keeping in mind the stability of the chosen model (Breiman, 1996b).

Apart from issues such as the applicability of a model for practical use and imprecision in the estimated effect of individual factors, experience of the instability problem prompted Sauerbrei (1999) to argue in favour of simplicity by building models which concentrate on including strong predictors (see also Hand (2006)).

Instability of the model selection process can be investigated by considering relationships between the inclusion of variables. Sauerbrei and Schumacher (1992) proposed a simple approach by analysing all pairwise relationships within a bootstrap result matrix. Log-linear modelling of the result matrix from bootstrap replications may be used for a more detailed analysis and is required for more complex problems such as establishing general relationships between the inclusion of variables and a simplified classification, for example FP1 versus FP2, of the functions chosen in the bootstrap replications. Details of one example have been given, and more can be found in Royston and Sauerbrei (2003). A key element of such an approach is the attempt to limit model instability by eliminating 'atypical' replications in which important variables were excluded or unimportant ones were included. Outliers or influential values which are repeated several times in some of the bootstrap samples are an important source of 'atypical' replications. Royston and Sauerbrei (2003) derived a stable subset comprising >60% of the replications in which the models did not include weakly influential variables but did include the strong predictors. In the stable subset in the GBSG example, three types of model dominated, models which seemed plausibly to account for the unknown complex relationship between the predictors and the outcome. A stability analysis can provide information not available in the original analysis. To take things further than this, subject-matter knowledge is required.

8.9.2   Stability of Functions

Breiman's (1996a) bagging method was used to compute an estimate of the function of a continuous predictor in subsets of bootstrap replications (see Figure 8.2). Although the bagged curve is composed of FP functions of degree ≤ 2, it is in fact almost always an FP function of degree > 2. The use of bagging in conjunction with the FP method generates extremely flexible functions from a low-dimensional basis. While such functions are useful for visualization and in the computation of instability indices, as they stand they are not the concise summaries which are needed when presenting models for practical use.

Royston and Sauerbrei (2003) proposed measures of model instability appropriate for continuous predictors, representing the variance within bootstrap samples (V) and the mean-squared deviation D^2 between the reference curve and the bagged curve. No attempt was made to amalgamate the measures across the predictors to form an overall measure for the whole model.

Handling replications in which a variable did not enter the model is an issue which needs further exploration. It is a generalization of the difficulties associated with selection bias of a regression parameter estimate (Miller, 2002). Royston and Sauerbrei (2003) considered the two extreme possibilities: either ignore these replications and use V_{cond}, or take $f(x) = 0$ in them and use V. The evidence for excluding a continuous variable from a reference model is conclusive if it enters in a few replications only. The decision to set $f(x) = 0$ hardly affects the instability measures if the variable enters in nearly all replications (e.g. pgr in the GBSG breast cancer example). By contrast, the difference between V and V_{cond} is much larger for 'weak' factors (e.g. size and er; see Table 8.4). If such a factor is of particular interest, then additional analyses, such as fixing the other parts of the model and estimating only the function for the factor of interest, may be informative. Deriving a 'stable' subset of bootstrap replications is a step in that direction.

CHAPTER 9

Some Comparisons of MFP with Splines

Summary

1. Spline functions provide flexible models for continuous covariates. Several 'flavours' of spline function are available. All require values for tuning parameters which control model fit.
2. No generally agreed procedures for multivariable model-building with splines exist.
3. Here, restricted cubic splines (RCSs) and smoothing splines (SSs) are compared with FPs in examples.
4. To allow 'like-for-like' comparisons, multivariable model-building procedures for the two types of spline were developed and are described here: MVRS for RCSs and MVSS for SSs. Both procedures adhere to the MFP philosophy of BE and function selection using a closed test procedure.
5. Selected spline and MFP models are compared in four datasets. With an adequate sample size, the selected variables and functions are broadly similar between FPs and splines.
6. Modelling using splines with many d.f. can generate uninterpretable fitted functions, particularly with smaller sample sizes.
7. Models that include different types of function (FPs or splines) nevertheless produce predictors that substantially agree.
8. Interpretation and transportability are easier with MFP models than with spline models. MFP is more generally useful for multivariable modelling than splines.

9.1 INTRODUCTION

In this chapter, MFP models are informally compared with the results from two model-building strategies that use spline functions for continuous predictors. We show that spline functions generate fitted curves that broadly resemble FP functions, albeit sometimes with artefactual features that defy sensible interpretation.

Multivariable Model-Building　　Patrick Royston, Willi Sauerbrei
© 2008 John Wiley & Sons, Ltd

9.2   BACKGROUND

In the final two sentences of their Discussion, Rosenberg et al. (2003) state that

> *non-parametric risk regression presents a valuable extension of the standard logistic regression model for case-control studies. However, it is our opinion that the most scientifically credible application of this approach requires a disciplined approach to model selection.*

The paper describes the application of several types of spline model to predict the risk of oral cancer from alcohol exposure in a case-control setting. It is clear from experience that, even in the case of a single continuous predictor (alcohol exposure), selecting a 'good' model and producing a 'convincing' interpretation of the results may prove challenging. There are several reasons for this difficulty. First, we never know the true model (if such a concept even means anything). Second, flexible ('nonparametric') regression methods present a locally detailed picture of the functional relationship in the data to hand, but such a picture is difficult to interpret. It may be hard to distinguish the signal from the noise. Third, the details are unlikely to be reproduced in other datasets. The multivariable setting complicates the situation considerably. See also Section 3.6 for further comments on nonlinear modelling.

Even in the simplest situation of a single predictor, model selection and interpretation of results are difficult issues with splines. Naturally, things are even more difficult with multivariable models. Reliable and generally accepted model-building procedures which allow for selection of variables are still lacking. It is hardly surprising that practitioners, faced with a bewildering variety of more or less complex methodologies with no clear guidance and no credible body of practical experience to draw on, are at a loss as to what to do.

Thoughtful analysts restrict their methodological toolkit to techniques they in principle understand. Differences between spline methods and how to work with them are comprehensible to experts but probably to few practitioners. Complex, unreliable and unconvincing functions published in the literature (e.g. Hastie et al., 1992) may discourage potential users of splines.

The MFP algorithm provides a principled approach for systematic selection of variables and FP functions. To increase the flexibility of the functions available, we have transferred the MFP philosophy to spline modelling. We describe two methods for selecting variables and spline functions of continuous predictors in a multivariable context. The first, MVRS (multivariable regression splines) combines backward elimination with a search for a suitable regression spline function (Royston and Sauerbrei, 2007b). The second, MVSS (multivariable smoothing splines), is based on a suggestion of Hastie and Tibshirani (1990); see Section 9.4. Cubic smoothing splines are the main competitor of regression splines, and have gained some popularity in applications. Both procedures are closely related to MFP and use a modified version of the FSP (see Section 4.10) to select a spline function while preserving the 'familywise' type I error probability for a given predictor.

The MVRS and MVSS procedures are just two possibilities for model-building with splines. We have chosen to illustrate them because (a) they are similar in spirit to MFP and, therefore, offer reasonable comparisons with MFP, (b) they can be used with generalized linear models and Cox models, and (c) they were not too difficult to program in Stata, utilizing existing tools. Other software platforms offer other procedures. For example, a recent method of multivariable modelling with penalized regression splines (Wood, 2006) is implemented in R.

Especially in a multivariable context, several problems remain with all spline methods. We restrict investigations to MVRS and MVSS.

Here, we compare the results of applying MFP, MVRS and MVSS to four datasets, with continuous, binary and time-to-event outcomes respectively. Reassuringly, we find that the results from these different approaches to modelling are generally compatible, albeit of course with some differences. The main motivation for splines is their ability to interpolate essentially any continuous function smoothly and accurately (de Boer, 2001; Wood, 2006). As discussed in Chapter 6, analysis of residuals is needed to assess the fit of a model. Judging whether additional features should be incorporated in a final model remains to some extent subjective, and depends also on subject-matter knowledge.

9.3   MVRS: A PROCEDURE FOR MODEL BUILDING WITH REGRESSION SPLINES

Before indicating how MVRS works, we outline restricted cubic spline functions, the building blocks of a multivariable additive spline model. The procedure for selecting a spline function of a single continuous predictor is then described. Further details and a software implementation (mvrs) in Stata are given by Royston and Sauerbrei (2007b).

9.3.1   Restricted Cubic Spline Functions

A spline function is created by joining several polynomials of the same degree at particular x values known as knots. A detailed account of splines is given by de Boer (2001). Joined-up polynomials have the potential for almost unlimited flexibility. When the polynomials have degree d, the spline function and all its derivatives up to order $d - 1$ are constructed to be continuous at all the knots. This condition ensures smoothness across the knots and limits the chance of overfitting the data by reducing the number of available parameters. A common choice is $d = 3$ (cubic splines), although linear and quadratic splines are also sometimes used.

It is recognized that cubic splines may show 'wild' behaviour at the extremes of x where the data are sparsest. Restricted cubic regression splines are designed to try to control this unsatisfactory behaviour. This type of spline has m interior knots $k_1 < \ldots < k_m$ and two 'boundary' knots $k_{min} < k_1$ and $k_{max} > k_m$, usually but not necessarily placed at the extremes of x. The function is linear in the tails, i.e. for $x < k_{min}$ and for $x > k_{max}$. The formula for such a spline is

$$s(x) = \gamma_{00} + \gamma_{10}x + \sum_{j=1}^{m} \gamma_j[(x - k_j)_+^3 - \lambda_j(x - k_{min})_+^3 - (1 - \lambda_j)(x - k_{max})_+^3]$$

where

$$(x - k)_+^3 = \begin{cases} (x - k)^3 & \text{if } x \geq k \\ 0 & \text{if } x < k \end{cases}$$

$$\lambda_j = \frac{k_{max} - k_j}{k_{max} - k_{min}}$$

If we write $\beta_0 = \gamma_{00}$, $\beta_1 = \gamma_{10}$ and for $j = 1, \ldots, m$:

$$\beta_{j+1} = \gamma_j$$
$$\beta_j(x) = (x - k_j)_+^3 - \lambda_j(x - k_{min})_+^3 - (1 - \lambda_j)(x - k_{max})_+^3$$

then $s(x) = \beta_0 + \beta_1 x + \beta_2 v_1(x) + \ldots + \beta_{m+1} v_m(x)$. This shows that the m knots give rise to a model with $m + 1$ 'basis functions' and hence $m + 1$ d.f. (β_0 is not counted). The number and positions of the interior knots are the key factors controlling the flexibility of the function. The two boundary knots are usually placed at the extreme values of x.

Purely for illustration, Figure 9.1 compares four spline functions, fitted to Rosenberg et al.'s (2003) oral cancer dataset. The first three splines have the same three interior knot positions, at the 25th, 50th and 75th centiles of the distribution of 1 oz drinks/week, i.e. 1.5, 15.75 and 48; the fourth spline has four additional knots. The restricted cubic splines have boundary knots at the extreme covariate values. Other choices of boundary knots could have changed the shape of the functions within the range of the data. Linear splines are discontinuous in the first derivative, generating the 'corners' seen in the upper left panel. The restricted cubic spline on four d.f. gives a very smooth function here. The unrestricted cubic spline on six d.f. has two additional d.f. due to two fewer constraints on the quadratic and cubic terms of the function. The restricted cubic spline on eight d.f. gives an unstable estimate of the function.

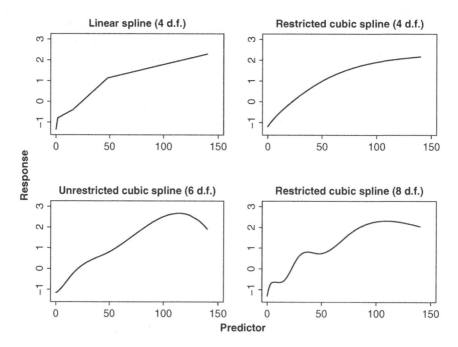

Figure 9.1 Oral cancer data. Four splines fitted to log odds of caseness as a function of the number of 1 oz drinks per week.

9.3.2   Function Selection Procedure for Restricted Cubic Splines

As with the FSP for FP functions (see Section 4.10), we constructed an approximate closed test procedure based on an ordered sequence of tests on spline functions designed to maintain the overall type I error probability at a prespecified nominal level α (Marcus et al.,1976). Initially, the most complex permitted spline model is chosen. Suppose it has $m + 1$ d.f., where m is the maximum number of knots to be considered and $m = 0$ means the linear function. As with the FSP for FP functions, the suggested default choice is four d.f.; therefore, $m = 3$.

The knot positions are chosen by default at predetermined percentiles of the distribution of x in the sample. Otherwise, they may be chosen according to subject-matter knowledge.

Let us call the most complex model with m knots M_m, the linear function M_{lin} and the null model (omitting x) M_0. First, M_m is compared with M_0 using a χ^2 test on $m + 1$ d.f. If the test is not significant at the α level, then x has no effect, and the FSP terminates. In a multivariable context, x would be eliminated. Otherwise, the algorithm next tests M_m against M_{lin} on m d.f. If the deviance difference is not significant at the α level, then M_{lin} is chosen and the algorithm stops.

Now consider the m possible spline models using just one of the m available knots. The best-fitting of these models, say M_1, is found and compared with M_m on $m - 1$ d.f. If M_m does not fit significantly better than M_1 at the α level, there is no evidence that a more complex model is needed, so M_1 is accepted and the algorithm stops. Otherwise, M_1 is augmented with each of the remaining $m - 1$ knots in turn, and the best fitting model, M_2, is found and compared with M_m on $m - 2$ d.f. The procedure continues in this fashion until either a test is non-significant and the procedure stops, or all the tests are significant, in which case model M_m is the final choice.

If x is to be 'forced' into the model, the first comparison, between M_m and M_0, is omitted.

All tests are based on χ^2 statistics from deviance ($-2 \times$ log-likelihood) differences. A more usual approach to selecting functions is to minimize an information criterion such as AIC, as for example in the S-plus routine `step.gam` (Chambers and Hastie, 1991, p. 282).

9.3.3   The MVRS Algorithm

Used with BE, the FSP for regression splines, as just described, is the building block for the multivariable model-building procedure, MVRS. The MVRS algorithm and the Stata program `mvrs` (Royston and Sauerbrei, 2007b) are identical in concept to MFP and `mfp` respectively. Predictors are considered in decreasing order of their statistical significance in a full linear model, with the most significant first. The algorithm repeatedly cycles over the predictors in the same order, if necessary updating the model in light of the results of FSP tests on each variable. The procedure terminates when there is no further change in the variables included in the model and in the spline functions (knots) chosen for each continuous variable.

9.4   MVSS: A PROCEDURE FOR MODEL BUILDING WITH CUBIC SMOOTHING SPLINES

9.4.1   Cubic Smoothing Splines

For normal-errors data, a cubic smoothing spline (CSS) function $f(x)$ on observations $\{x_i, y_i\}$ ($i = 1, \ldots, n$), with each x_i lying in an interval $[a, b]$, minimizes a penalized residual sum of squares defined by

$$\text{RSS}_{\text{pen}} = \sum_{i=1}^{n} \{y_i - f(x_i)\}^2 + \lambda \int_a^b \{f''(x)\}^2 \, dx \qquad (9.1)$$

where $f''(x)$ is the second derivative of $f(x)$. For a given, nonzero value of λ, the solution $\widehat{f}(x)$ which minimizes RSS_{pen} is a cubic spline with a knot at every distinct value of x and with regression coefficients shrunk towards zero. With $\lambda = 0$, the solution interpolates the data; with $\lambda = \infty$, it is a straight line; and with intermediate values of λ, it is a more or less 'wiggly' curve. The 'wiggliness' (complexity) is controlled by the equivalent degrees of freedom (e.d.f.), a monotonic function of λ (see Green and Silverman (1994, p. 37) for details).

In generalized linear and Cox models, Equation (9.1) is generalized to a penalized (partial) likelihood function. The solution $\widehat{f}(x)$ is the additive predictor or index of the resulting model.

In software implementations, the complexity of the function is specified through the e.d.f. A CSS function with four e.d.f. is approximately equivalent in complexity to a restricted cubic spline with three knots.

9.4.2   Function Selection Procedure for Cubic Smoothing Splines

Here, we adopt essentially the method suggested by Hastie and Tibshirani (1990, section 9.4, pp. 259–260) to decide whether x is significant, and if so, how complex its spline function should be. A 'coarse' regimen of three possible fits is considered, defined by e.d.f. 0 (drop x), 1 (linear function) or ν (most complex allowed function). As the primary method, we take Hastie and Tibshirani's suggestion of $\nu = 4$. To allow greater flexibility and, thereby, to demonstrate problems that may arise with overcomplex functions, we also illustrate some results with $\nu = 8$. Note that restricting ν to be 0, 1 or 8 is not intended as a serious modelling proposal, since the chosen function (with $\nu = 8$ e.d.f.) is almost certain to be overfitted if nonlinearity is detected.

The closed-test approach embodied in the FSP for FPs and regression splines applies straightforwardly to selecting a CSS function. Only two tests are required: the model with ν e.d.f. versus the null model, to determine if x should be eliminated, and the model with ν e.d.f. versus the linear model, to check if a nonlinear function is required. The two tests are based on an assumed χ^2 null distribution for the difference in penalized deviances, requiring ν and $\nu - 1$ d.f. respectively.

9.4.3   The MVSS Algorithm

Used with BE, the FSP for CSS functions, as just described, is the building block for a multivariable model-building procedure, MVSS, which is conceptually similar to MVRS. MVSS actually builds generalized additive models (GAMs) (Hastie and Tibshirani, 1990). The main difference from MVRS is that MVSS considers only three possible models for each continuous predictor, with 0, 1 or ν e.d.f., representing omission of the variable, a linear function or, depending on ν, a more or less wiggly function. In contrast, MVRS has access to a richer set of possible models. Predictors are considered in decreasing order of their statistical significance in a full linear model, with the most significant first. The algorithm cycles over the predictors repeatedly in the same order. If necessary, the model is updated in light of the results of FSP tests for CSS functions on each individual variable. The procedure terminates when there is no further change in the variables included in the model and in the spline functions (identified by their e.d.f.) chosen for each continuous variable.

To fit CSS models, we used the Stata program gam (Royston and Ambler, 1998), which calls a Fortran program (gamfit) written by T. Hastie and R. Tibshirani (see http:// lib.stat.cmu.edu/general/gamfit for the source code). MVSS is implemented in an unpublished Stata program, mvss, which is available from the first author on request. mvss repeatedly calls gam to fit the required GAMs.

9.5   EXAMPLE 1: BOSTON HOUSING DATA

We obtain results with the MFP, MVRS and MVSS model-building procedures, with the following aims:

- to compare the models with respect to the selected variables and the complexity of the selected functions;
- to study the effects of varying the sample size on the above;
- to examine what happens if more complex spline functions (with eight d.f. rather than our preferred default of four d.f.) are used as the starting point for function selection;
- to compare predictors (indexes) from MFP and spline models.

The procedures are applied to the Boston housing data in this section, and to three additional datasets in subsequent sections.

As reported by Harrison and Rubinfeld (1978), the data were assembled for an analysis of the possible effect of air pollution on median house prices in 506 census tracts of the Boston Standard Metropolitan Area. In addition to nox (concentration of airborne nitrogen oxide pollutants), 12 other variables (11 of them continuous) were recorded, all considered likely to influence house prices. The correlation matrix of the predictors (not shown) clearly indicates a high degree of interdependence, e.g. three absolute Spearman correlations are > 0.8. The outcome variable y is $\log_e$ of the median value of owner-occupied houses in thousands of dollars. As discussed in Chapter 2, we must assume that the variables in the published dataset were rigorously preselected. Practically all of them are significant in a multivariable model. The dataset has been used by several authors (Samarov, 1993; Pena, 2005; Yuan and Lin, 2005) e.g. to illustrate methodological issues, including selecting influential covariates and modelling continuous predictors.

Table 9.1 shows the models selected by the five procedures. There is a fair degree of consistency among the results. For example, zn, indus and age are clearly uninfluential in a multivariable setting, except for indus with MVRS(8), where the function suggests overfitting (not shown). Predictors chas, ptratio, crim, rm, dis, lstat are always selected, the last four with a nonlinear function and ptratio as linear (chas is binary). Results for nox, rad, tax and bk are more mixed. The four variables are always selected, sometimes with linear and sometimes with non-linear functions.

Figure 9.2 shows the estimated functions for the nine continuous predictors that were always selected.

All functions are plotted on the same vertical scale to allow one to compare the strength of the effects. The predominant impression is of similarity of results from the different model-building procedures. The strongest predictor appears to be lstat. Closer inspection reveals that some implausibly 'wiggly' functions have been chosen by MVRS(8) and MVSS(8), notably for nox, dis, tax and bk. The FP function for crim differs from the spline functions

Table 9.1   Boston housing data. Comparison of predictors and functions selected at the 5% nominal significance level by five different model-building methods. MFP, MFP(0.05, 0.05); MVRS(4) or MVRS(8), multivariable regression spline procedure with four or eight default degrees of freedom per continuous predictor; MVSS(4) or MVSS(8), multivariable smoothing spline procedure with four or eight default degrees of freedom per continuous predictor.[a]

Variable	MFP	MVRS(4)	MVSS(4)	MVRS(8)	MVSS(8)
crim	1, 2	4	4	2	8
zn	–	–	–	–	–
indus	–	–	–	5	–
chas[b]	✓	✓	✓	✓	✓
nox	lin	lin	4	8	8
rm	0.5, 0.5	3	4	2	8
age	–	–	–	–	–
dis	−2, 1	4	4	4	8
rad	lin	lin	4	lin	lin
tax	lin	3	4	lin	lin
ptratio	lin	lin	lin	lin	lin
bk	lin	lin	4	lin	8
lstat	0.5	4	4	4	8
R^2	0.834	0.848	0.857	0.872	0.879

[a] Key: for MFP, numbers indicate FP powers selected; for MVRS/MVSS, numbers indicate d.f. of selected spline function; lin = linear; ✓ = binary variable selected; – = variable eliminated.
[b] Binary predictor.

for values >25. The proportion of variance explained R^2 is broadly similar across the models, with the more complex eight-d.f. models not surprisingly yielding the largest values (see Table 9.1).

9.5.1   Effect of Reducing the Sample Size

A sample size of 506 and explained variation exceeding 0.80 constitutes a dataset–model combination with a lot of information. To see how the procedures would compare with smaller amounts of information, we divided the sample at random into four nearly equal subsets. The models were rebuilt using one half of the sample, and again with half of that subset (i.e. 25% of the original observations). The results are given in Table 9.2.

With a reduced sample size, greater variation occurs among the variables and functions selected. Overall, a smaller number of variables and a smaller number of nonlinear functions are selected. With half of the data used, MFP and MVRS eliminate bk, and some variables are included as a linear instead of a nonlinear function. The latter can also be seen for the MVSS procedures. With only a quarter of the data used, the differences are much more pronounced. Using half of the data, 10 variables are selected by the two spline procedures with eight d.f., which reduces to five selected variables in the quarter of the data.

With the smallest sample size, the considerable loss of power for variable inclusion when allowing eight d.f. instead of four d.f. is not surprising. The first step of the FSP concerns the

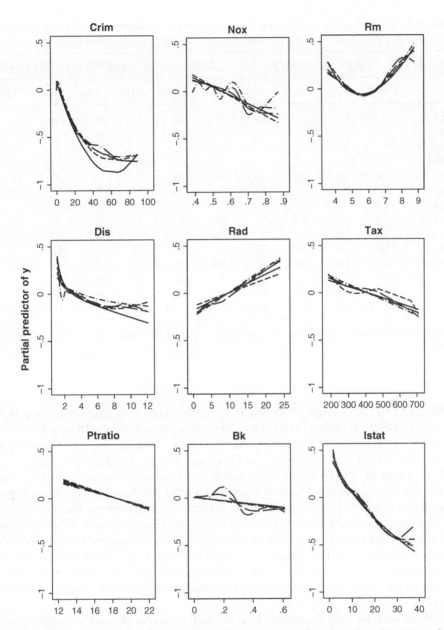

Figure 9.2 Boston housing data. Partial predictors (centred on zero) for selected continuous covariates, according to five multivariable models: MFP (solid lines), MVRS(4) (short dashes), MVSS(4) (long dashes), MVRS(8) (short dashes and dots), MVSS(8) (long dashes and dots).

inclusion or exclusion of a variable. This test uses a critical value from a χ^2 test on either four or eight d.f. The improvement in fit achieved by the more complex eight d.f. function is less marked than with the four d.f. function. MVRS(4) includes four variables not selected by MVRS(8). The latter adds age with a nonlinear function. This is surprising, since age

Table 9.2 Boston housing data. Comparison of predictors and functions selected at the 5% nominal significance level by five different model-building methods using reduced sample sizes.[a]

Variable	MFP		MVRS(4)		MVSS(4)		MVRS(8)		MVSS(8)	
	Half	Qtr.	Half	Qtr.	Half	Qtr.	Half	Qtr.	Half	Qtr.
crim	0, 0.5	lin	2	lin	lin	lin	2	lin	lin	8
zn	–	–	–	–	–	–	–	–	–	–
indus	–	–	–	–	–	–	–	–	–	–
chas	in	–	in	–	in	in	in	–	in	–
nox	lin	lin	lin	lin	4	lin	4	–	8	lin
rm	1, 1	3	2	lin	4	lin	2	lin	8	lin
age	–	–	–	–	–	–	–	5	–	–
dis	−1	−1	2	3	lin	4	3	3	lin	8
rad	lin	lin	lin	lin	lin	lin	lin	–	lin	–
tax	lin	lin	lin	lin	4	lin	lin	–	lin	–
ptratio	lin	lin	lin	lin	lin	lin	lin	–	lin	–
bk	–	–	–	–	4	lin	–	–	8	–
lstat	3, 3	lin	lin	4	4	4	4	lin	8	8
R^2	0.830	0.825	0.817	0.826	0.849	0.837	0.854	0.815	0.873	0.839

[a] Half: random half of the data, $n = 253$; Qtr.: quarter (random half of the random half), $n = 127$. Other details as for Table 9.1.

is included in no other model. With MVSS the difference is even larger. Starting with four d.f., five variables (all linear or binary) are additionally included compared with the eight-d.f. version. With the smallest sample size, the number of nonlinear functions selected is also much reduced. Only dis is selected with a nonlinear function by all five procedures.

Partial predictors for the five variables with the most markedly nonlinear effects are shown for the three sample sizes in Figure 9.3. Results for MVSS(8) have been omitted because of space constraints. The plots further illustrate the tendency to postulate a linear function or a function with only a small amount of curvature for the smallest sample size. Results from MFP and MVRS(4) are remarkably similar for all sample sizes, and agreement with the MVSS(4) functions is also acceptable. Clearly, some of the MVRS(8) and MVSS(8) functions (the latter not shown) are too 'wiggly' and are hardly interpretable. The wide pointwise CIs (not shown) for the spline curves caution against overinterpretation of minor features of the fitted functions.

Residual plots for several variables from the MFP model for the smallest sample size (see Table 9.2) are given in Figure 9.4. In general, the fit seems reasonable, but some of the curvature in the data has been missed, probably because the power is too low to detect more complex nonlinearity.

Despite the smaller number of variables and nonlinear functions selected with the reduced sample sizes, R^2 values decrease only slightly. Compared with the full data, R^2 decreases in the quarter of the data by 0.009 for MFP, 0.022 for MVRS(4), 0.020 for MVSS(4), 0.057 for MVRS(8) and 0.040 for MVSS(8). This example illustrates the general point that the explained variation may be increased only slightly by adding factors with a weak effect or incorporating weak nonlinearity. See also the discussion in Section 2.9.4.

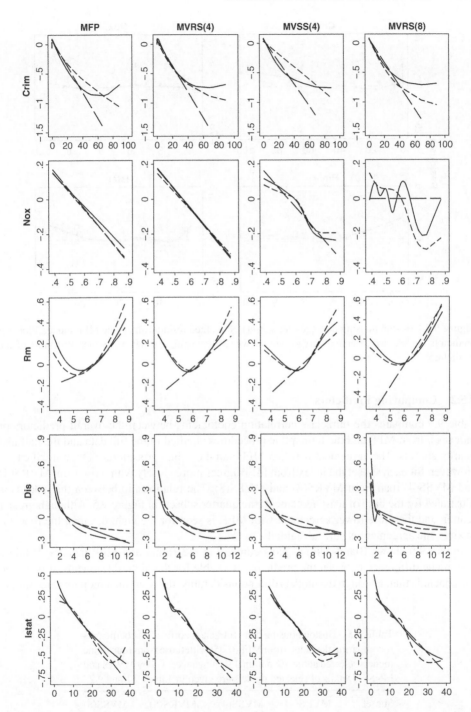

Figure 9.3 Boston housing data. Partial predictors (centred on zero) for five continuous covariates, displayed in rows of the graph, according to MFP, MVRS(4), MVSS(4) and MVRS(8) model-building procedures. The three curves in each panel correspond to the full data ($n = 506$, solid lines), a random 50% ($n = 253$, short dashes), and a further random 50% (quarter subset, $n = 127$, long dashes). See Tables 9.1 and 9.2 for details of the selected models.

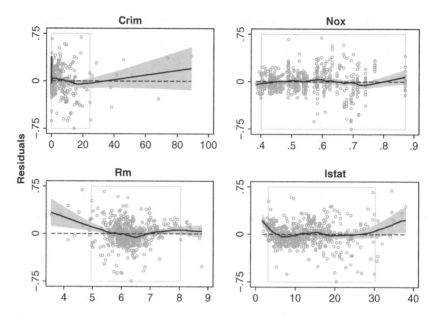

Figure 9.4 Boston housing data (quarter subset). Smoothed residuals from the MFP model. Plots for predictors crim, nox, rm and lstat are shown. Minor nonlinearities are evident for most of the predictors.

9.5.2 Comparing Predictors

Table 9.3 compares the intraclass correlation coefficients between the overall predictors or indexes $\hat{\eta}$ from MFP and the four spline procedures, applied to the full data and the half and quarter subsets. The agreement between MFP and the spline procedures is high in all cases. However, for each subset the index from MFP agrees more closely with those from MVRS(4) and MVSS(4) than from MVRS(8) and MVSS(8). The relationship between the indexes is illustrated for the case of least agreement (the quarter subset) in Figure 9.5. Although greater scatter, and therefore lower agreement with MFP, is seen for the eight d.f. than the four d.f. models, the agreement is still substantial.

Figure 9.6 shows Bland-Altman plots comparing the spline and MFP indexes. While a systematic difference between the predictors is visible for the eight-d.f. models, it is minor. We conclude that, although the models differ considerably, the resulting predictors are similar.

Table 9.3 Boston housing data. Intraclass correlation coefficients r_1 comparing the index from MFP(0.05) with those from four spline models selected at the 5% nominal significance level by MVRS and MVSS. Details of selected models are given in Tables 9.1 and 9.2.

Subset	MVRS(4)	MVSS(4)	MVRS(8)	MVSS(8)
Full data	0.991	0.991	0.978	0.983
Half	0.988	0.983	0.983	0.974
Quarter	0.992	0.986	0.954	0.967

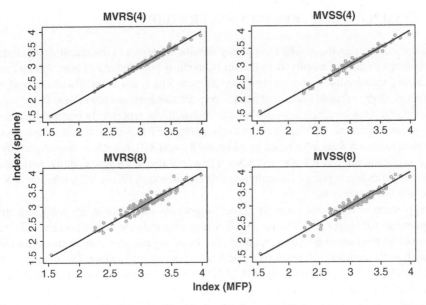

Figure 9.5 Boston housing data (quarter subset). Comparison between the indexes $\hat{\eta}$ from MFP and the four spline procedures.

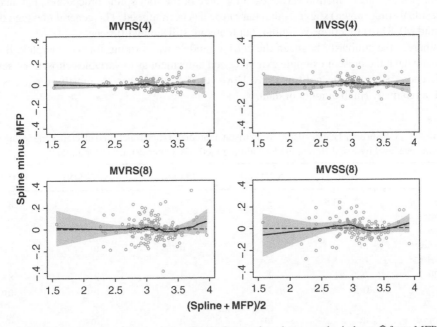

Figure 9.6 Boston housing data (quarter subset). Comparison between the indexes $\hat{\eta}$ from MFP and the four spline procedures. Bland–Altman plot of difference (spline minus MFP) against mean (spline + MFP)/2, with smoothing and 95% pointwise CI.

9.6   EXAMPLE 2: GBSG BREAST CANCER STUDY

Table 9.4 summarizes the results of applying the five strategies to the breast cancer dataset.

In contrast to the main analysis of the study, medical knowledge has been ignored and the preliminary transformation for `nodes` (see Section 5.6) is not used. There are only minor differences in the variables selected: `meno`, `gradd3` and `er` are always excluded, whereas `age`, `nodes`, `pgr` and `hormon` are always included. As reported in our stability analysis (Royston and Sauerbrei, 2003), `size` was selected by MFP in 40.2% of bootstrap replications and seems to have a weak effect (see also Figures 8.1 and 8.2), but it was not selected by MFP on the original data. MVRS also excludes `size`, but the MVSS procedures both select it. MVSS(8) yields a complicated, uninterpretable function (see Figure 9.7), whereas MVSS(4) suggests a linear function.

The postulated functions for `age` agree for patients up to about 55 years, but there is disagreement for older ages. The two MVRS functions indicate a small decrease in risk for woman older than about 65 years. As mortality from any cause is considered as an event in this analysis of recurrence-free survival, such behaviour is implausible. The function selected by MVSS(8) is also rather unlikely. MVSS(4) shows a little more curvature than MFP but agrees well in general.

The selected FP2 function for `nodes` has an inexplicable 'hook' at low values (see also Figure 6.9), but is otherwise plausible. Apart from the hook, the spline functions agree up to about 15 nodes, but the wiggles and the decrease in risk from about 20 nodes are definitely contrary to medical knowledge. The downturn in the spline functions appears to be driven by the single extreme observation at 51 nodes.

For `pgr`, a linear function was chosen by three of the four spline procedures, but analysis of residuals (not shown) suggests that curvature has been missed. The general message of the FP and MVRS(4) functions is similar, but there are still some differences.

Whereas the similarity between the chosen models was striking for the complete Boston housing data, the present example exhibits good agreement as to variables chosen but several differences between selected functions. However, the differences are mostly in regions of x with a relatively small amount of data.

Table 9.4   GBSG breast cancer data. Comparison of predictors and functions selected at the 5% nominal significance level by five different model-building methods. Other details as for Table 9.1.

Var	MFP	MVRS(4)	MVSS(4)	MVRS(8)	MVSS(8)
age	−2, −0.5	3	4	3	8
meno[a]	−	−	−	−	−
size	−	−	lin	−	8
gradd1[a]	in	−	in	in	in
gradd2[a]	−	−	−	−	−
nodes	−2, −1	2	4	2	8
pgr	0.5	3	lin	lin	lin
er	−	−	−	−	−
hormon[a]	in	in	in	in	in
R_D^2	0.276	0.285	0.288	0.278	0.327

[a] Binary predictor.

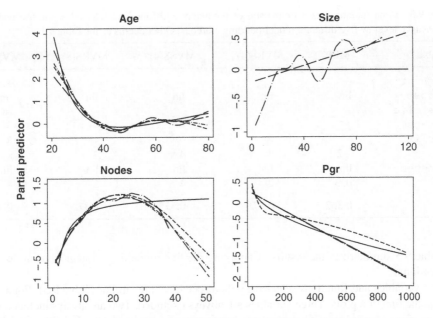

Figure 9.7   GBSG breast cancer data. Partial predictors (log relative hazards scale, centred on zero) for variables selected by the model-building procedures MFP (solid lines), MVRS(4) (short dashes), MVSS(4) (long dashes), MVRS(8) (short dashes and dots), MVSS(8) (long dashes and dots). size is not selected by MFP and its effect is plotted as zero. For pgr fits from three of the four spline models are indistinguishable.

As with the Boston housing data, R^2 is similar for all models, only MVSS(8) giving a substantially higher value (0.327). However, the MVSS(8) model includes three variables fitted with eight-d.f. spline functions. The improbable nature of these fitted functions suggests that the higher R^2 value arises through overfitting the data.

9.7   EXAMPLE 3: PIMA INDIANS

The Pima Indians dataset is available at http://archive.ics.uci.edu/beta/ datasets/Pima+Indians+Diabetes. It has been analysed as a classification problem many times in the literature. The dataset arises from an investigation of potential predictors of the onset of diabetes in a cohort of 768 female Pima Indians (a group peculiarly susceptible to the disease), of whom 268 developed diabetes. It comprises a binary outcome y (diabetes, yes/no) and the following eight ordinal or continuous measurements: pregnant (number of times pregnant), glucose (plasma glucose concentration at 2 h in an oral glucose tolerance test), diastolic (diastolic blood pressure, mmHg), triceps (triceps skin fold thickness, mm), insulin (2-h serum insulin, μU ml^{-1}), bmi (body mass index, weight in kg/height in m^2), diabetes (diabetes pedigree function), and age (years). Substantial missingness was present for triceps (30%) and insulin (49%), much less (< 5%) for glucose, diastolic and bmi. To be able to work with complete data, we imputed the missing values once by using the ice procedure for Stata (Royston, 2005).

Table 9.5 Pima Indians data. Comparison of predictors and functions selected at the 5% nominal significance level by five different model-building methods. Other details as for Table 9.1.

Variable	MFP	MVRS(4)	MVSS(4)	MVRS(8)	MVSS(8)
pregnant	–	–	–	–	–
glucose	lin	lin	lin	lin	lin
diastolic	–	–	–	–	–
triceps	–	–	–	–	–
insulin	–	–	–	–	8
bmi	−2	3	4	3	8
diabetes	lin	lin	lin	lin	lin
age	0, 3	2	4	2	8
R^2	0.362	0.371	0.375	0.372	0.395

Table 9.5 summarizes the results of applying the five modelling strategies, using logistic regression.

The variables and functions selected by the procedures are similar (Table 9.5). Figure 9.8 shows the function plots. Three variables are always excluded. Two are always included with a linear function and two others with a nonlinear function. insulin is included only by MVSS(8), and the plot of the function shows wild, uninterpretable behaviour, indicative of

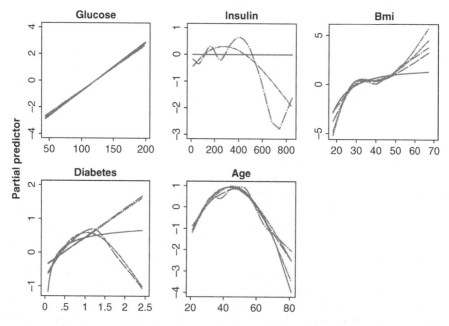

Figure 9.8 Pima Indians data. Partial predictors (log odds scale, centred on zero) for variables selected by the model-building procedures MFP (solid lines), MVRS(4) (short dashes), MVSS(4) (long dashes), MVRS(8) (short dashes and dots), MVSS(8) (long dashes and dots). Note that insulin was selected only by MVRS(8) and MVSS(8).

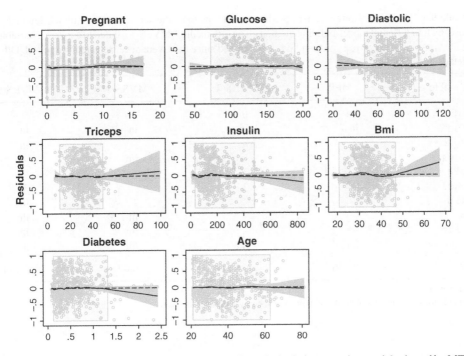

Figure 9.9   Pima Indians data. Smoothed residuals from the logistic regression model selected by MFP, comprising `glucose`, FP1(bmi;−2), `diabetes`, FP2(age; 0, 3) (see Table 9.5).

major overfitting. The functions for `glucose`, `diabetes` and `age` agree remarkably well. The MFP function for `bmi` differs from the four spline functions (see comments below). As discussed in Section 6.5, model criticism should be an integral part of every analysis, whether by MFP or another technique.

Plots of the smoothed residuals from the MFP model (see Figure 9.9) show a good fit for all predictors except possibly `bmi`. The FP fit is less good in the central portion (`bmi` about 30 to 40) where the splines indicate a plateau, also for the 2% of obese individuals with `bmi` > 47 kg m^{-2} where the splines show an increased probability of diabetes in these 16 individuals (see Figure 9.8), 13 of whom actually developed diabetes. As in previous examples, R^2 is higher for MVSS(8) than for the other procedures.

9.8   EXAMPLE 4: PBC

Table 9.6 shows model-building results using Cox regression for the PBC dataset, for which the outcome is time to death (312 patients, 125 events).

Results for MVSS(8) are not available since the GAM models with eight d.f. could not be fitted by the software (convergence failure). The 10 variables selected by at least one procedure are shown; the remaining seven predictors were never chosen. Greater variation in the selected models is seen in this example than in the previous three. MVRS(8) selected nine variables, compared with only five for MFP. Four variables were selected by all procedures, of which two (`age`, `bil`) are continuous. `chol` was selected by the MVRS procedures but not by MFP or MVSS(4). Figure 9.10 shows, once again, that implausible functions are sometimes found

Table 9.6   PBC data. Comparison of predictors and functions selected at the 5% nominal significance level by four different model-building methods. The fifth method, MVSS(8), failed to converge. Variables trt, sex, hep, ap, sgot, plt and pro are omitted since they were never selected in a model. Other details as for Table 9.1.

Variable	MFP	MVRS(4)	MVSS(4)	MVRS(8)
age	lin	lin	4	5
asc[a]	in	in	in	in
spider[a]	–	–	–	in
edema[b]	lin	lin	lin	–
bil	−2, −1	3	4	4
chol	–	3	–	3
alb	–	–	–	lin
cu	–	–	lin	lin
trig	–	–	lin	lin
stage[b]	lin	lin	lin	lin
R_D^2	0.646	0.669	0.673	0.744

[a] Binary predictor.
[b] Predictor with small number of different values, considered only as linear.

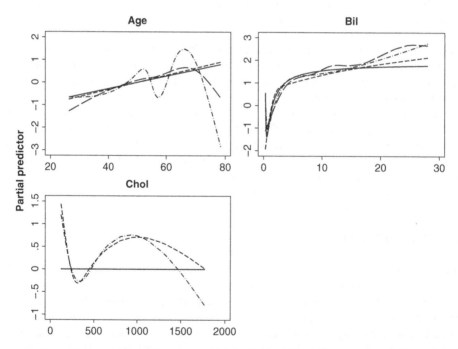

Figure 9.10   PBC data. Partial predictors (log relative hazard scale, centred on zero) for age, bil and chol, selected by MFP (solid lines), MVRS(4) (short dashes), MVSS(4) (long dashes) and MVRS(8) (short dashes and dots). chol was selected only by MVRS(4) and MVRS(8).

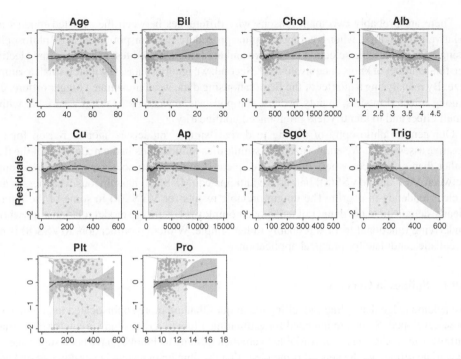

Figure 9.11   PBC data. Smoothed martingale residuals from the Cox regression model selected by MFP, comprising `age`, `edema`, FP2(`bil`;−2, −1), `stage` (see Table 9.6). Residuals and confidence intervals have been truncated to the interval [−2, 1], but the original values were used to create the smooths.

by the spline-based procedures; this is particularly evident for `age` and `chol`. Modification of the FP function for `bil` to remove the unsatisfactory 'hook' present in the FP2 model was discussed in Section 5.5.2.

Figure 9.11 shows smoothed martingale residuals from the MFP model for all 10 continuous predictors. There is some evidence of lack of fit for some of the predictors. However, the functions indicated by the patterns of the smoothed residuals are complex and not readily interpretable. The fitted functions from MVRS(8) (see Figure 9.10) suggest substantial overfitting, presumably accounting for the large R^2. The values of R^2 for the other models are similar.

9.9 DISCUSSION

We have shown in several examples that the MFP and spline model-building procedures give roughly comparable multivariable models. Some of the weaker predictors may be included by one procedure but not by another, and estimated functions often show some differences. In regions with sparse data the differences can be major, but where there is enough data the functions and their interpretation are similar. If interest lies mainly in the simpler task of deriving a good predictor, the differences between the selection procedures become smaller. These initial results are promising in the context of multivariable model-building with continuous variables.

There are probably two main reasons why differences between the selected models are relatively small. First, when creating the spline procedures we transported the MFP philosophy, using BE as a variable selection procedure and applying a closed test procedure for selecting the final number of knots or equivalent d.f. Second, we ignored datasets with too small a sample size. By reanalysing a quarter of the Boston housing data, we demonstrated bigger differences among selected models, mainly because the power is inadequate to select a variable with a small effect and to identify less pronounced nonlinearity.

Our general philosophy of aiming to derive 'simple' models is another reason for our positive assessment. We emphasized the similarities of model fits in the main body of the data rather than differences where the data are sparse. Larger differences were sometimes found between MFP and MVSS(8); but, as we anticipated, MVSS(8) has obvious weaknesses and a clear tendency to overfit. The uninterpretable results for MVSS(8) in some of the figures clearly demonstrate the danger of using overcomplex predictors to model relationships whose underlying functional form is probably rather simple. Without modification, MVSS(8) is not a realistic candidate for practical applications.

9.9.1 Splines in General

We acknowledge that spline-modelling techniques that are more sophisticated than ours have been developed. Some are intended for estimating a function of a single predictor (univariate fitting), while others are multivariable in character. Many techniques have been suggested for univariate fitting, but the general principle is that the domain of a covariate is covered by knots and local polynomial pieces are anchored at these knots. The techniques differ in the number of knots used, the method of determining knot positions, and how the parameters for the polynomial pieces are estimated. Smoothing splines essentially place one knot at each unique covariate value and use a roughness penalty for parameter estimation (Green and Silverman, 1994). Regression spline approaches use only a small number of knots, with the advantage that parameter estimation can be performed by standard regression techniques. If a small number of knots is used, then the shape of the fitted function depends to a greater extent on their precise positions. Alternatively, regression splines can be extended by using a large number of knots (e.g. equally spaced) in combination with penalized parameter estimation (Eilers and Marx, 1996) to shrink and, hence, stabilize the fitted functions.

We developed algorithms in Stata for multivariable model-building using regression splines (MVRS; see Royston and Sauerbrei (2007b)) and smoothing splines (MVSS, unpublished) to conform with the MFP strategy. Using generally available software (R packages design from F. Harrell, mgcv from S. Wood and survival from Therneau and Grambsch), Sauerbrei et al. (2007a) presented functions for the GBSG breast cancer and the oral cancer studies derived using restricted cubic splines with four knots and from several models fitted by penalized B-splines. As in the present analyses, some of the functions are very wiggly and hardly interpretable. Rosenberg et al. (2003) showed that multiple minima in an optimization criterion (e.g. AIC) can pose an additional problem for model selection with splines. A practical disadvantage of some of the spline-fitting techniques is the difficulty in adapting them to more complex data structures, such as censored survival times.

Holländer and Schumacher (2006) compared several methods for estimating the functional form for one continuous covariate in the Cox model. In addition to linear, quadratic and cutpoint functions, they considered FPs and restricted cubic splines with four knots. They summarized their results as follows:

Taking all models into account the method based on fractional polynomials performed best: If the given risk function is linear FP reduced to [a linear function] in nearly all replications, whereas this rate was lower for the other methods. Furthermore, FP was the sole method holding the type I error rate for all three model selection strategies.

They studied as selection strategies two of the constituent tests in our FSP, namely FP2 versus linear and FP2 versus FP1.

9.9.2   Complexity of Functions

Although splines are more flexible than FPs for modelling continuous covariates, their use brings the danger of potential instability. We have seen no published stability analysis of spline functions in a multivariable context. In our examples we used splines with four and eight d.f. Whereas results from MVRS(4) and MVSS(4) seemed to make sense in general, use of eight d.f. often resulted in uninterpretable functions. We doubt whether such a large number of d.f. should be used in a multivariable context with selection of variables and of functions. Further, with our closed test procedure, the power to select the variable is reduced if the d.f. is large (e.g. with the reduced Boston data; see Section 9.5).

Extensive simulation studies are needed for a detailed assessment of the strengths and weaknesses of different spline procedures. Comparisons of the better-performing spline techniques with MFP are also required.

9.9.3   Optimal Fit or Transferability?

Some of the functions derived in our examples clearly demonstrate substantial overfitting of the data. If R^2 is taken as a measure of model performance, then MVSS(8) seems to have selected the best model in several cases, but some of its functions are wiggly and uninterpretable. The model is clearly overfitted and, therefore, its results are not of much practical use. Such functions are not reproducible and are not transferable to other settings. In the settings considered, there is no clear winner with respect to R^2 among MFP, MVRS(4), MVRS(8) and MVSS(4). However, MFP selects the model with the better transferability and general usefulness.

9.9.4   Reporting of Selected Models

In medicine, it has been recognized that, despite intensive research and many papers, progress in understanding the role of prognostic factors, risk factors and, more generally, explanatory variables is slow. Developing guidelines for reporting studies is considered an important step to improve research and to make results more accessible to the scientific community and other stake-holders (e.g. decision-makers and members of the public). For example, McShane et al. (2005) and von Elm et al. (2007) developed guidelines for reporting prognostic marker studies and epidemiological studies respectively. The main goal of such guidelines is to encourage transparent and complete reporting. The guidelines recommend that details of the statistical analysis, estimated effects of a variable on the outcome and much more should be provided.

Evidence-based summary assessments of prognostic markers, risk factors, etc. are becoming increasingly important. Adequate reporting of the original studies and use of appropriate techniques for building reliable and accurate models are prerequisites for constructing the evidence base (Sauerbrei et al., 2006a).

Such reporting of analyses and results from MFP should not cause any difficulty. However, reporting key details of the analysis and the results from multivariable spline modelling is not straightforward. For example, the common practice of showing graphs as a means of presenting the main results of a spline analysis is in this sense unsatisfactory. Presenting sufficient details of the analysis and results using complex spline models, e.g. with many knots, is extremely challenging.

9.9.5   Conclusion

We present a limited comparison of MFP with two spline procedures developed according to the MFP philosophy. If the sample size is sufficiently large, then the procedures give similar results. For weakly influential factors or in regions with sparse data, differences sometimes appear. Results from MFP analyses are more stable, are easier to interpret and simpler to transfer to other environments; therefore, MFP is our preferred approach. As is clear from the reporting guidelines, transportability and general usefulness are more important properties than minor advantages in goodness of fit.

CHAPTER 10

How To Work with MFP

Summary

1. The chapter describes in some detail univariate and multivariable analyses of an artificial but realistic dataset (the 'ART' study), created by simulation.
2. Various approaches to model criticism that have been introduced in our book are further illustrated in an integrated context. Methods include function plots, residual plots to indicate lack of fit, analyses for increased robustness, diagnostic plots for influential observations, and the detection of interactions.
3. A simple stability analysis of the selection of variables and FP functions is described.
4. Several issues to be aware of in multivariable modelling are discussed and summarized.

10.1 INTRODUCTION

Chapters 4 to 6 have laid out the fundamentals of the FP approach to the analysis of the effects of one or more continuous covariates on an outcome variable. Chapter 6 focuses specifically on building a multivariable model using the MFP algorithm. In this chapter, we suggest a guide to the steps required in a detailed multivariable analysis of a single dataset using MFP. The data are assumed to be complete (no missing values). We also point out pitfalls that could be hiding within the observed data, of which the user needs to be aware.

10.2 THE DATASET

Tables 10.1 and 10.2 summarize an artificial dataset that is used for illustration purposes throughout this chapter. It is called the 'ART study' (ART denoting 'artificial'). The distribution of the predictors and their correlation structure was informed by the GBSG study. It has $n = 250$

Multivariable Model-Building   Patrick Royston, Willi Sauerbrei
© 2008 John Wiley & Sons, Ltd

Table 10.1   ART study. Distribution of continuous variables.

Variable	Mean	SD	Min.	Max.	Skewness	Kurtosis	Missing (%)
y [response]	12.2	1.0	8.3	15.2	−0.0	3.7	0
x1	54.2	9.7	24	85	0.2	3.0	0
x3	20.8	9.3	5	81	1.9	10.4	0
x5	6.9	16.9	0.5	234.3	10.3	132.5	0
x6	147.9	184.2	0	1200	2.4	10.4	0
x7	106.0	169.8	0	1727	5.3	43.4	0
x10	17.0	8.3	0.6	43.5	0.7	3.3	0

Table 10.2   ART study. Distribution of categorical variables, which have scores $\{0, 1\}$ or $\{1, 2, 3\}$.

Variable	0	1	2	3	Missing (%)
x2	62 (25%)	188 (75%)			0
x4[a]		30 (12%)	164 (66%)	56 (22%)	0
x8	165 (66%)	85 (34%)			0
x9[b]		176 (70%)	58 (23%)	16 (6%)	0

[a] Ordered.
[b] Unordered.

observations, a continuous, uncensored response variable y, and 10 covariates x1, ..., x10, a mixture of continuous, binary, ordinal and unordered categorical variables. The distributions of the continuous and the categorical variables are shown in Tables 10.1 and 10.2 respectively. The columns for percentage of missing values serve as a reminder that in real studies some data are usually missing. (Handling missing data in regression models is beyond the scope of our book.) For continuous variables, the minimum and maximum are important when FP modelling is contemplated (see Sections 4.7, 5.4 and 5.5). We strongly recommend producing the equivalent of Tables 10.1 and 10.2 before beginning any multivariable analysis. It is well worth spending time 'getting a feel' for the data first.

The response variable y is approximately normally distributed. There are two binary variables (x2 and x8) and two three-level categorical variables (x4 and x9), of which x4 is ordered and x9 is unordered. Note that the category x9 = 3 is sparse (only 16 observations, 6%); it may be worth combining x9 = 3 with x9 = 1 or x9 = 2, whichever is more sensible from a subject-matter point of view.

Turning to the six continuous explanatory variables, the large values of skewness and kurtosis for x3, x5, x6 and x7 show that their distributions are markedly nonnormal, whereas the others (x1 and x10) are approximately normal. To supplement the summary statistics, the distributions of the continuous variables may be presented as histograms, box plots, normal probability plots, etc. – not shown here. The distributions of x3, x5, x6 and x7 may include extreme observations, some of which may be influential in an analysis. It is important to be aware of this possibility. Furthermore, x6 and x7 have a minimum value of zero (see Table 10.1); x6 +1 and x7 +1 are used for FP analysis (see Section 4.7).

Figure 10.1 shows the distributions of all pairs of continuous variables. While extreme observations are visible for x5 and x7, there are no obvious bivariate outliers. The relationships between the response and the continuous predictors individually are not strong.

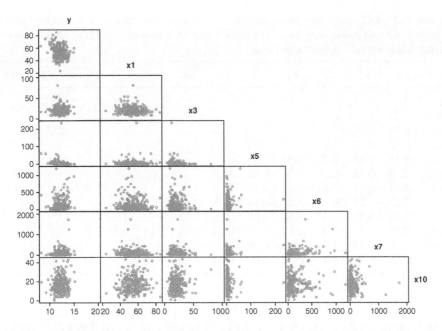

Figure 10.1   ART study. Matrix scatter plot of the response and six continuous predictors.

The matrix scatter plot is rather dominated by one large outlier in $x5$ and two less extreme values of $x7$. Figure 10.2, the same plot without these values, gives a somewhat clearer impression of the relationships among the variables.

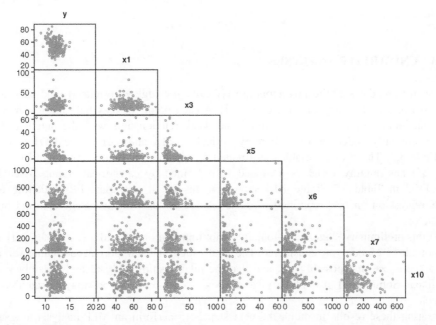

Figure 10.2   ART study. Matrix scatter plot of the response and six continuous predictors, with three extreme values of $x5$ and $x7$ removed.

Table 10.3   ART study. Spearman correlation coefficients. Note that x4 is an ordinal variable and so is not split into dummy variables here, whereas x9 is unordered and is represented by two separate dummy variables, x9a and x9b.

	y	x1	x2	x3	x4	x5	x6	x7	x8	x9a	x9b	x10
y	1											
x1	−0.21	1										
x2	0.16	−0.59	1									
x3	0.03	−0.10	0.03	1								
x4	−0.18	0.04	−0.02	−0.03	1							
x5	−0.45	0.06	−0.06	−0.30	−0.05	1						
x6	0.21	−0.05	0.06	−0.03	−0.38	0.29	1					
x7	0.22	−0.11	0.15	−0.04	−0.14	0.05	0.40	1				
x8	0.11	−0.02	0.00	−0.06	−0.01	0.10	−0.01	−0.06	1			
x9a	0.02	0.18	−0.06	−0.29	0.07	0.11	0.29	−0.01	0.05	1		
x9b	0.10	0.15	−0.00	−0.28	0.04	−0.08	0.07	−0.09	0.12	−0.14	1	
x10	0.09	0.17	−0.07	0.08	0.01	0.00	0.09	−0.02	−0.02	0.10	0.02	1

The Spearman rank correlation matrix for all 11 variables is given in Table 10.3. The largest absolute correlation (0.59) is between x1 and x2. The covariate most highly correlated with y is x5 (−0.45). No other correlation exceeds 0.40.

We suppose that subject-matter knowledge requires monotonic functional forms for predictors x5 and x6, but not much is known about the functional form for x1, x3, x7 or x10. Different types of function are possible for the latter variables.

10.3   UNIVARIATE ANALYSES

Since in most datasets the predictors are typically not highly correlated (and such is the case here), a preliminary univariate analysis of each predictor may help the researcher to gain an initial impression of the strong and weak relationships with the outcome. First, x4 is ordinally coded into two dummy variables, x4a and x4b – see DO1 and DO2 in Table 3.2. These two variables are treated as separate predictors. The unordered variable x9 has dummy variables x9a and x9b formed by categorical coding – see DC1 and DC2 in Table 3.2. Table 10.4 shows the results of univariate FP modelling using OLS regression on each continuous predictor, and of tests of the categorical and binary predictors.

At this preliminary stage, monotonicity is not imposed on x5 and x6, so potentially nonmonotonic FP2 models are considered. The results suggest that important continuous variables are x1, x5, x6 and possibly x7, and that the relationships with x5 and possibly x6 are nonlinear. Some or all of the binary variables x2, x4a, x4b and x8 may be required in a model.

Bearing these results in mind, the next step is to perform an MFP analysis to select a multivariable model.

Table 10.4 ART study. Univariate analyses.

Type	Variable	*P*-value for testing			Variables and powers selected[b]
		Inclusion[a]	FP2 vs linear	FP2 vs FP1	
Continuous	x1	0.001	0.2	0.4	1
	x3	0.9	0.8	0.7	out
	x5	< 0.001	< 0.001	0.03	0, 3
	x6[c]	0.002	0.05	0.3	0
	x7[c]	0.01	0.4	0.7	1
	x10	0.1	0.3	0.3	out
Categorical	x9a, x9b	0.4			out
Binary	x2	0.02			in
	x4a	0.006			in
	x4b	0.06			out
	x8	0.09			out

[a] Denotes test of FP2 vs null for continuous variables, or test of dummy variable(s) for categorical or binary variables.
[b] The 'in' and 'out' denote that a variable was selected or not selected at the 5% significance level. [c] The '1' is added to avoid zeros in FP analysis; see Section 4.7.

10.4 MFP ANALYSIS

Since x4 is ordinal, ordinal coding, with x4a and x4b representing x4 $\geq$ 2 and x4 = 3 respectively, is chosen (see Section 3.3). Standard dummy-variable coding is used for x9, with the two dummies tested jointly on two d.f. Alternatively, the dummies may be tested separately. Table 10.5 shows the results of an analysis with MFP(0.05, 0.05). Seven variables

Table 10.5 ART study. Multivariable model selected by MFP(0.05, 0.05). For further details, see Table 10.4.

Type	Variable	*P*-value for testing			Model A
		Inclusion[a]	FP2 vs linear	FP2 vs FP1	
Continuous	x1	< 0.001	0.2	–	1
	x3	< 0.001	0.6	–	1
	x5	< 0.001	< 0.001	0.005	0, 3
	x6[b]	< 0.001	< 0.001	0.4	0
	x7[b]	0.2	–	–	out
	x10	0.003	0.1	–	1
Categorical	x9a, x9b	0.3			out
Binary	x2	0.5			out
	x4a	0.01			in
	x4b	0.8			out
	x8	0.001			in

[a] Denotes test of FP2 vs null for continuous variables, or test of dummy variable(s) for categorical or binary variables.
[b] The '1' added to avoid zeros in FP analysis; see Section 4.7.

are selected, five of which are continuous and two binary. The selected model, called in what follows model A, comprises x1, x3 and x10 linear, $\varphi_1(x6 + 1; 0)$, $\varphi_2(x5; 0, 3)$, and binary predictors x4a and x8. The variance explained by model A is 51.6%.

10.5   MODEL CRITICISM

Having created model A, we now illustrate some important aspects of model criticism.

10.5.1   Function Plots

Figure 10.3 shows function plots with 95% pointwise CIs for x1, x3, x5, x6 and x10. The fitted function for x5 is nonmonotonic and passes close to the large outlier in x5 already noted in Figure 10.1. The function is unsatisfactory, since it is strongly affected by the outlier and since subject-matter knowledge calls for a monotonic function. We return to this issue in Section 10.5.3. The other variable with outliers, x7, is not selected – see Section 10.5.2. All the other functions appear reasonable.

10.5.2   Residuals and Lack of Fit

Figure 10.4 shows residuals smoothed as a function of each covariate (including the eliminated variable x7) using running-line smoothers. There is a hint of lack of fit for several

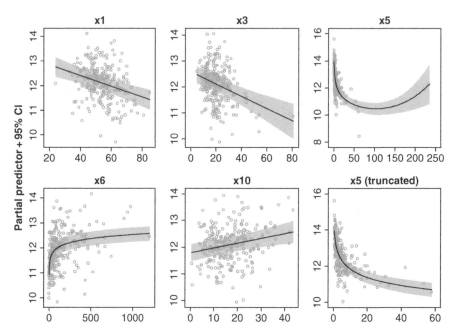

Figure 10.3   ART study. Function plots for continuous predictors in the selected MFP model. Note that the plot for x5 is repeated.

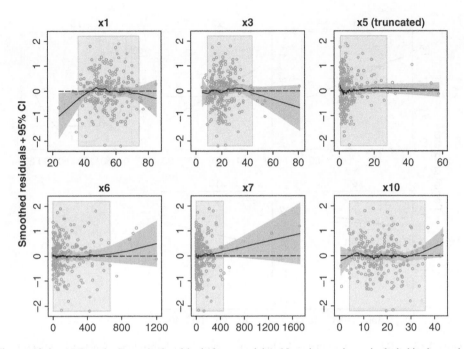

Figure 10.4 ART study. Smoothed residuals from model A. Note that x7 is not included in the model.

covariates, mainly outside the shaded areas, i.e. beyond the body of the observations. The lack of fit is not serious enough to motivate altering model A. A nonmonotonic function may be needed for x1. Other departures from the line $y = 0$ appear to be driven by single observations, an issue that is considered later.

10.5.3   Robustness Transformation and Subject-Matter Knowledge

Some extreme values are present in the data. They may influence the model that is chosen. To check the robustness of the functions shown in Figure 10.3, the preliminary transformation $g_\delta(x)$ (see Section 5.5) was applied to each continuous covariate. On rerunning MFP(0.05, 0.05) with the transformed variables, the same variables are selected as before, but all with FP1 or linear functions of the continuous variables. Figure 10.5 compares the two sets of fitted curves visually. The pairs of functions are similar for the bulk of the observations. The greatest differences are for x3 and x5. The previously nonmonotonic function of x5 is now monotonic and, therefore, is much more satisfactory. The functions for x3 are identical up to about 45. The upper truncation point ($\bar{x} + 2.8 \times$ SD) for x3 with respect to $g_\delta(x3)$ is 46.7, so the fitted function of $g_\delta(x3)$ is forced to be constant for x3 > 46.7.

The explained variation for the revised model is less than before (45.9% vs 51.6%). Since all the other functions are satisfactory (see Figure 10.3), robustification may be damaging the overall model fit and may be needed only for x5. To guarantee a monotonic function for x5, MFP(0.05, 0.05) was run again, now allowing only FP1 functions for x5. The resulting model

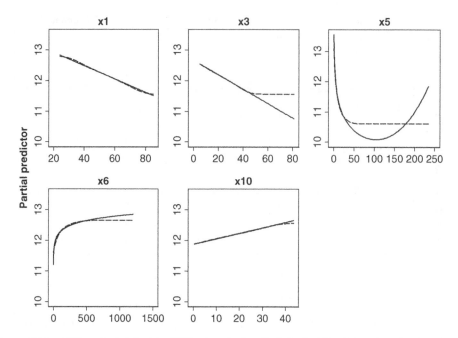

Figure 10.5   ART study. Original (solid lines) and robustified (dashed lines) functions for the selected continuous variables.

has an FP1 function with power −0.5 for x5, but is otherwise identical to the original. The explained variation is 49.4%, which is a little lower than for model A.

In summary, the robustness analysis may suggest that an FP1 function is preferable for x5, but no other changes to model A.

10.5.4   Diagnostic Plot for Influential Observations

The FP2 function chosen for x5 may be further scrutinized by examining the FP diagnostic plot described in Section 5.3. The plot of deviance differences for FP2 versus linear for x5 (not shown) indicates that there is no single observation whose deletion would make the test of FP2 versus linear nonsignificant. It follows that a linear function is ruled out, and that at least an FP1 function is needed. The left panel of Figure 10.6 shows the deviance difference plot for FP2 versus FP1 for x5, adjusted for the other variables in the selected multivariable model. With case 175 removed, the FP2 – FP1 difference is no longer significant ($X^2 = 3.71$, $P = 0.16$), so an FP1 function would be chosen.

Interestingly, case 175 is not the most extreme value of x5, but the second most extreme. The most extreme value belongs to case 151, which is not indicated as influential in the left panel of Figure 10.6. The right-hand panel of Figure 10.6 suggests that a fitted curve might pass near to cases 175 and 151, meaning that a nonmonotonic function would be required.

This example shows that the diagnostic plot has value beyond simply omitting the most extreme covariate observation. Figure 10.6 shows that an FP1 function should be used for x5, conforming with the need for a monotonic function.

These types of plot should be used for all continuous variables.

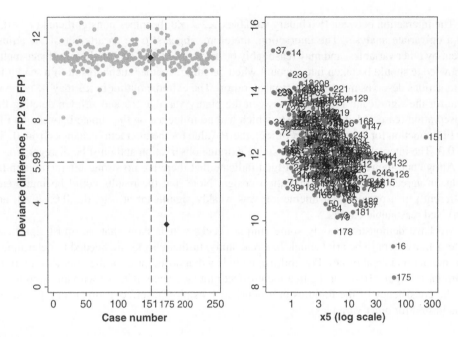

Figure 10.6   ART study. Influence of observations of x5 on fit of FP functions. Left panel: FP diagnostic plot for choosing between FP2 and FP1 functions. A single influential observation (case 175) is apparent. The most extreme value of x5 (case 151) is seen to be uninfluential. Cases 151 and 175 are shown as diamonds. Right panel: scatter plot of y versus x5 (note log scale) with case numbers shown.

10.5.5   Refined Model

Subject-matter knowledge, the robustness analysis and the diagnostic plot suggested that it was desirable to change model A by restricting the function for x5 to be FP1. On rerunning MFP with this restriction, and not omitting any observations from the analysis, the only difference from model A is the selection of an FP1 function for x5 with power −0.5. Call this model B. In principle, model B should be checked again in the same way.

Another option that an analyst might consider is to remove observation 151, which is an extreme outlier for x5. MFP again selects model A, in agreement with Figure 10.6. A further option is to use a preliminary transformation for heavily skewed variables (e.g. x5, x7), such as a negative exponential (see Section 5.6.2). The option is not explored here.

10.5.6   Interactions

As explained in Section 7.11.2, we use the MFPIgen algorithm to look at all possible two-way interactions between variables in the ART dataset, whether their main effects are significant or not. Of the 45 tests done, three were significant at the 0.05 level and an additional two (x2 × x8 and x5 × x8) at the 0.01 level. Since, owing to multiple testing, the interactions significant at the 0.05 level are less likely to be real, we check only whether those significant at the 0.01 level are reliable.

The interaction between two binary variables, $x2 \times x8$, becomes nonsignificant ($P = 0.3$) in a univariate analysis. The interaction, therefore, appears to be an artefact of the adjustment by other variables, and may reasonably be disregarded. In a real analysis, subject-matter knowledge should be taken into account when deciding whether such a result of a search for interactions deserves more serious consideration. The extreme outlier in $x5$ may be responsible for the significant interaction between the binary variable $x8$ and $x5$. On deleting this observation (case 151 in Figure 10.6, which had no influence on the choice between an FP1 or FP2 function for the main effect of $x5$), the P-value for the interaction changes from 0.009 to 0.3. The interaction is driven by a single extreme observation and must be disregarded.

Altogether, the results from MFPIgen indicate that there are no strong interactions in the data, in agreement with the simulation design. Note that the results could be interpreted differently if a prespecified interaction was weakly significant at, say, the 0.05 level, and survived reasonable checks.

We have demonstrated only some simple checks which show that the most significant interactions seem to be artefactual. In a real study, further checks are needed to determine if an interaction is real or not. Possibilities include a diagnostic plot for the interaction term(s) (similar to Figure 10.6), or the treatment-effect plot (see Section 7.3.6) with an examination of the consistency of the findings in groups, such as plots or effect estimates (adjusted and unadjusted for other variables).

10.6 STABILITY ANALYSIS

A basic stability analysis is given here. The principles of more detailed investigations are described in Chapter 8 and by Royston and Sauerbrei (2003).

The key elements of a bootstrap stability analysis are the BIFs (see Section 8.3.2) for individual variables and the issue of linearity/nonlinearity for continuous variables (see Section 8.7).

As a working minimum, 200 bootstrap replications were taken from the ART dataset. Table 10.6 gives the BIFs and percentage of nonlinear functions selected with MFP(0.05, 0.05). It is clear that $x1$, $x3$, $x5$, $x6$, $x8$ and $x10$ must be included in the model and that $x4b$, $x9a/b$ and probably $x2$ must be excluded. Decisions regarding $x4a$ and $x7$ are less clear-cut. As seen in Figure 10.1, $x7$ has two extreme values which influence the selection of this variable and which is also reflected in the bootstrap replications. $x5$ and $x6$ clearly require a nonlinear function and $x3$ a linear function.

The position with $x1$, $x7$ and $x10$ is uncertain. Is $x7$ required at all? Do the three variables require a nonlinear function or not? To clarify the issues, we consider $x1$ as an example. A plot of 20 individual curves and the 'bagged' (bootstrap aggregated) estimate for $x1$ may be helpful (see Figure 10.7). The plot of functions in individual bootstrap samples (left panel) shows considerable variation, being a mixture of straight lines and curves. In replications in which straight lines are selected, the power to detect nonlinearity may be inadequate. The solid and dashed lines in the right panel show the mean, 10 th and 90th centiles of the fitted function for $x1$ when $x1$ was selected (i.e. in 96% of replications). There is some evidence here favouring a nonmonotonic function, but the very wide interval for $x1 < 35$ suggests considerable uncertainty. The dash-dotted line in Figure 10.7 shows the best FP2 function for $x1$ from the original data, adjusting for the other selected variables and functions in model A. This has powers (0.5, 0.5). Despite the bagged estimate including a substantial proportion of

Table 10.6   ART study. Percentage BIFs for inclusion of variables and nonlinearity of continuous functions. The numerator for the former and the denominator for the latter is the number of bootstrap replicates in which the variable was selected.

Type	Variable	BIF (%)	
		Selection	Nonlinear
Continuous	x1	96	41
	x3	97	9
	x5	100	100
	x6	100	100
	x7	48	27
	x10	92	38
Categorical	x9a, x9b	11	–
Binary	x2	32	–
	x4a	67	–
	x4b	10	–
	x8	95	–

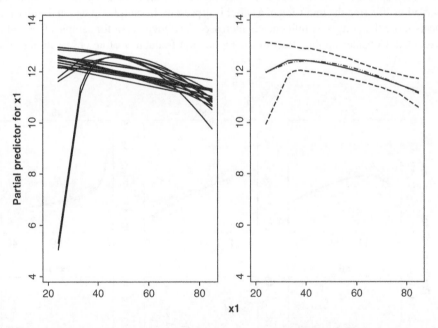

Figure 10.7   ART study. Estimated functions for x1 within multivariable models. Left panel: estimates of the partial predictor for x1 in 20 bootstrap samples. Right panel: bootstrap average curve (solid line) with 95% pointwise CI (dashed lines); best FP2 curve from the original data (dash-dots).

linear functions, the agreement between the FP2 function and the bagged estimate is striking. All in all, and considering the graphical evidence of departure from linearity offered by the smoothed residuals (see Figure 10.4), the FP2 function is a reasonable alternative to a linear function here.

Figure 10.8 shows the bagged functions for all six continuous variables, taken from bootstrap samples in which each variable was selected. Nonlinearity of the functions for x1, x5 and x6 is evident, whereas the others are approximately linear. Also shown, as dashed lines, is the true function for each variable. Apart from x7, the bagged and true functions are in remarkable agreement. The function for x7 is subject to marked selection bias, since only the 48% of bootstrap samples in which x7 was selected as 'significant' are represented in the bagged function. If x7 was taken seriously as a predictor, then it might be preferable to study its bagged function with x7 forced into the model in all samples.

When bagging, the question of whether to exclude replications in which the variable was not selected, or to set the selected function to zero and include it in the bagged estimate, is difficult. When the BIF is large, (say, > 90%), the results from the two approaches will be similar. However, had zero estimates of the function been included, the bagged estimate for x7 (see Figure 10.8) would have been much nearer to zero.

In general, an observation that affects model selection in the original data may have its influence magnified or reduced in individual bootstrap samples. In cases of doubt (as with x7 above), an analysis such as the one shown in Figure 10.6 may clarify whether a moderately large BIF is due to a small number of influential observations. The probability of an observation appearing at least once in a given bootstrap sample is about 63%. Assume that an observation affecting the inclusion of a variable is present. In samples in which the observation appears more than once, its influence is increased, and vice versa for samples from which it is excluded. Considering the effect of the replication frequency of an observation (i.e. zero,

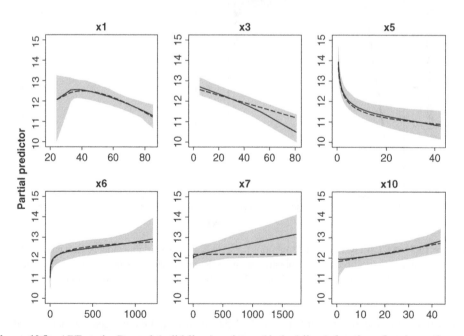

Figure 10.8 ART study. Bagged (solid lines) and true (dashed lines) functions for six continuous predictors from 200 bootstrap samples. The horizontal axis for the function for x5 has been truncated for clarity. Note that the BIF for x7 is only 48%.

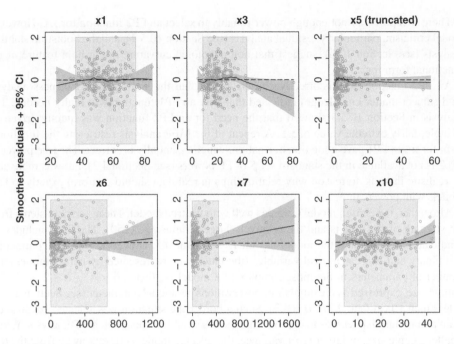

Figure 10.9 ART study. Smoothed residuals from model C. Note that x7 is not included in the model.

one or more than one time) on the selected models suggests a further way of checking for influential observations.

10.7   FINAL MODEL

We have illustrated the development of model A using MFP. Several checks, carried out for particular variables, suggested that some modifications would improve the model. The linear function for x1 may be better replaced with an FP2 function, and the FP2 for x5 with an FP1 function. Let the resulting model be model C. Compared with the residuals from model A (see Figure 10.4), the smoothed residuals from model C show some improvement (see Figure 10.9). Most of the patterns in the residuals are driven by small numbers of observations. In a real study, further checks might lead to minor changes to model C.

10.8   ISSUES TO BE AWARE OF

10.8.1   Selecting the Main-Effects Model

The true model used to generate the ART data included the following variables and transformations: $\varphi_2(x1; 0.5, 1)$, x3, x4a, $\varphi_1(x5; -0.2)$, $\varphi_1(x6 + 1; 0)$, x8, x10. As might be expected from the results of the bootstrap analysis (see Table 10.6), the MFP model reported in Table 10.5 included all the influential variables and did not include any 'noise' variables. To this extent it was very successful, but the FP degree chosen for x1 and x5 was wrong.

There was probably not enough power reliably to select an FP2 function for x1. However, model criticism, particularly residual analysis (see Section 10.5.2), and the bootstrap stability analysis (see Section 10.6), suggest that nonmonotonic curvature might be a feature of the functional form for x1.

An FP2 function was incorrectly chosen for x5, but the subsequent robustness analysis (and subject-matter knowledge) indicated that only an FP1 function was really needed. The analysis in Section 10.5.4 showed that the need for an FP2 function was mainly driven by a single, fairly extreme value of x5. A repeat of the MFP analysis restricting the functional form for x5 to FP1 gave a log transformation, which essentially is correct. The true power of -0.2 is not available in the standard set S of FP powers (see Section 4.3), but was motivated by realism: there is no reason why relationships in real data should conform exactly to FP1 models with powers in S.

Altogether, the chosen model C agrees well with the true model. The agreement stems from the strength of the effects and the relatively large sample size (250). The data includes 12 candidate predictors and the model has an R^2 of about 0.5. However, with six continuous, two binary and two categorical variables (the latter each represented by two dummies), the number of parameters in the most complex FP2 model is 30 (i.e. $6 \times 4 + 2 + 2 \times 2$). The sample size, expressed as the number of observations per model parameter (see Section 2.9.1), is less than 10 (in fact, $250/30 = 8.3$). Such a borderline situation was selected because the main aim in this chapter was to demonstrate the essential features of an MFP analysis. With a smaller sample size, or larger error variance, the selected model is further away from the true model and instability increases. The sample-size effect is well known from stability analyses in less demanding scenarios (Sauerbrei and Schumacher, 1992) (model selection limited to inclusion or exclusion of variables), or as reported by Royston and Sauerbrei (2003) for the GBSG study. We illustrate the effect of varying the sample size more systematically in an example (briefly in Section 10.8.2 and more fully in Section 9.5). The summary of BIFs from the ART study (Table 10.6) shows that there are just three critical issues: a linear or nonlinear function for x1 or x10, and whether to include x7; in the latter case, whether x7 should be modelled as linear or nonlinear. In addition, x2 may be selected if a model including weak factors is preferred.

10.8.2   Further Comments on Stability

Evidence of how difficult it is to select the correct model for a dataset of modest size is provided by Table 10.7. The table shows the results of applying MFP to 10 simulated replicates of the ART data, each with $n = 250$. The ART dataset used in this chapter is the first replicate. The true model includes x1, x3, x4a, x5, x6, x8 and x10. Instability is seen for selection of all variables except x5 and x6, but even for them there is instability in the FP function that is chosen. Larger samples are needed to have a reasonable chance of getting the 'correct' model in most replications.

The true function for x1, an FP2 with powers (0.5, 1), is not selected in any of the 10 replicates. Figure 10.10 (left-hand panel) shows the true function and the estimated partial predictor for x1 in the eight replicates in which MFP selected this variable. Although the six selected FP2 models differ, the curves are broadly similar and resemble the true function quite well. When the power is too low, a linear function is selected (in two replicates), and the estimates are clearly biased. When assessing the functions estimated for x1, it should be borne in mind that other variables are also mismodelled. The correlation between x1 and other

Table 10.7 ART study. Variables and FP functions selected by MFP(0.05) in 10 simulated replicates of the artificial dataset.[a]

Variable	True model	1	2	3	4	5	6	7	8	9	10
						Replicate					
x1	0.5, 1	1	3, 3		−2, −1	−2, 3	−0.5, 3		1	1, 2	1, 2
x2				✓					✓		
x3	1	1	1	1	1		1		1	1	1
x4a	✓	✓	✓		✓	✓		✓	✓	✓	✓
x4b											
x5	−0.2	0, 3	0	0	−0.5, −0.5	0	−1, −0.5	−0.5	0	0	−0.5
x6	0	0	0	0	0.5, 0.5	0	−0.5		0	0	0
x7											
x8	✓	✓	✓	✓	✓		✓		✓	✓	✓
x9						✓					
x10	1	1	1	1					1	1	1

[a] $n = 250$ in each replicate. 'True model' denotes the FP powers or binary variables that were used to generate the data. ✓ denotes binary variable included.

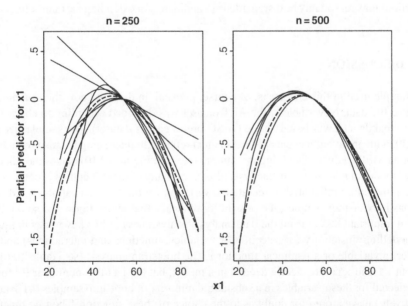

Figure 10.10 ART study. True function (dashed black line) and estimated partial predictors (solid lines) for x1, adjusted to the mean age in replicate 1. Left panel: estimated functions in the eight replicates each with $n = 250$ in which x1 was selected. In two replicates a linear function was found; the others are all FP2 functions with different powers (see Table 10.7). Right panel: estimated functions in five replicates each with $n = 500$. An FP2 function for x1 was selected in all five replicates.

variables will affect the function chosen for x1. In two replicates x1 is not selected at all, but is replaced by the highly correlated variable x2.

The right-hand panel of Figure 10.10 repeats the analysis, but in five replicates each with $n = 500$ (formed by merging neighbouring pairs with $n = 250$). The events-per-parameter ratio is now about 17 and the power is sufficient to select an FP2 function every time. The correct powers of (0.5, 1) are selected in one replicate. The improvement in stability is striking, and a consequence of better identification of the important variables and functions for x_2 to x_{10}.

Despite the instability, MFP can still find a *useful* model that summarizes the relationships in the data quite well.

10.8.3   Searching for Interactions

In the same way that nonlinearity is not considered in many analyses, interactions are perhaps even less often looked for. Note that if a variable has a nonlinear (main) effect that is incorrectly modelled as linear, then apparently significant linear-by-linear interactions of that variable with others may be spurious. In the ART data, MFPIgen identified several significant interactions which were not designed. Therefore, all are artefacts, either of the data (by chance) or of the modelling. Covariates with heavily skewed distributions may be particularly prone to generating artefactual interactions. In a hypothesis generation mode, it must be expected that some apparently significant interactions are not real. Highly significant interactions should be taken seriously and checked for consistency. In the example, neither of the interactions significant at the 0.01 level survived simple checks. Because of multiple testing, weaker interactions may reasonably be disregarded as candidates for extending the main-effects model.

10.9   DISCUSSION

The example used in this chapter is, of course, unusual, in that we know the true model that generated the data. Nevertheless, the ART dataset was designed according to characteristics found in real data (it was based on the GBSG breast cancer dataset). The sample size chosen ($n = 250$, with about eight events per parameter) reflects an interesting but also realistic scenario for an MFP analysis. Doubling the sample size (see Figure 10.10) makes the selection of a 'good' model much easier. The analysis illustrates many of the difficulties encountered with real data, including influential observations, variations in the selected variables and selected functions in bootstrap samples, the possibility of artefactual interactions, and so on. Perhaps the most important lesson from the data analysis is awareness of two key issues in multivariable modelling: instability (of selection of variables, functions and interactions) and power (to detect a variable or a nonlinear function). The bootstrap analysis (see Table 10.6) shows that both x1 and x10 are clearly needed in a model, but the fact that nonlinear FP functions were selected for these variables in a substantial minority of bootstrap samples (41% and 38% respectively) leaves room for doubt as to the choice of 'best' function. The true function for x1 was nonmonotonic and that for x10 was linear. Similarly, x7 was selected in 48% of samples, suggesting that it may be influential. However, x7 is quite highly correlated (Spearman $r_S = 0.4$) with a true predictor, x6, and may be selected by chance if the function for x6 is mismodelled, leaving residual confounding. Relationships between BIFs for different variables may be studied by inspection of two-way frequency tables, or in a more sophisticated way by log-linear modelling (see Royston and Sauerbrei (2003)). Our analysis is mainly for demonstration. In reality, at least 1000 bootstrap replications are needed.

Besides stability analysis, several methods of model checking are emphasized. If carefully done, they can uncover major discrepancies between the data and the model, if such discrepancies exist. It is well known that outliers and influential observations can easily result in poor models. Because heavily skewed distributions of covariates are often found in practice, they are a feature of the data in the ART study. Several tools have been demonstrated to cope with such situations. The analyst should aim to arrive at a model which survives these checks. Such a model would adequately summarize the relationships in the data. It goes without saying that the model should also be consistent with subject-matter knowledge. This aspect could not be discussed realistically in the ART study, but it is important in real studies. MFP has some options allowing integration of subject-matter knowledge, including complete flexibility in the choice of the 'tuning' parameters (nominal significance levels) α_1 and α_2, the ability to pre-define the maximum FP degree for each continuous predictor, and the possibility of changing the set of powers available to transform any continuous predictor (see Section 6.9.2).

In the present chapter, a nominal significance level of 0.05 was used for selecting variables and functional forms. Depending on the aim of a study, it may be preferable to choose a different value. For example, to develop a diagnostic index, the 0.01 level may be chosen, because keeping the model simple is important. In contrast, when selecting confounders in an epidemiological study, it may be more appropriate to use a large value, such as 0.20.

It is also important to investigate whether important interactions exist in the data. However, one should be aware that, partly because of multiple testing, MFPIgen can often generate artefacts, particularly when small numbers of potentially influential observations are present. The checks that have been described here and in Chapter 7 should usually enable one to distinguish between interactions 'truly in the data' and artefacts.

Once transportability and practical usefulness have been added to the list of criteria for a good model, the analyst should use a model-building procedure capable of detecting all strong factors, nonlinearity for continuous variables, important interactions between variables and, for survival data, nonproportionality of hazards (Sauerbrei et al., 2007a). Provided the (effective) sample size is adequate, then MFP and its extensions (for survival data, see Section 11.1.2) are useful procedures to fulfil these aims.

CHAPTER 11

Special Topics Involving Fractional Polynomials

11.1 TIME-VARYING HAZARD RATIOS IN THE COX MODEL

The semiparametric proportional hazards model of Cox (1972) has become the standard for the analysis of survival time data. The assumption of proportional hazards (PH) implies that the effect on the hazard function of each covariate measured at the beginning of a study is unchanged throughout the whole observation period. However, this assumption is questionable with long-term follow-up.

Methods for checking the PH assumption have been available for some time. A popular test is Grambsch and Therneau's (1994) test, based on the association between a covariate and its standardized Schoenfeld residuals. For an overview and some comparisons of properties of test statistics, see Hess (1994), Hess (1995), Ng'Andu (1997) and Berger et al. (2003).

Cox (1972) suggested relaxing the PH assumption by including an interaction between a covariate and a prespecified parametric function of time. Typically, linear or logarithmic functions have been used. Since then, other methods have been proposed for incorporating such a time-dependent function of a covariate, including a step-function model based on cutpoints on the time axis, smoothing splines (Hastie and Tibshirani, 1993), penalized regression splines (Gray, 1992), regression splines (Hess, 1994; Heinzl and Kaider, 1997) and FPs (Berger et al., 2003). Estimation of locally time-varying coefficients has been suggested (Verweij and van Houwelingen, 1995; Martinussen et al., 2002; Cai and Sun, 2003; Tian et al., 2005). However, there is no consensus with respect to advantages, disadvantages and practical usefulness of the many methods available.

Further to the diminishing prognostic effect of a covariate over time, it has been shown that a time-varying regression coefficient may result from model misspecification. PH may break down in a number of ways, including gross failures such as crossing hazard functions, converging hazards due to the presence of a fraction of 'cured' patients, or frailty effects whereby the sickest patients succumb to an event early on. More subtly, as Abrahamowicz and MacKenzie (2007), Therneau and Grambsch (2000) and others have discussed, omission of an important covariate or misspecification of the functional form for a time-fixed covariate may induce non-PH, even when the PH assumption is satisfied by a more appropriate model.

Multivariable Model-Building   Patrick Royston, Willi Sauerbrei
© 2008 John Wiley & Sons, Ltd

Methods for assessing possible time-varying covariate effects usually take a time-fixed model as their starting point. The model may already include nonlinear transformations of predictors. Only violation of the PH assumption is considered as mismodelling. However, in most analysis of observational studies, selection of important variables and of the functional relationships for continuous predictors is also necessary.

Sauerbrei et al. (2007c) proposed a new procedure, MFP time (MFPT), which transfers the FP concept for the selection of nonlinear functions to the investigation and modelling of time-varying effects of a covariate. In order to avoid dubious time-varying effects caused by major misspecification of other parts of a Cox model, it combines selection of variables, functional form for continuous variables and time-varying covariate effects.

Assuming PH initially, MFPT starts with a detailed model-building step, including a search for possible nonlinear functions for continuous covariates. A variable with a strong short-term effect may appear weak or noninfluential when 'averaged' over time under the PH assumption. To protect against omitting such variables, the procedure repeats the analysis over a restricted time interval. Any additional prognostic variables identified by this second analysis are added to create a final time-fixed multivariable model. Using a forward-selection algorithm, MFPT determines whether including time-varying effects for any of the covariates significantly improves model fit. The first step, creating a time-fixed model, does not require the use of MFP; any prespecified model may be used.

11.1.1 The Fractional Polynomial Time Procedure

The aim of the FP-time (FPT) procedure is to detect and model the time-varying effect of a single covariate. By allowing the parameter vector β to vary in time, the Cox model may be extended to a non-PH model

$$h(t|\mathbf{x}) = h_0(t) \exp[\mathbf{x}\beta(t)] \tag{11.1}$$

The relative hazard $h(t|\mathbf{x})/h_0(t) = \exp[\mathbf{x}\beta(t)]$ is now a function of $\mathbf{x}$ and t. For a single variable x, Cox (1972) proposed adding an unspecified transformation $\omega(t)$ of time, giving

$$h(t|x) = h_0(t) \exp[\beta_0 x + \beta_1 \omega(t)x] \tag{11.2}$$

Implicitly, $\beta(t) = \beta_0 + \beta_1\omega(t)$. Model (11.2) may also be viewed as a usual Cox model extended with a time-dependent covariate $z(t) = \omega(t)x$ constructed as a multiplicative interaction term.

The key practical issue is how to specify $\omega(t)$. Sauerbrei et al. (2007c) suggested estimating $\omega(t)$ for a single covariate in data-dependent fashion as an FP function $\varphi(t)$, determined by a systematic search. Because it is often used and allows short-term effects to be modelled, they took $\varphi_1(t; 0) = \log t$ as a default. A better-fitting FP function is chosen if strongly indicated by the data. The procedure is called the FPT algorithm.

FPT searches for a possible improvement in fit compared with the default function, $\log t$, by considering FP1 and FP2 functions of t. For example, a time-varying Cox model with covariate x and an FP2 function of t with powers $\mathbf{q} = (q_1, q_2)$ (to distinguish them from powers $\mathbf{p}$ used for covariates) is

$$h(t|x) = h_0(t) \exp[x\varphi_2(t; \mathbf{q})]$$
$$= \begin{cases} h_0(t) \exp[x(\beta_0 + \beta_1 t^{q_1} + \beta_2 t^{q_2})], & q_1 \neq q_2 \\ h_0(t) \exp[x(\beta_0 + \beta_1 t^q + \beta_2 t^q \log t)], & q_1 = q_2 = q \end{cases}$$

The covariates are multiplicative interactions between x and FP2 transformations t^q of t, i.e. products of x with $(1, t^{q_1}, t^{q_2})$ or $(1, t^q, t^q \log t)$. The best-fitting FP1 or FP2 function of t is defined as that which minimizes the deviance within its class (FP1 or FP2), the deviance being defined as minus twice the maximized partial log likelihood for the interaction model. The best power(s) $\mathbf{q}$ are estimated by a systematic search of the standard set S (see Section 4.2.1), with coefficients β estimated in the usual way, conditional on $\mathbf{q}$. Apart from chosing the log function as default for $\varphi(t)$ instead of the linear function, the approach to function selection exactly mimics the FSP (see Section 4.10). The 'familywise' error rate for incorrectly selecting a time-varying effect when none is present is controlled at a nominal level, say α_3.

11.1.2 The MFP Time Procedure

The aim of the the MFPT procedure is to detect and model time-varying effects of several covariates in a multivariable model. Figure 11.1 illustrates the motivation, gives the FPT procedure for one covariate and shows the three stages of the algorithm. MFP is used to determine a time-fixed model in two stages. If the second stage identifies no variable with only a short-term effect, then the 'final' PH model M_1 is the MFP model M_0 from stage 1. Sometimes, covariates with a short-term effect are added to M_0 to give M_1. Alternatively, M_1 may be provided externally, e.g. prespecified in the statistical analysis plan of a randomized trial, restricting the use of MFPT to stage 3.

Next, the nominal P-value (α_3) for testing time-varying effects (often 0.05 or 0.01) and the maximum degree of FP functions (we propose to use FP2) are chosen. The FPT algorithm is then used to investigate whether the effect of each variable in M_1 varies in time, adjusting for the other variables in M_1.

Addition of time-varying effects to M_1 is done as in the usual FS procedure for variable selection. For further details, see Sauerbrei et al. (2007c).

11.1.3 Prognostic Model with Time-Varying Effects for Patients with Breast Cancer

Sauerbrei et al. (2007c) derived a model for 2982 primary breast cancer patients whose records were included in the Rotterdam tumour bank. Follow-up time ranged from 1 to 231 months (median 107 months). Event-free survival time was defined as time from primary surgery to disease recurrence or death from breast cancer. Times to death from other causes were treated as censored. With this outcome measure, 1518 events were observed. To work with complete data, a single imputation of the small number of missing values was made using the MICE algorithm (van Buuren et al., 1999), implemented in the ice program for Stata (Royston, 2005). Nine prognostic variables were considered. The results from the MFPT procedure are summarized in Table 11.1.

Model M_0 is the result of applying MFP(0.05, 0.05). The continuous variables age (linear), squared transformed number of positive lymph nodes (enodes = $\exp(-0.12 \times \text{nodes})$),

(a) MOTIVATION

Multivariable modelling strategy required to determine

- which variables influence the outcome;
- functional form for continuous variables (is usual linearity assumption justified?);
- whether the proportional hazards assumption is fulfilled, or one or more time-varying functions fit the data better.

(b) TIME-VARYING FUNCTION FOR ONE COVARIATE

FPT algorithm to select $\varphi(t)$, compare deviances between models

FPT2 ($\varphi_2(t)$) to null (i.e. **time-fixed effect**, $\varphi_c(t)$)	four d.f.
FPT2 ($\varphi_2(t)$) to $\log(t)$ ($\varphi_{\log}(t)$)	three d.f.
FPT2 ($\varphi_2(t)$) to FPT1 ($\varphi_1(t)$)	two d.f.

(c) STAGES OF THE MFPT ALGORITHM

STAGE 1: Determine time-fixed model M_0

- Select M_0 using MFP, assuming PH, over the full time period.

STAGE 2: If necessary, add covariates with short-term effect

- Start with model M_0, keep variables and functions from M_0.
- Restrict the time period to $(0, \tilde{t})$, e.g. $\tilde{t}$ defined by the first half of events.
- Rerun MFP for times in $(0, \tilde{t})$; if necessary, add significant covariates to M_0. This gives M_1.

STAGE 3: Investigate time-varying effects of variables in M_1

- For each covariate (with selected FP transformation) in M_1, run the FPT algorithm to investigate time-varying effect adjusting for all other covariates in M_1.
- Use an FS procedure to add significant time-varying effects to M_1. This gives the final model M_2.

Figure 11.1 The MFPT algorithm. (Adapted from Sauerbrei et al. (2007c) with permission from John Wiley & Sons Ltd.)

the binary variables grad1 (grade 2,3 versus 1) and sized1 (tumour size $\leq$ 20 mm versus > 20 mm) and the two treatment variables hormon (hormonal therapy) and chemo (chemotherapy) were selected. The variables meno (menopausal status) and er (oestrogen receptors) were eliminated.

Using M_0 with the powers as given, but allowing re-estimation of the regression parameters, MFP(0.05, 0.05) was rerun for the restricted time period $\leq$ 2.54 years (the median uncensored event time). Predictors sized2 (tumour size $\leq$ 50 mm versus > 50 mm) and pgr were additionally selected, both significant at $P < 0.001$ over the restricted time period. An FP1 model with power 0 was selected for pgr, the actual transformation being $\log(\text{pgr} + 1)$ (see Section 4.7).

Adjusting for the variables and transformations from model M_1, the FPT algorithm identified univariately significant ($P < 0.01$) time-varying effects for the four variables sized1,

Table 11.1   Rotterdam breast cancer data. Estimates of parameters of models M_0 (step 1; MFP, time fixed), M_1 (step 2; added variables to M_0, time fixed) and M_2 (step 3; adding time-varying functions to M_1).

Variable	Model M_0		Model M_1		Model M_2	
	$\widehat{\beta}$	SE	$\widehat{\beta}$	SE	$\widehat{\beta}$	SE
age	−0.013	0.002	−0.013	0.002	−0.013	0.002
sized2	–	–	0.171	0.080	0.150	0.081
grade	0.39	0.064	0.354	0.065	0.375	0.065
enodes^2	−1.71	0.081	−1.681	0.083	−1.696	0.084
hormon	−0.39	0.085	−0.389	0.085	−0.411	0.085
chemo	−0.45	0.073	−0.443	0.073	−0.446	0.073
sized1	0.29	0.057	0.249	0.059	−0.112	0.107
log(pgr + 1)	–	–	−0.032	0.012	0.137	0.024
sized1 × log t	–	–	–	–	−0.298	0.073
log(pgr + 1) × log t	–	–	–	–	0.128	0.016

(Adapted from Sauerbrei et al. (2007c) with permission from John Wiley & Sons Ltd.)

enodes^2, log(pgr + 1) and chemo. Since sized2 had no significant time-fixed effect over the full time period but was significant in the restricted time period, it had a short-term effect only. Nevertheless, it just missed having a significant time-varying effect ($P = 0.012$) over the full time period in a model adjusting for M_1. Instead, the effect of sized1 was highly significant ($P < 0.001$). The largest time-varying effect, which was identified in the first step, was for log(pgr + 1). The time-varying effect of sized1 was still significant in the second step. After including these two time-varying effects, no further such effects were significant at the chosen 1% level. This gives the final model M_2. For further details, see Sauerbrei et al. (2007c).

Estimated time-varying β values for log(pgr + 1) and sized1 are shown in Figure 11.2. A logarithmic function of t (the default in FPT) was chosen for both variables. The effect of log(pgr + 1) actually changes sign over time. In the early follow-up phase, the sign is negative (indicating a 'protective' factor), whereas after about 3 years the sign is positive. By contrast, the effect of sized1 diminishes over time.

11.1.4   Categorization of Survival Time

To estimate time-varying functions, the dataset must be enlarged considerably. The standard approach is to split all episodes at each event time. In the Rotterdam dataset, about 2.2 million additional observations would be produced. Model-fitting and exploration become impracticably slow, even with a fast computer. Instead, time was categorized into half-year intervals up to year 15 and a single period of > 15 years. The dataset was split at the 31 categorized times, giving a manageable 35 698 observations. Sensitivity analyses showed negligible information loss through this approach. Categorization of time helps to overcome computational difficulties in larger studies and reduces computational costs when many models are considered. Therefore, it increases the practical usefulness of model-building strategies that include a time-varying component, and should make it possible to carry out simulation studies within a reasonable computation time (Buchholz et al., 2007).

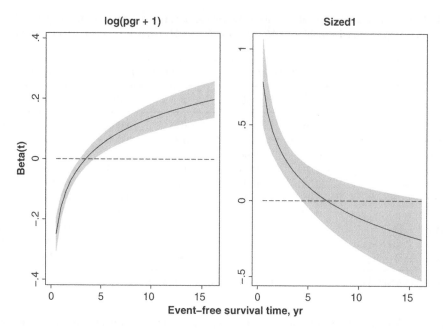

Figure 11.2   Rotterdam breast cancer data. Time-varying coefficients with 95% pointwise CIs are plotted against time for log(pgr + 1) (transformed progesterone receptors, left panel) and sized1 (tumour size ≥ 20 mm versus < 20 mm, right panel), adjusted for predictors in Table 11.1. (Adapted from Sauerbrei et al. (2007c) with permission from John Wiley & Sons Ltd.)

11.1.5   Discussion

With long term follow-up, time-varying effects, particularly an effect diminishing in time, are biologically plausible. It is well known that mismodelling a part of a model can introduce spurious time-varying effects. Many approaches are available to investigate and model time-varying effects, but none is generally accepted and results often differ. Procedures usually concentrate on the time-varying aspect, ignoring the possibility of a nonlinear effect for a continuous variable and not considering sufficiently the importance of variable selection. It is assumed that a PH model has been developed somehow, and only time-varying effects need further attention.

MFPT combines variable selection, selection of functional relationships and selection of time-varying functions. The procedure is an attempt to minimize the risk of spurious time-varying effects caused by mismodelling one component. Abrahamowicz and MacKenzie (2007) propose a model in which time-varying and nonlinear covariate effects are modelled simultaneously using regression splines. In contrast, MFPT works hierarchically by first considering the linearity or otherwise of the main (covariate) effects, while permitting the removal of variables with a nonsignificant influence on the outcome. Time-varying effects are considered in a second step. Overfitting is an issue for both procedures. Although no results are available, we believe that the greater flexibility of Abrahamowicz and MacKenzie's (2007) method of simultaneously considering functional forms of covariates and time-varying effects is more likely to produce spurious associations than is MFPT.

11.2 AGE-SPECIFIC REFERENCE INTERVALS

The construction of age-specific reference intervals (sometimes confusingly called 'normal ranges') is required in several branches of medicine. It is particularly important in clinical laboratory measurements and in studies of human growth. The idea is that certain extreme centiles of a distribution, known as 'normal limits' or 'reference limits', define a range of values outside which some abnormality may be suspected. In the case of a blood measurement, for example, an unusually high or low value, beyond the reference limits, may indicate a haematological problem. In the case of growth, a body weight below a lower reference limit may indicate a failure to thrive.

The scenario is as follows. To derive the limits, a large reference sample of 'normal' individuals must be assembled. (The concept of reference limits has certain philosophical issues, such as what a 'normal' individual is, and normative versus descriptive approaches, but we do not engage with them here.) Predetermined centiles of the reference distribution, typically the 2.5th and 97.5th, are estimated from the data. In many cases, and the main topic here, the distribution depends on the person's age. The challenge is to estimate the age-dependent limits in such a way that the resulting age-specific reference interval (RI) is accurate and is easy to apply in clinical settings. Let y be the measurement of interest and t be age. The statistical task is to estimate centiles $c_\alpha(y|t)$ of the conditional distribution of y for specified values of α such as 0.025, 0.975. Usually, $c_\alpha(y|t)$ is required to be a smooth function of t.

Because of the importance of the clinical problem, much statistical literature on the estimation of age-specific RIs has appeared, going back at least to 1942 (Count, 1942). There are two main issues: approximating the distribution of $y|t$ and smoothing the parameters or estimated centiles of the latter on t. Quantile regression (Koenker and Bassett, 1978) and kernel density estimation (Rossiter, 1991) have been used to obtain $c_\alpha(y|t)$ for given values of α, without regard for the remainder of the distribution. Such methods have the severe disadvantage that, because no constraint is imposed to ensure separation at all t, centile curves may cross each other at some t. Attempts (e.g. He, 1992) have been made to remedy this difficulty. Here, we concentrate on methods that assume a parametric form for the density, $f(y|t)$. The resulting centile curves cannot cross.

We do not attempt to describe or exemplify the many available methods. A good summary is given in Borghi et al. (2006, table 1). Interested readers are directed to this excellent paper for further background, comments on RI estimation and bibliography. The aim here is to outline relatively straightforward methodology to estimate age-specific RIs, focusing on simple distributions and mainly using FP functions as smoothers on t. A real example, introduced next, provides motivation.

11.2.1 Example: Fetal growth

Figure 11.3 (upper left panel) shows the relationship between $y = \texttt{hemi}$, the hemisphere diameter (a measure of fetal brain size) and $t = \texttt{gawks}$ (gestational age in weeks) in 574 fetuses. The measurements were made by ultrasound scanning of the woman's abdomen and were part of a large cross-sectional study of *in utero* fetal growth performed by Lyn S. Chitty (Altman and Chitty, 1993). If we suppose that $y|t$ is normally distributed $N(\mu(t), \sigma^2(t))$ then the running-line smooth in the upper left panel of Figure 11.3 is an estimate of $\mu(t)$. However, to construct an RI we also need an estimate of $\sigma(t)$. A symmetrical RI with coverage $1 - \alpha$ is then given by $\mu(t) \pm z_{1-\alpha/2}\sigma(t)$. How is $\sigma(t)$ to be estimated?

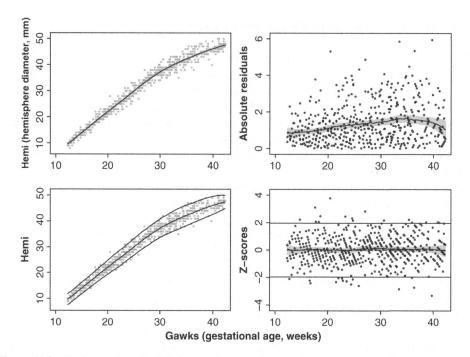

Figure 11.3 Fetal growth study. Relationship between hemi and gawks. Upper left: raw data; upper right: absolute residuals; lower left: 95% age-specific RI (central line shows the estimated mean); lower right: Z-scores. In all plots except the lower left, solid lines and grey areas show running-line smooths and 95% pointwise CIs respectively. $\sigma(t)$ is estimated by the absolute residuals $\times 1.253$.

A useful result, exploited by Altman (1993), is that the mean of the absolute values of a sample from a normal distribution $N(0, \sigma^2)$ is $(2/\pi)^{1/2}\sigma$. Thus, $\sigma(t)$ may be estimated regressing (smoothing) $|y - \mu(t)|$ on t and multiplying the predicted values $E(|y - \mu(t)|)$ by $(\pi/2)^{1/2}$ or about 1.253. The upper right panel of Figure 11.3 shows the smoothed absolute residuals, computed using a running-line smooth of y on t. $\sigma(t)$ appears to increase up to about 33 weeks' gestation and then to decrease. The lower left panel shows the estimated 95% RI, i.e. taking $\alpha = 0.05$ and $z_{1-\alpha/2} = 1.96$. The lower right panel shows the Z-scores (standardized residuals), defined as $(y - \mu(t))/\sigma(t)$. If the model is correct then the Z-scores should have a standard normal distribution, independent of t. This appears (reasonably) true in the plot. The overall skewness and kurtosis of the Z-scores is 0.01 and 3.22, compared with normal values of 0 and 3.

11.2.2   Using FP Functions as Smoothers

'Nonparametric' smoothers (e.g. running line, lowess, kernel methods) are not simple and transportable; therefore, they are difficult for practical use. Smoothing with FPs provides simple functions which are easy to use in practice (Altman and Chitty, 1993).

The normal-based methodology just described was implemented in the xrigls command for Stata, and uses FP functions to smooth $\mu(t)$ and $\sigma(t)$ by Royston and Wright (1997). A version of the FSP is used to select the best-fitting FP functions. The regression of y on x

is weighted by $\sigma^{-2}(t)$ in a second cycle of an iterative algorithm, following an initial cycle in which an unweighted estimate of $\mu(t)$ is made. In the fetal example, FP2 functions were selected for $\mu(t)$ and $\sigma(t)$, with powers (2, 2) and (3, 3) respectively. The results (not shown) are similar to those in the lower panels of Figure 11.3.

11.2.3 More Sophisticated Distributional Assumptions

When the normality assumption applies, estimation of age-specific RIs is not in principle diffi-cult, as we have just seen. It is, however, critical to get a good approximation to the distribution of $y|t$. Failing this, RIs, by definition in the tail areas of the distribution, are poorly estimated. When the normality assumption is wrong, age-related transformation of y to normality is the technique favoured by most researchers. The Box–Cox–normal three-parameter distribution (Cole, 1988; Cole and Green, 1992) was the first proposal and, more recently, four-parameter distributions have been used. For example, Royston and Wright (1998) used the modulus–exponential–normal distribution and Rigby and Stasinopoulos (2004) the Box–Cox power exponential distribution. The use of three parameters allows (possibly age-specific) skewness to be modelled, whereas using four parameters allows (possibly age-specific) kurtosis to be modelled also. These methodologies appear to be current 'state of the art' for RI estimation. Parameter estimation is by maximum likelihood. Smoothing of parameters on t may be done using any class of functions suited to the task, including linear functions, polynomials, FPs, restricted cubic regression splines, smoothing splines, and so on. The choice depends on the complexity of the curves to be modelled.

As a second example, in which the normality assumption is untenable, consider cholesterol as a function of age. Serum total cholesterol and age were observed in a random sample of 502 normal men and 553 normal women aged between 26 and 64 years in a study investigating cardiovascular risk factors (Mann et al., 1988). Following preliminary analysis of the data for women with `xrigls` along the lines described in Section 11.2.2, FP1 functions with powers 3 and 1 (i.e. linear) in age were chosen for $\mu(t)$ and $\sigma(t)$ respectively. The Z-scores from this model have significant positive skewness ($\sqrt{b_1} = 0.44$, $P < 0.001$), showing that normality does not hold. An exponential–normal three-parameter model (Royston and Wright, 1998) shows a much improved fit, with a deviance about 20 lower then the normal model ($P < 0.001$); the skewness and kurtosis of the Z-scores are now 0.02 and 2.72 respectively. The model was fit by maximum likelihood using the Stata command `xriml` (Wright and Royston, 1996). Adding a parameter to model kurtosis, i.e. fitting a modulus–exponential–normal model, does not significantly improve the fit ($P = 0.12$). Figure 11.4 shows a 95% RI and smoothed Z-scores from the exponential–normal model. The results appear satisfactory. The positive skewness can be easily seen, since the reference limits are asymmetrically disposed around the central (median) line. The Z-scores that exceed the reference limits are evenly spread across age. The smoothed mean curve does show some systematic departure from the zero line, but it is minor.

11.2.4 Discussion

While FP functions are clearly useful for modelling age-specific RIs in fetal medicine, where the functions are generally simple, this is not the case in all fields. For example, human growth curves are more complex, showing 'spurts' near puberty and levelling off near adulthood. For

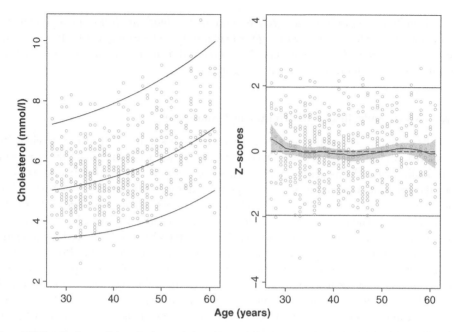

Figure 11.4   Cholesterol data. Left panel: 95% RI derived using FP1 functions and a three-parameter distribution; right panel: Z-scores with a running-line smooth and 95% pointwise CI.

these applications, additional flexibility, e.g. using spline functions (Cole and Green, 1992), is required. However, FPs are still useful for a wide variety of types of measurement.

11.3   OTHER TOPICS

11.3.1   Quantitative Risk Assessment in Developmental Toxicity Studies

An important problem impacting public health is the determination of safe dose levels for exposures to drugs, harmful chemicals, and other environmental hazards. Benchmark doses must be defined. Developmental toxicity studies are required. For example, pregnant dams (mice, rats) are exposed to different dose levels of a compound of interest to determine a dose–response model. The use of FPs instead of classical polynomial predictors for quantitative risk assessment (QRA) has been investigated and seems to be beneficial in such a setting (Geys et al., 1999; Faes et al., 2003; Faes et al., 2007).

The standard approach to QRA based on dose–response modelling requires the specification of an adverse event, along with its risk expressed as a function of dose. In contrast to situations considered in our book so far, analysis of developmental toxicity data is based on clustered data, e.g. offspring are clustered within litters. Faes et al. (2003) focus on the risk function $r(d)$, representing the probability of observing a malformation at dose level d for at least one fetus within a litter. Based on this probability, a common measure for the excess risk over background is given by $r^*(d) = (r(d) - r(0))/(1 - r(0))$. Greater weight is given to outcomes with larger background risk. This definition of the excess risk measures the relative increase in risk above background. A benchmark dose is then defined as the dose corresponding to a very small increase in risk over background.

A prerequisite for risk assessment based on a dose–response model is that the model should fit the data well. This has implications for both the model family chosen (the probability model), as well as for the form of the dose–response model. While the probability model can take special features of the data into account, the dose–response model must be sufficiently flexible. Because the standard polynomial models often give poor results, especially when low dose extrapolation is envisaged, there is a clear need for alternative specifications. FPs provide much more flexible curve shapes than conventional polynomials, but reduce to polynomials when such extension is unnecessary. Their use is strongly recommended in an important public health context, such as the determination of safe limits for human exposure to potentially hazardous agents (Faes et al., 2003).

This recommendation is supported by asymptotic and small sample simulations. Faes et al. (2003) consider several models and conclude that the results from FP modelling are much closer to the true benchmark dose than those from polynomials. This seems to indicate that FP functions can attain the correct benchmark dose better than conventional linear polynomials. Differences between dose–response models can produce large variations in benchmark doses. For example, in a study on ethylene glycol in mice, Faes et al. (2003) use a Molenberghs and Ryan (1999) probability model, and compare polynomial dose–response models with FPs. They show that the fit of the FP models is much better, particularly in the very low dose range. For example, the estimated benchmark dose from FP modelling was $84 \text{ mg kg}^{-1} \text{ day}^{-1}$, compared with $382 \text{ mg kg}^{-1} \text{ day}^{-1}$ from a conventional polynomial. Faes et al. (2003) conclude that FP models not only can (partly) correct for a misspecification of the dose–response model, but also of the probability model.

Other flexible parametric models could be considered too, such as models based on nonlinear predictors and penalized splines. In contrast to FPs (which are easy to handle), these methods pose nontrivial methodological challenges (Faes et al., 2003).

11.3.2 Model Uncertainty for Functions

The standard statistical approach to parameter estimation assumes that the model was pre-specified, i.e. that no data-dependent model building was done. For selected models, ignoring that the uncertainty from the selection process may lead to unjustifiable confidence in these estimates (Chatfield, 1995; Draper, 1995; Hoeting et al., 1999). For example, CIs around predictions may be too narrow and may not yield the nominal coverage. One way to try to get more realistic estimates incorporating model selection uncertainty is to average over a set of considered models. SEs may be calculated, allowing for the between-model variance of parameter estimates (Buckland et al., 1997).

The topic of model uncertainty has received much attention in the literature. Suggestions on how to improve a predictor by model averaging (MA) have been published for many types of outcome variable. However, MA of nonlinear functions of continuous variables has seldom been considered. The MA estimate is a weighted average of a set of parameter estimates, one from each model considered. Most of the work has been done in a Bayesian MA framework. Typically, poorly fitting models are excluded according to the principle of Occam's window before averaging is done (Hoeting et al., 1999). An alternative to Bayesian MA is bootstrap MA, first proposed by Buckland et al. (1997), who emphasized the case of a small set of candidate models. With many variables, and therefore a large model class, Augustin et al. (2005)

suggested modifying the bootstrap MA procedure by including a variable screening step done before averaging across selected models.

With a single continuous predictor, the FSP (see Section 4.10) selects the best-fitting function from the FP class in data-dependent fashion. The powers are determined and the corresponding parameters are estimated as though the powers were prespecified. The variance of the resulting predictor, being conditional on selection of the powers, is underestimated (see also Section 4.9.2). For determining a safe dose-level from a dose–response curve, Faes et al. (2007) propose MA on a set of FP functions. If only FP1 and FP2 functions are considered, then the model class is restricted to 44 functions. The additional step of eliminating poorly fitting curves before the MA step is not required. Weights for the MA step are based on the AIC values of each function. Faes et al. (2007) show that model averaging reduces bias and results in better precision in estimating a safe level of exposure compared with an estimator from the selected best model. They suggest using the bootstrap to estimate the variance of an MA function.

MA of FP functions can easily be transferred to many other relevant situations, such as estimating a dose–response function for a continuous exposure factor in a meta-analysis of several studies. Some research along this line is under way.

11.3.3 Relative Survival

Relative survival concerns the estimation of survival probabilities for patients with a particular disease such as breast cancer, excluding causes of death unrelated to the disease (Cutler and Aztell, 1969). Crude survival probabilities do not allow for mortality in the general population. In relative survival studies, the observed survival is compared with that expected in an age- and sex-matched group from the population (other matching factors may be used as well).

The following treatment is essentially that of Lambert et al. (2005). Suppose t is the time since some starting point (e.g. diagnosis of breast cancer), x_1 is a covariate pattern of interest for a patient whose survival function is $S(t; x_1)$ and x_2 a matched covariate pattern in the general population, with survival function $S^*(t; x_2)$. Covariates in x_2 usually comprise age, sex and epoch of diagnosis. The covariates in x_1 may be the same as in x_2, but may also include clinically relevant information such as disease severity (e.g. in breast cancer, prognostic group or stage). The relative survival for x_1 is a function of t, x_1 and x_2, defined as

$$R(t; x_1) = \frac{S(t; x_1)}{S^*(t; x_2)} \tag{11.3}$$

After a little algebra, Equation (11.3) may be re-expressed on the hazard scale as

$$\lambda(t; x_1) = \lambda^*(t; x_2) + \lambda_d(t; x_1) \tag{11.4}$$

i.e. the overall hazard of death (mortality) is composed of an underlying, known population hazard $\lambda^*(t; x_2)$ and an excess hazard $\lambda_d(t; x_1)$, which is attributable to the disease and to the additional covariate(s) in x_1, if any. Interest then lies in estimating $\lambda_d(t; x_1)$.

Lambert et al. (2005) approach estimation of $\lambda_d(t; x_1)$ in Equation (11.4) in a flexible way, (a) by dividing the time axis into a large number of periods, with a Poisson model for mortality in each period, and (b) by expressing the bivariate function $\lambda_d(t; x_1)$ as the sum of an FP function of time, a function of x_1 and possibly an interaction between the two. They found that FP3 functions of t were needed to model the time component (i.e. the baseline excess hazard function) adequately, but that only FP1 or FP2 functions of t were needed to model the interaction with x_1 (time-dependent covariate effects on the excess hazard).

The analysis nicely showed that breast cancer patients from a deprived background had worse relative survival than affluent patients, but that the excess mortality difference between these groups disappeared after about 5 years. FP3 models were shown to fit the baseline excess hazard function better than FP2 functions, so provide another example of when higher order FPs may be useful (see Section 5.7).

The authors commented that 'the fractional polynomial models used here provide a better fit to the data than the piecewise models and they also have the advantage of using fewer parameters'. They further noted that 'when using spline functions there is the problem of subjectivity in the choice of the number and location of knots and we found that fitted values could sometime be fairly sensitive to the choice of knots'.

All models presented in their paper can be estimated within a GLM framework and, thus, are easily implemented using standard software.

11.3.4 Approximating Smooth Functions

It is common in statistics to require a simple approximating equation for a 'function', i.e. a smooth relationship between a y and an x. Such 'functions' are either completely smooth or have minimal noise. The exact mathematical form may be known but complex (e.g. a difficult-to-evaluate integral), or unknown. In several examples, Royston and Altman (1997) showed that FPs are particularly good at providing concise and often sufficiently accurate formulae for representing such relationships.

As a 'toy' example, consider approximating the quantile function $y = G(3, x)$ of the chi-square distribution on three d.f., where x is the P-value (upper tail area) and y is the corresponding quantile. For example, $G(3, 0.05) = 7.815$. Table 11.2 shows the residual SD (i.e. root-mean-square error) for least-squares regression of y on x over the range $0.001 \leq x \leq 0.9$, using a grid on x with 1000 equally spaced values. FP functions of degree 1, 2, 3, 4 and 5 were used. The error decreases rapidly up to FP3, but much more slowly for FP4 and FP5 functions. FP3 seems to be a good choice in this example. The FP3 function has powers $(0, 0.5, 3)$. The formula to predict y from x is

$$y = 1.7536 - 2.1070 \ln x - 1.1198 x^{0.5} - 0.4313 x^3$$

Table 11.2 Approximating the quantile function of $\chi^2(3)$ by FP functions of different degree. Goodness of fit (residual SD) from least squares regression on P-values.

Degree of FP	Residual SD
1	0.1394
2	0.0165
3	0.0028
4	0.0022
5	0.0011

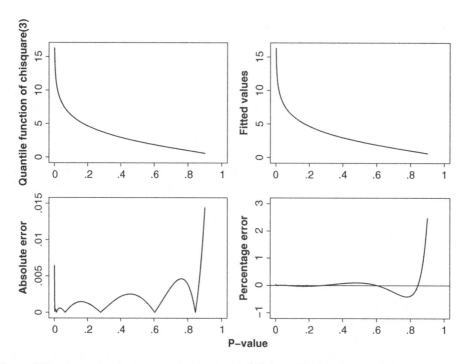

Figure 11.5 Approximating the quantile function of $\chi^2(3)$ by an FP3 function. Top left: actual function; top right: fitted values of FP3 function fitted by least squares to 1000 'observations'; bottom left: absolute error of approximation; bottom right: percentage error.

Figure 11.5 shows various aspects of the FP3 approximation. The observed and fitted functions are indistinguishable (upper panels). The percentage error for $P \leq 0.5$ is tiny ($< 0.02\%$; see lower right panel). Since low P-values are more important than high ones, the approximation is sufficient for practical use. Obviously, it would not be accurate enough for detailed mathematical work, nor should it be extrapolated beyond the defined P-value range.

Getting a reasonable functional approximation is usually quite straightforward using FPs. Functions with degree higher than four should probably be avoided, since they are cumbersome to use, defeating the object of the exercise, to obtain a simple approximation. Sometimes it is sensible to transform x or y or both before applying FP regression. If x has nonpositive values, then the negative exponential pre-transformation (see Section 5.6) may be useful.

11.3.5 Miscellaneous Applications

Faes et al. (2006a) used FPs to determine dose–response models in hierarchical and other complex models in a Bayesian framework. FPs are also used to estimate herd-specific force of infection by using a random-effects model for clustered binary data (Faes et al., 2006b).

In another example, Shkedy et al. (2006) argue that the set S of powers (see Section 4.3) offers insufficient flexibility for estimating dose–response functions in their models. They extend the power terms to allow a finer grid search.

CHAPTER 12

Epilogue

12.1 INTRODUCTION

In this chapter, we summarize several themes that run through our book and present them as recommendations for practice. We briefly discuss some themes omitted from our book, thus pointing the way towards topics requiring further research. Bayesian methodology for multivariable modelling using spline and FP functions is under development, and may become more important in the near future. It is not considered here.

12.2 TOWARDS RECOMMENDATIONS FOR PRACTICE

We started writing our book with a belief that MFP is useful for deriving a 'good' model in many applications and, therefore, is helpful for practitioners who urgently need recommendations for selecting important variables and determining functional form in multivariable model-building. In Section 1.6 (see Table 1.3), we discussed the scenarios in which we expected the methods to be useful. Briefly, these were no subject matter knowledge, about 5 to 30 candidate predictors, no strong correlations among predictors, at least 10 observations per variable, no missing data, and no interactions. Allowing for the possibility of interactions (see Chapter 7), our recommendations for analysis, a slightly modified version of Sauerbrei et al.'s (2007a) table 3, are given in Table 12.1.

We are well aware that some of the recommendations need more supporting evidence from simulations, which may result in modifications. We hope that more researchers use the methods to gain practical experience, to investigate methodological aspects systematically, and to improve knowledge of multivariable model-building on issues discussed in our book. We concur with Harrell's (2001, p. 79) statement 'At the least these default strategies are concrete enough to be criticized so that statisticians can devise better ones'.

12.2.1 Variable Selection Procedure

Even when the model-building task is restricted to selecting influential variables assuming a linear effect for continuous variables, statements in the literature conflict and are unhelpful.

Multivariable Model-Building   Patrick Royston, Willi Sauerbrei
© 2008 John Wiley & Sons, Ltd

Table 12.1   Towards recommendations for model building by selection of variables and functional forms for continuous predictors in observational studies, under the assumptions of Table 1.3.

Issue	Recommendation	Explanation
Variable selection procedure	BE with significance level as key tuning parameter. The choice depends on the aim of the study	BE seems to be the best of all stepwise procedures. Although BE does not optimize any specific criteria, it will select models which are at least very similar to AIC or BIC (provided the appropriate significance level is chosen). Programs for all types of regression model are available. Combining BE with other tasks, e.g. FSP, is often straightforward
Functional form for continuous covariates	Linear function as the 'default', check improvement in model fit by FPs. Check derived function for overfitting and undetected local features	FP functions are sufficiently flexible for most purposes. A well-defined procedure (MFP) for the multivariable situation is available. Spline models or smoothers with a local character lack stability and transportability. Disagreement about the most sensible spline approaches, useful only in the hands of experienced practitioners
Extreme values or influential points	Check at least univariately for outliers and influential points in continuous variables. A preliminary transformation may improve the model selected. For a proposal, see Section 5.5	Outliers or influential points often affect the functions selected, particularly at the left or right end of a covariate distribution. They may even influence the selection of a variable in a model. Many methods have been developed to identify outliers and influential points, but none seems to be generally accepted in multivariable model-building
Sensitivity analysis	Important assumptions should be checked by a sensitivity analysis. Highly context dependent	In multivariable model-building, many assumptions are required. Sensitivity analyses are important for checking whether main results are consistent when assumptions are varied
Check for model stability	The bootstrap is a suitable approach to check for model stability	Data-dependent model building requires many decisions and models selected may depend on a small fraction of the data. Repeating the model selection process in many (say ≥ 1000) bootstrap replications allows an assessment of model stability. Most important variables and a reasonable functional form can be identified
Complexity of a predictor	A predictor should be 'as parsimonious as possible'	Regarding statistical criteria such as MSE, complex predictors based on many variables or including wiggly functions may be only slightly better than simple predictors including the strong variables and strong nonlinearity for continuous variables. However, complex predictors are more prone to overfitting, are less stable and less likely to be reproducible in new data (external validation). Predictors based on many variables are unlikely to be transportable to other settings (general usefulness)
Check for interactions	Testing for binary × binary interactions can be done by standard methods. Continuous × binary and continuous × continuous interactions can be checked by applying the MFPI and MFPIgen methods respectively	Multivariable models should be checked for interactions, including the relevant main effects. Interactions detected should be subjected to consistency checks, as described in Chapter 7. Since models with interactions are rather complicated, we suggest including only stronger effects, e.g. by using $P < 0.01$ for significance testing

(Adapted from Sauerbrei et al. (2007a) with permission from John Wiley & Sons Ltd.)

From theoretical points of view, every procedure and every single step of a model building strategy can be easily criticized. Nevertheless, researchers expect and need answers about the important variables and about appropriate functional forms for continuous variables. Practical experience and our own research lead us to believe that, within the limitations imposed by Table 1.3, BE is a useful method for selecting variables.

12.2.2   Functional Form for Continuous Covariates

In many applications a linear function is acceptable. However, sometimes the data contradict linearity, so a check for possible nonlinearity is always required. This can best be done using a 'global' function. A systematic search for better-fitting FP functions is a good approach. The MFP procedure combines BE with FP function selection. The nominal significance level can be chosen separately for the BE and FP components, depending on the aim of the study. For many real analyses, the choice of significance level should allow one to choose the complexity of the model according to the problem at hand. Nevertheless, the variable selection and function selection process is still comprehensible and reproducible. In contrast to spline modelling, MFP is less dependent on careful 'tuning' by an expert.

Even with a smaller number of variables, say 10, of which five are continuous, several million MFP models are possible. Naturally, good statistical practice should be followed. Before starting the model building process, the data must be checked for extreme values or influential points. Functions selected in an MFP model should be checked for important overlooked local features which cannot be detected by MFP. For example, Sauerbrei and Royston (1999) checked the functions in the breast cancer model by a GAM with four equivalent d.f.

Data-dependent model building introduces several biases, including selection bias. For the variable selection problem alone, combining selection with shrinkage seems to be useful. Sauerbrei and Royston (1999) present post-estimation shrinkage factors for the breast cancer MFP model. These shrinkage factors may reveal that a function is too steep. Combining variable selection with shrinkage in the context of function selection seems promising, but more research is needed.

12.2.3   Extreme Values or Influential Points

Despite the existence of books on the subject of influential observations, guidance from the literature is confusing. Should outliers and influential observations be checked univariately or in a multivariable model? Different methods of detecting them give different results. Furthermore, it is unclear whether such observations should be modified, included or excluded from an analysis. For example, truncating a few extreme observations may or may not affect the model that is selected (Royston and Sauerbrei, 2007a).

12.2.4   Sensitivity Analysis

Sensitivity analyses are required to check whether the selected model depends (strongly) on particular assumptions. Sensitivity analyses may reveal that the proposed model is dubious and the search for a 'good' model may have to start again with some assumptions modified.

12.2.5   Check for Model Stability

Whenever data-dependent modelling is used, the analyst should be aware that the selected model is partly due to chance and can have many weaknesses. Investigation of its stability should become standard practice (Sauerbrei, 1999; Royston and Sauerbrei, 2003). Often, stability investigation reveals that weakly influential variables or implausible functions have been selected. Such a type of internal validation may suggest that certain features of the model will not be reproduced in independent data. However, external validation is the ultimate check of a data-driven model.

12.2.6   Complexity of a Predictor

Good agreement between simple and complex predictors, and experiences with checks of model stability and also with external validation result in the recommendation that a predictor should be 'as parsimonious as possible'.

12.2.7   Check for Interactions

The possibility of interactions between covariates is often ignored. If interactions are found, then it is essential to check their consistency in the data by confirmatory analyses and plots. Interactions detected by data-driven methods should be regarded with suspicion unless they are highly significant ($P < 0.01$) and confirmed by consistency checks. For interpretation purposes, it is crucial to distinguish prespecified interactions from those found in hypothesis-generation mode.

12.3   OMITTED TOPICS AND FUTURE DIRECTIONS

Of necessity, many topics pertaining to univariate and multivariable model-building with continuous covariates have been omitted from our book. In this section, we make brief comments on some of them. All of these topics are ripe for further research. Many are discussed in textbooks and are not specific to FP modelling. For some of them, FP methodology offers possible improvements. However, further research is required.

12.3.1   Measurement Error in Covariates

Measurement error could cause apparent curvature in what is really an underlying linear association. For example, that would happen if the variance of the measurement error increased with the level of the covariate. Conversely, a truly curved relationship may be made to appear more linear by homogeneous measurement error. Hence, curvature detected by FP modelling may not accurately reflect the underlying relationship if measurement error is substantial.

12.3.2   Meta-analysis

One interesting application of FPs is to combine evidence of continuous dose–response relationships across studies. However, several challenges must be overcome. For example, it is likely that the relationships will differ among datasets; therefore, combining them will present a challenge. Also, there may be heterogeneity between studies in many aspects (such as types of measurement technique or choice of confounders) that must be accommodated in any overall statistical model. The latter requires a function averaged over the studies.

12.3.3   Multi-level (Hierarchical) Models

A paper using FPs in hierarchical models is mentioned in Section 11.3.1. In principle, FP modelling may be applied in a wider context to hierarchically structured datasets at any level of the model. For example, Royston (1995) used FPs to describe fetal growth patterns in the context of two-level models for longitudinal data. The continuous covariate of interest was gestational age. As with meta-analysis, a family of FP functions with the same degree and powers would have to be able adequately to represent the relationship of interest across the subunits of a given level of the hierarchy. Then, models with the regression coefficients for the FP terms as random effects could be fitted using standard methods. We would expect FP2 functions with a single pair of best-fitting powers to be flexible enough to do a good job in many applications.

With more levels containing continuous covariates, computational problems and instability would likely increase, because more combinations of FP powers would have to be considered.

12.3.4   Missing Covariate Data

The use of multiple imputation methodology (e.g. Schafer, 1997) for coping with missing covariate data is an active area of research. The idea is that, once the imputations have been created, complete-data methods are applied to each imputed dataset and parameter estimates from a model of interest are combined across datasets using Rubin's (1987) rules. Within this paradigm, many practical issues await a solution. For example, to build a multivariable model, selection of variables and functions needs to be applied across several datasets, and somehow combined to give a single model. Selection of variables assuming linear covariate effects has been considered by Wood et al. (2008). Extension of MFP to this scenario is currently under investigation.

12.3.5   Other Types of Model

Although many types of model have not been explicitly mentioned, MFP still applies to them. Provided a model includes (i) a linear combination of the covariates and (ii) a defined likelihood, then MFP can in principle be used with it. Excluded from consideration are other multivariable techniques, such as classification and regression trees, neural networks, machine learning techniques (random forests, boosting, support vector machines, etc.), and time-series models. We also do not consider the special problems of modelling high-dimensional data with more predictors than samples – microarrays, '-omics' data, and so on.

12.4   CONCLUSION

In a fascinating and thought-provoking paper, with four eminent discussants, that we strongly recommend to readers, Breiman (2001) stated that

There are two cultures in the use of statistical modeling to reach conclusions from data. One assumes that the data are generated by a given stochastic data model. The other uses algorithmic models and treats the data mechanism as unknown. The statistical community has been committed to the almost exclusive use of data models. This commitment has led to irrelevant theory, questionable conclusions, and has kept statisticians from working on a large range of interesting current problems. . . . If our goal as a field is to use data to solve problems, then we need to move away from exclusive dependence on data models and adopt a more diverse set of tools.

(Reproduced with permission from International Journal of
Epidemiology, copyright 1999 by the Oxford University Press.)

In a thoughtful and detailed response, the first discussant, Sir David Cox, disagreed with Breiman's focus on algorithmic rather than model-based solutions as being too limited and focusing too much on data and not enough on the questions of substantive interest, remarked that 'Formal models are useful and often almost, if not quite, essential for incisive thinking'. We consider that our approach lies in between these extremes. Whereas FP models rarely have formal justification in subject-matter terms as 'data models', nevertheless they often lead to interpretable, relevant conclusions. FP modelling has elements of the algorithmic approach, in that a number of different functions are considered in the search for a well-fitting descriptor of the data to hand. Once the model has been selected, inference follows conventional lines.

Model development is still a controversial subject. We cannot hope to address in a single book all issues in the context of the many types of model that are available in today's technique- and software-rich statistical environment. We have had to be very selective in our attempt to put over a single approach in a way that is comprehensible and helpful to the user. Nevertheless, we hope and believe that the techniques described in our book will enable practitioners to build and critically assess multivariable regression models with understanding, confidence and reliability. We recommend readers to try out the methods using the software mentioned in Appendix A.3. We also encourage teachers of statistical methodology who are interested in introducing MFP to students to visit the book's website, where they will find material designed to make a start in instructing others in the area. See the Preface for further information.

APPENDIX A

Data and Software Resources

A.1 SUMMARIES OF DATASETS

In total, analyses of 23 real datasets are used in our book to illustrate methodological principles and practice. The datasets have been divided into two classes according to whether they have been used just once or more than once. The former are listed in Table A.1. Descriptions of the data are provided where the analysis is presented. Datasets used more than once are listed in Table A.2 and described in Appendix A.2.

Table A.1 Datasets used once in our book. N/A = not applicable. Further details accompany the example in the relevant section.

Name	Outcome	Obs.	Events	Variables[a]	Section reference
Myeloma	Survival	65	48	16	2.7.1
Freiburg DNA breast cancer	Survival	109	56	1	3.4.1
Cervix cancer	Binary	899	141	21	3.5
Nerve conduction	Cont.	406	N/A	1	5.7.1
Triceps skinfold thickness	Cont.	892	N/A	1	5.7.2
Diabetes	Cont.	43	N/A	2	6.7.2
Advanced prostate cancer	Survival	475	338	13	7.5
Quit smoking study	Cont.	250	N/A	2	7.11.3
Breast cancer diagnosis	Binary	458	133	6	8.6
Boston housing	Cont.	506	N/A	13	9.5
Pima Indians	Binary	768	268	8	9.7
Rotterdam breast cancer	Survival	2982	1518	10	11.1.3
Fetal growth	Cont.	574	N/A	1	11.2.1
Cholesterol	Cont.	553	N/A	1	11.2.3

[a] Maximum number of predictors used in analyses. Categorical variables count as >1 predictor, if modelled using several dummy variables.

Multivariable Model-Building   Patrick Royston, Willi Sauerbrei
© 2008 John Wiley & Sons, Ltd

Table A.2   Datasets used more than once in our book. N/A = not applicable. Further details are given in Appendix A.2.

Name	Outcome	Obs.	Events	Variables[a]	Section reference
Research body fat	Cont.	326	N/A	1	1.1.3, 4.2.1, 4.9.1, 4.9.2, 4.10.3, 4.12
GBSG breast cancer	Survival	686	299	9	1.1.4, 3.6.2, 5.6.2, 6.5.2, 6.5.3, 6.5.4, 6.6.5, 6.6.6, 6.8.2, 7.6, 7.7.2, 8.8, 9.6
Educational body fat	Cont.	252	N/A	13	2.7.2, 2.8.6, 5.2, 5.3.1, 5.5.1, 8.5
Glioma	Survival	411	274	15	2.7.3, 8.4
Prostate cancer	Cont.	97	N/A	7	3.6.2, 3.6.3, 4.15, 6.2, 6.3.2, 6.4.2, 6.4.3, 6.5.1, 6.5.3, 6.6.1, 6.6.2, 6.6.3, 6.6.4, 7.11.3
Whitehall I	Survival	17 260	2576	11	6.7.3
	Binary	17 260	1670	11	4.13.1, 4.13.2, 4.14, 7.11.1, 7.11.3
PBC	Survival	418	161	17	5.3.2, 5.4, 5.5.2, 9.8
Oral cancer	Binary	397	194	4	6.7.1, 9.3.1
Kidney cancer	Survival	347	322	11	5.8.2, 7.9

[a] Maximum number of predictors used in analyses. Categorical variables count as >1 predictor, if modelled using several dummy variables.

A.2   DATASETS USED MORE THAN ONCE

Note that a variable indicated in a table as (Y/N) is binary and is coded as 0 = no, 1 = yes. Sex is coded 0 = male, 1 = female, and age is always in years.

A.2.1   Research Body Fat

Luke et al. (1997) described the relationship between percentage body fat content pbfm and body-mass index bmi in samples of black people from three countries (Nigeria, Jamaica and the USA). The authors aimed to find out how well bmi predicted pbfm. The latter was estimated by a relatively complex bioelectrical impedance analysis, whereas bmi was calculated as the ratio of weight (in kilograms) to squared height (in square metres), and is of course relatively straightforward to determine. For illustration purposes, we restrict attention to the subsample of 326 females from the USA.

A.2.2   GBSG Breast Cancer

From July 1984 to December 1989, the GBSG recruited 720 patients with primary node positive breast cancer into a factorial 2×2 trial investigating the effectiveness of three versus six cycles of chemotherapy and of additional hormonal treatment with tamoxifen. The number

Table A.3   GBSG breast cancer data.

Variable	Name	Details
x_1	age	Age (years)
x_2	meno	Menopausal status (0 = premeno, 1 = postmeno)
x_3	size	Tumour size (mm)
x_{4a}	gradd1	1 = tumour grade 2 or 3, 0 = grade 1
x_{4b}	gradd2	1 = tumour grade 3, 0 = grade 1 or 2
x_5	nodes	Number of positive lymph nodes
x_{5e}	enodes	$\exp(-0.12 \times \text{nodes})$
x_6	pgr	Progesterone receptor status (fmol l^{-1})
x_7	er	Oestrogen receptor status (fmol l^{-1})
x_8	hormon	Tamoxifen treatment (0 = no, 1 = yes)
t	rectime	Time (days) to death or cancer recurrence[a]
d	censrec	Censoring (0 = censored, 1 = event)[a]

[a] Response variable.

of cycles of chemotherapy had no effect on RFS and is not considered further. The dataset we use comprises RFS time of the 686 patients (with 299 events) who had complete data on the variables shown in Table A.3. Tumour grade is represented by two ordinally coded dummy variables gradd1 and gradd2. The variable hormon had an effect, and analyses are often adjusted for it. Sometimes this variable is treated as a prognostic factor, which does not reflect the fact that it is a treatment variable. Stratification by this variable is an alternative approach which is not considered in our book. enodes is a derived variable that is often used in analyses of this dataset. For further information on the study, see Sauerbrei and Royston (1999) and the literature references there.

A.2.3   Educational Body Fat

To exemplify some issues in multiple regression analysis, Johnson (1996) explored a dataset comprising a response variable (the estimated percentage of body fat) and 13 continuous covariates (age, weight, height and 10 body circumference measurements) in 252 men (see Table A.4). The mixture of imperial and metric measurement units is how the data were provided. The aim was to predict percentage body fat from the covariates.

Table A.4   Educational body-fat data.

Variable	Name	Details	Variable	Name	Details
x_1	age	Age (years)	x_8	thigh	Circumference (cm)
x_2	weight	Weight (lb)	x_9	knee	Circumference (cm)
x_3	height	Height (in)	x_{10}	ankle	Circumference (cm)
x_4	neck	Circumference (cm)	x_{11}	biceps	Circumference (cm)
x_5	chest	Circumference (cm)	x_{12}	forearm	Circumference (cm)
x_6	ab	Circumference (cm)	x_{13}	wrist	Circumference (cm)
x_7	hip	Circumference (cm)	y	pcfat	Body fat[a] (%)

[a] Response variable.

The study was reported in the sports medicine literature. The dataset is available at http://lib.stat.cmu.edu/datasets/bodyfat and elsewhere on the Web. As suggested by Johnson (1996), we have corrected an apparent error in the height of case 42 by replacing the value of 29.5 in with 69.5 in.

A.2.4 Glioma

A randomized trial to compare two chemotherapy regimes included 447 patients with malignant glioma, an aggressive type of brain cancer. At the time of the analysis 293 patients had died and the median survival time from the date of randomization was about 11 months. Survival times are analysed with the Cox model. Apart from therapy, 12 variables (age, three ordinal and eight binary variables) which might affect survival time were considered (see Table A.5).

The three variables measured on an ordinal scale (the Karnofsky index, the type of surgical resection and the grade of malignancy) were each represented by two dummy variables, giving 15 predictors in all. Complete data on these predictors was available for 411 patients (274 events) and is used here. The study has previously been used in methodological investigations (e.g. Ulm et al., 1989; Sauerbrei and Schumacher, 1992).

A.2.5 Prostate Cancer

Stamey et al. (1989) studied potential predictors of prostate-specific antigen (PSA) in 97 patients with adenocarcinoma of the prostate. The aim was to see which factors or combinations of factors were associated with a raised PSA level. All observations were made around the time of radical prostatectomy, a surgical operation in which the entire prostate gland is

Table A.5   Glioma data.

Variable	Name	Details	Variable	Name	Details
x_1	sex	Sex	x_{10}	convul	Convulsions (Y/N)[c]
x_2	time	Interval to diagnosis (1 = short, 2 = long)	x_{11}	cort	Cortisone (Y/N)[c]
			x_{12}	epi	Epilepsy (Y/N)[c]
x_3	gradd1	Malignancy grade[a]	x_{13}	amnesia	Amnesia (Y/N)[c]
x_4	gradd2	Malignancy grade[a]	x_{14}	ops	Organic psycho-syndrome (Y/N)[c]
x_5	age	Age (years)	x_{15}	aph	Aphasia (Y/N)[c]
x_6	kard1	Karnofsky index[a]	x_{16}	trt	Treatment (0 = std, 1 = new)
x_7	kard2	Karnofsky index[a]			
x_8	surgd1	Resection type[a]	t	survtime	Time to death[b] (days)
x_9	surgd2	Resection type[a]	d	cens	Censoring[b] (0 = censored, 1 = died)

[a] Categorical predictor represented by 2 dummy variables.
[b] Response variable.
[c] Coded N(o) = 0, Y(es) = 1.

Table A.6 Prostate cancer data.

Variable	Name	Details
x_1	age	Age (years)
x_2	svi	Seminal vessel invasion (Y/N)
x_3	pgg45	Percentage Gleason score 4 or 5
x_4	cavol	Cancer volume (ml)
x_5	weight	Prostate weight (g)
x_6	bph	Amount of benign prostatic hyperplasia (g)
x_7	cp	Amount of capsular penetration (g)
y	lpsa	log PSA concentration[a]

[a] Response variable.

removed. A Box–Cox regression analysis of PSA on all seven predictors gave an optimal power transformation of 0.08 (95% CI, −0.03 to 0.18), supporting a log transformation (i.e. power 0) of the response. In all analyses, therefore, we use log PSA (lpsa) as the outcome variable. Furthermore, one recorded value of 449 g for the variable weight (weight of the prostate gland) seems so unlikely that we have regarded it as an outlier (possibly a transcription error). We divided it by 10 to give a more realistic value of 44.9 g. Gleason score was also available, but since it was highly correlated with percentage Gleason score 4 or 5 (pgg45) it was discarded. See Table A.6.

A.2.6   Whitehall I

Whitehall I is a prospective, cross-sectional cohort study of 18 403 male British Civil Servants employed in London. Its aim was to examine factors, particularly socio-economic features, which influence death rates in a defined population. Identified causes of death included CHD, stroke and cancer. As in Royston et al. (1999) we focus on a cohort of 17 260 male Civil Servants aged 40–64 years with complete 10-year follow-up. Two outcomes are analysed. First, 10-year all-cause mortality is considered as a binary outcome and modelled using logistic regression. Second, time to death from CHD is treated as a censored survival-time outcome and analysed using Cox regression. Only age, cigarette smoking, systolic blood pressure, cholesterol, height, weight and job grade are available for analyses, a highly selected list of variables. Except for job grade, all are continuous (see Table A.7). More details can be found in Royston et al. (1999) and in the original papers referenced there.

A.2.7   PBC

PBC is an abbreviation of primary biliary cirrhosis, a serious liver disease which usually results in liver failure and death. The dataset was used by Fleming and Harrington (1991) to illustrate certain aspects of survival analysis. Data on the first 312 patients (125 deaths) were obtained from a randomized controlled trial of two treatments for PBC, performed at the Mayo Clinic between 1974 and 1984. A further 106 patients (36 deaths) did not take part in the trial, but consented to have six variables measured and to be followed up for survival (the 'cohort study'). Randomized treatment (D-penicillamine versus placebo) and 16 prognostic factors were recorded in the trial, including the subset of six recorded in the cohort study (see Table A.8). The outcome measure was overall survival time.

Table A.7   Whitehall I data.

Variable	Name	Details	Variable	Name	Details
x_1	cigs	Daily cigarette consumption	x_9	gradd1	Job grade[a]
x_2	sysbp	Systolic blood pressure (mmHg)	x_{10}	gradd2	Job grade[a]
x_3	diasbp	Diatolic blood pressure (mmHg)	x_{11}	gradd3	Job grade[a]
x_4	map	Mean arterial pressure	y	all10	10-year mortality[b] (0 = alive, 1 = dead)
x_5	age	Age (years)			
x_6	ht	Height (cm)	t	pyar	Years of follow-up[c]
x_7	wt	Weight (kg)	d	chd	Censoring[c] (0 = censored, 1 = died of CHD)
x_8	chol	Cholesterol (mmol/l)			

[a] Unordered categorical predictor represented by three dummy variables.
[b] Response variable for binary outcome.
[c] Response variable for survival outcome.

Table A.8   PBC data.

Variable	Name	Details	Variable	Name	Details
x_1	trt	Treatment (1 = D-penicillamine, 2 = placebo)	x_{10}	alb	Albumin[a] (g dl^{-1})
x_2	age	Age[a]	x_{11}	cu	Urinary copper (mg day^{-1})
x_3	sex	Sex (0 = male, 1= female)	x_{12}	ap	Alkaline phosphatase (U l^{-1})
x_4	asc	Ascites (Y/N)[d]	x_{13}	sgot	SGOT (U ml^{-1})
x_5	hep	Hepatitis (Y/N)[d]	x_{14}	trig	Triglyceride (mg dl^{-1})
x_6	spider	Spiders (Y/N)[d]	x_{15}	plt	Platelets[a] (number/ml, $\times 10^{-3}$)
x_7	edema	Oedema[a,b] (0/0.5/1)	x_{16}	pro	Prothrombin time[a] (s)
x_8	bil	Bilirubin[a] (mg dl^{-1})	x_{17}	stage	Histological stage[b] (1, 2, 3, 4)
x_9	chol	Cholesterol (mg dl^{-1})	t	survtime	Time to death, days[c]
			d	cens	Censoring[c] (0 = censored, 1 = died)

[a] Measured in trial and cohort study.   [c] Response variable.
[b] Categorical predictor, modelled as linear.   [d] Coded N(o) = 0, Y(es) = 1.

A.2.8   Oral Cancer

The oral cancer data are described in Section 6.7.1.

A.2.9   Kidney Cancer

The MRC RE01 trial compared interferon-α (IFN) with medroxyprogesterone acetate (MPA) on the overall survival of patients with advanced kidney cancer that had metastasized

Table A.9   Kidney cancer data.

Variable	Name	Details	Variable	Name	Details
x_1	age	Age (years)	x_6	rem	Kidney removed (0 = no, 1 = yes)
x_2	sex	Sex (0 = male, 1 = female)	x_7	mets	Metastatic sites (0 = one, 1 = multiple)
x_{3a}	whod1	0 = WHO 0, 1 = WHO 1 or 2	x_8	haem	Haemoglobin (g dl^{-1})
x_{3b}	whod2	0 = WHO 0 or 1, 1 = WHO 2	x_9	wcc	White cell count (l^{-1} $\times 10^{-9}$)
x_4	t_dt	Days from original diagnosis to treatment for advanced disease	x_{10}	trt	Treatment (0 = MPA, 1 = IFN)
x_5	t_mt	Days from metastasis to treatment for advanced disease	t	survtime	Time to death[a] (days)
			d	cens	Censoring[a] (0 = censored, 1 = died)

[a] Response variable.

(i.e. spread from the original cancer site to other organs). The study is a randomized trial recruiting between 1992 and 1997. A 28% reduction in the risk of death in the interferon-α group was reported (Medical Research Council Renal Cancer Collaborators, 1999). Using data updated to June 2001, with 322 deaths in 347 patients, Royston et al. (2004) analysed interactions between treatment and 10 prognostic factors, of which six were continuous. Because of the high proportion of missing data, one of the factors (serum calcium) is not considered here. Otherwise, we use the same dataset in our book. The variables are listed in Table A.9. A small proportion of missing values in some of the covariates were imputed once using the ice program (Royston, 2005) for Stata.

A.3   SOFTWARE

We used Stata (StataCorp, 2007) for essentially all the analyses and graphics presented in our book. The principal analysis procedures are fracpoly (FPs for a single predictor) and mfp (MFPs). fracpoly and mfp are standard components of Stata. We have also developed programs to perform MFP modelling in SAS and R (Sauerbrei et al., 2006b). The mfp package in R is freely available at http://cran.r-project.org. It was originally written for S-plus by Gareth Ambler (University College London) and converted to R by Axel Benner (University of Heidelberg). The SAS macro may be downloaded from http://www.imbi.uni-freiburg.de/biom/mfp/.

Examples of how some of the analyses in our book may be done in Stata are supplied on the book's website. Also provided are (links to) Stata routines that implement extensions to MFP, including MFPI, MFPIgen, MFPT, and the spline procedure MVRS.

Glossary of Abbreviations

Table B.1 Abbreviations used in our book.

Abbreviation	Meaning	Abbreviation	Meaning
AIC	Akaike information criterion	ICC	Intraclass correlation coefficient
BE-hier	Hierarchical backward elimination	LRT	Likelihood ratio test
BE	Backward elimination	MA	Model averaging
BIC	Bayes information criterion	MFP	Multivariable FP
BIF	Bootstrap inclusion fraction	MFPI	MFP (interaction) – see MFP
BMI	Body mass index	MFPIgen	MFP (general interaction) – see MFP
Bagging	Bootstrap aggregating	MFPT	MFP (time) – see MFP
CDF	Cumulative distributive function	ML	Maximum likelihood
CHD	Coronary heart disease	MLE	Maximum likelihood estimate
CI	Confidence interval	MRC	Medical Research Council
CSS	Cubic smoothing spline	MVRS	Multivariable regression spline
d.f.	Degrees of freedom	MVSS	Multivariable smoothing spline
e.d.f.	Equivalent degrees of freedom	OLS	Ordinary least squares
EPV	Events per variable	OR	Odds ratio
FP	Fractional polynomial	PBC	Primary biliary (i.e. liver) cirrhosis
FP1	First-degree FP	PH	Proportional hazards
FP2	Second-degree FP	PSA	Prostate-specific antigen
FPm	FP of degree m	PWSF	Parameterwise shrinkage factor
FS	Forward selection	RI	Reference interval
FSP	Function selection procedure	SD	Standard deviation
GAM	Generalized additive model	SE	Standard error
GBSG	German Breast Cancer Study Group	STEPP	Subpopulation treatment effect pattern plot
GLM	Generalized linear model	StS	Stepwise selection
HR	Hazard ratio		

Multivariable Model-Building   Patrick Royston, Willi Sauerbrei
© 2008 John Wiley & Sons, Ltd

References

Abrahamowicz, M. and MacKenzie, T. (2007). Joint estimation of time-dependent and non-linear effects of continuous covariates on survival, *Statistics in Medicine* **26**: 392–408. [Cited on pp. 241, 246]

Aiken, L. S. and West, S. G. (1991). *Multiple regression: testing and interpreting interactions*, Age Publications, Newbury Park, Ca. [Cited on p. 182]

Akaike, H. (1973). Information theory as an extension of the maximum likelihood principle, *in* B. N. Petrov and F. Csaki (eds), *Second International Symposium on Information Theory*, Akademiai Kiado, Budapest, pp. 267–281. [Cited on p. 30]

Allen, D. M. and Cady, F. B. (1982). *Analyzing Experimental Data by Regression*, Wadsworth, Belmont, CA. [Cited on p. 34]

Altman, D. G. (1993). Construction of age-related reference centiles using absolute residuals, *Statistics in Medicine* **12**: 917–924. [Cited on p. 248]

Altman, D. G. and Andersen, P. K. (1989). Bootstrap investigation of the stability of a Cox regression model, *Statistics in Medicine* **8**: 771–783. [Cited on pp. 50, 184, 197]

Altman, D. G. and Chitty, L. S. (1993). Design and analysis of studies to derive charts of fetal size, *Ultrasound in Obstetrics and Gynecology* **3**: 378–384. [Cited on pp. 247, 248]

Altman, D. G., Lausen, B., Sauerbrei, W. and Schumacher, M. (1994). The dangers of using 'optimal' cut-points in the evaluation of prognostic factors, *Journal of the National Cancer Institute* **86**: 829–835. [Cited on pp. 2, 58, 59, 60]

Altman, D. G. and Royston, P. (2000). What do we mean by validating a prognostic model?, *Statistics in Medicine* **19**: 453–473. [Cited on p. 67]

Ambler, G., Brady, A. R. and Royston, P. (2002). Simplifying a prognostic model: a simulation study based on clinical data, *Statistics in Medicine* **21**: 3803–3822. [Cited on p. 135]

Ambler, G. and Royston, P. (2001). Fractional polynomial model selection procedures: investigation of Type I error rate, *Journal of Statistical Computation and Simulation* **69**: 89–108. [Cited on pp. 19, 80, 82, 83, 84, 104]

Andrews, D. F. and Herzberg, A. M. (1985). *Data*, Springer-Verlag, Berlin. [Cited on p. 157]

Armitage, P., Berry, G. and Matthews, J. N. S. (2002). *Statistical Methods in Medical Research*, fourth edn, Blackwell, Oxford. [Cited on p. 68]

Assmann, S. F., Pocock, S. J., Enos, L. E. and Kasten, L. E. (2000). Subgroup analysis and other (mis)uses of baseline data in clinical trials, *Lancet* **355**: 1064–1069. [Cited on p. 165]

Augustin, N., Sauerbrei, W. and Schumacher, M. (2005). The practical utility of incorporating model selection uncertainty into prognostic models for survival data, *Statistical Modelling* **5**: 95–118. [Cited on pp. 28, 51, 185, 251]

Austin, P. C. and Brunner, L. J. (2004). Inflation of the Type I error rate when a continuous confounding variable is categorized in logistic regression analyses, *Statistics in Medicine* **23**: 1159–1178. [Cited on p. 58]

Austin, P. C. and Tu, J. V. (2004a). Automated variable selection methods for logistic regression produced unstable models for predicting acute myocardial infarction mortality, *Journal of Clinical Epidemiology* **57**: 1138–1146. [Cited on p. 184]

Austin, P. C. and Tu, J. V. (2004b). Bootstrap methods for developing predictive models, *The American Statistician* **58**: 131–137. [Cited on p. 197]

Balslev, I., Axelsson, C. K., Zedeler, K., Rasmussen, B. B., Carstensen, B. and Mouridsen, H. T. (1994). The nottingham prognostic index applied to 9,149 patients from the studies of the danish breast cancer cooperative group (DBCG), *Breast Cancer Research and Treatment* **32**: 281–290. [Cited on p. 137]

Becher, H. (1992). The concept of residual confounding in regression models and some applications, *Statistics in Medicine* **11**: 1747–1758. [Cited on pp. 60, 138]

Becher, H. (2005). General principles of data analysis: continuous covariables in epidemiological studies, *in* W. Ahrens and I. Pigeot (eds), *Handbook of Epidemiology*, Springer, Berlin, pp. 595–624. [Cited on p. 94]

Belsley, D. A., Kuh, E. and Welsch, R. E. (1980). *Regression diagnostics*, John Wiley & Sons, New York. [Cited on pp. 15, 176]

Berger, U., Schäfer, J. and Ulm, K. (2003). Dynamic Cox modelling based on fractional polynomials: time variations in gastric cancer prognosis, *Statistics in Medicine* **22**: 1163–1180. [Cited on pp. 84, 241]

Bishop, Y. M., Fienberg, S. E. and Holland, P. W. (1975). *Discrete Multivariate Analysis: Theory and Practice*, The MIT Press, Cambridge, MA. [Cited on pp. 153, 172]

Bland, M. J. and Altman, D. G. (1986). Statistical methods for assessing agreement between two methods of clinical measurement, *The Lancet* **8**: 307–310. [Cited on pp. 29, 37]

Blettner, M. and Sauerbrei, W. (1993). Influence of model-building strategies on the results of a case-control study, *Statistics in Medicine* **12**: 1325–1338. [Cited on pp. 19, 24, 25, 26, 55]

Bonetti, M. and Gelber, R. D. (2000). A graphical method to assess treatment–covariate interactions using the Cox model on subsets of the data, *Statistics in Medicine* **19**: 2595–2609. [Cited on p. 167]

Bonetti, M. and Gelber, R. D. (2004). Patterns of treatment effects in subsets of patients in clinical trials, *Biostatistics* **5**: 465–481. [Cited on pp. 167, 168]

Borghi, E., de Onis, M., Garza, C., van den Broeck, J., Frongillo, E. A., Grummer-Strawn, L., van Buuren, S., Pan, H., Molinari, L., Martorell, R., Onyango, A. W. and Martines, J. C. (2006). Construction of the World Health Organization child growth standards: selection of methods for attained growth curves, *Statistics in Medicine* **25**:247–265. [Cited on p. 247]

Box, G. E. P. and Cox, D. R. (1964). An analysis of transformations, *Journal of the Royal Statistical Society, Series B* **26**: 211–252. [Cited on p. 11]

Box, G. E. P. and Jenkins, G. M. (1970). *Time series analysis: forecasting and control*, Holden-Day, London. [Cited on p. 49]

Box, G. E. P. and Tidwell, P. W. (1962). Transformation of the independent variables, *Technometrics* **4**: 531–550. [Cited on pp. 2, 73]

Breiman, L. (1992). The little bootstrap and other methods for dimensionality selection in regression: X-fixed prediction error, *Journal of the American Statistical Association* **87**: 738–754. [Cited on p. 48]

Breiman, L. (1995). Better subset regression using the nonnegative garotte, *Technometrics* **37**: 373–384. [Cited on p. 43]

Breiman, L. (1996a). Bagging predictors, *Machine Learning* **24**: 123–140. [Cited on pp. 192, 198]

Breiman, L. (1996b). The heuristics of instability in model selection, *Annals of Statistics* **24**: 2350–2381. [Cited on pp. 168, 198]

Breiman, L. (2001). Statistical modeling: The two cultures, *Statistical Science* **16**: 199–231. [Cited on pp. 260]

Brenner, H. and Blettner, M. (1997). Controlling for continuous confounders in epidemiologic research, *Epidemiology* **8**: 429–434. [Cited on pp. 60, 138]

Breslow, N. E. and Day, N. E. (1980). *Statistical methods in cancer research*, Vol. 1, IARC Scientific Publications, Lyon. [Cited on p. 60]

Buchholz, A., Holländer, N. and Sauerbrei, W. (2008). On properties of predictors derived with a two-step bootstrap model averaging approach – a simulation study in the linear regression model, *Computational Statistics and Data Analysis*. To appear. [Cited on p. 51]

Buchholz, A., Sauerbrei, W. and Royston, P. (2007). Investigation of time-varying effects in survival analysis may require categorisation of time – does it matter? Submitted. [Cited on p. 245]

Buckland, S. T., Burnham, K. P. and Augustin, N. H. (1997). Model selection: an integral part of inference, *Biometrics* **53**: 603–618. [Cited on pp. 28, 51, 185, 251]

Burnham, K. P. and Anderson, D. R. (2002). *Model selection and multimodel inference*, second edn, Springer, New York. [Cited on pp. 28, 30, 49]

Burnham, K. P. and Anderson, D. R. (2004). Multimodel inference – understanding AIC and BIC in model selection, *Sociological Methods & Research* **33**: 261–304. [Cited on pp. 30, 33]

Byar, D. P. (1984). Identification of prognostic factors, *in* M. E. Buyse, M. J. Staquet and R. J. Sylvester (eds), *Cancer clinical trials – methods and practice*, Oxford University Press, Oxford, pp. 423–443. [Cited on p. 57]

Byar, D. P. and Green, S. B. (1980). The choice of treatment for cancer patients based on covariate information: application to prostate cancer, *Bulletin du Cancer* **67**: 477–490. [Cited on pp. 157, 158, 162]

Cai, Z. and Sun, Y. (2003). Local linear estimation of time-dependent coefficients in Cox's regression models based on covariate information: application to prostate cancer, *Scandinavian Journal of Statistics* **30**: 93–111. [Cited on p. 241]

Chambers, J. M. and Hastie, T. J. (1991). *Statistical models in S*, Chapman & Hall/CRC, Boca Raton. [Cited on p. 205]

Chatfield, C. (1995). Model uncertainty, data mining and statistical inference (with discussion), *Journal of the Royal Statistical Society, Series B* **158**: 419–466. [Cited on pp. 51, 80, 251]

Chatfield, C. (2002). Confessions of a pragmatic statistician, *The Statistician* **51**: 1–20. [Cited on pp. xv, 19, 24, 26]

Chen, C. and George, S. L. (1985). The bootstrap and identification of prognostic factors via Cox's proportional hazards regression model, *Statistics in Medicine* **4**: 39–46. [Cited on pp. 50, 184, 197]

Chernick, M. R. (1999). *Bootstrap Methods: A Practitioner's Guide*, John Wiley & Sons, Ltd/Inc., New York. [Cited on p. 50]

Cleveland, W. S. and Devlin, S. J. (1988). Locally weighted regression: an approach to regression analysis ny local fitting, *Journal of the American Statistical Association* **83**: 596–610. [Cited on pp. 9, 11, 67]

Cohen, J., Cohen, P., West, S. G. and Aiken, L. S. (2003). *Applied Multiple Regression/Correlation Analysis for the Behavioral Sciences*, third edn, Lawrence Erlbaum Associates, New Jersey. [Cited on pp. 10, 57, 152, 172, 175, 180, 182]

Cole, T. J. (1988). Fitting smoothed centile curves to reference data, *Journal of the Royal Statistical Society, Series A* **151**: 385–418. [Cited on p. 249]

Cole, T. J. and Green, P. J. (1992). Smoothing reference centile curves: the LMS method and penalized likelihood, *Statistics in Medicine* **11**: 1305–1319. [Cited on pp. 110, 249, 250]

Collett, D. (2003a). *Modelling binary data*, second edn, Chapman & Hall/CRC, Boca Raton. [Cited on pp. 12, 60]

Collett, D. (2003b). *Modelling survival data in medical research*, second edn, John Wiley & Sons, London. [Cited on p. 13]

Concato, J., Peduzzi, P., Holford, T. R. and Feinstein, A. R. (1995). Importance of events per independent variable in proportional hazards analysis: II. Background, gaols and general strategy, *Journal of Clinical Epidemiology* **48**: 1495–1501. [Cited on p. 47]

Copas, J. B. (1983a). Plotting *p* against *x*, *Applied Statistics* **32**: 25–31. [Cited on p. 12]

Copas, J. B. (1983b). Regression, prediction and shrinkage (with discussion), *Journal of the Royal Statistical Society, Series B* **45**: 311–354. [Cited on pp. 19, 27]

Copas, J. B. and Long, T. (1991). Estimating the residual variance in orthogonal regression with variable selection, *Statistician* **40**: 51–59. [Cited on p. 40]

Count, E. W. (1942). A quantitative analysis of growth in certain human skull dimensions, *Human Biology* **14**: 143–165. [Cited on p. 247]

Cox, D. R. (1972). Regression models and life-tables (with discussion), *Journal of the Royal Statistical Society, Series B* **34**: 187–220. [Cited on pp. 12, 241, 242]

Cutler, S. J. and Aztell, L. M. (1969). Adjustments of long-term survival rates due to intercurrent disease, *Journal of Chronic Diseases* **22**: 485–495. [Cited on p. 252]

Dales, L. G. and Ury, H. K. (1978). An improper use of statistical significance testing in studying covariables, *International Journal of Epidemiology* **4**: 373–375. [Cited on p. 133]

Davison, A. C. and Hinkley, D. V. (1997). *Bootstrap methods and their application*, Cambridge University Press, Cambridge. [Cited on pp. 50, 185]

de Boer, C. (2001). *A practical guide to splines*, revised edn, Springer, New York. [Cited on pp. 9, 14, 67, 127]

DeMaris, A. (2004). *Regression with Social Data: modeling continuous and limited response variables*, John Wiley & Sons, New Jersey. [Cited on pp. 10, 152, 180, 182]

Draper, D. (1995). Assessment and propagation of model selection uncertainty (with) discussion, *Journal of the Royal Statistical Society, Series B* **57**: 45–97. [Cited on pp. 51, 251]

Draper, N. R. and Smith, H. (1998). *Applied regression analysis*, third edn, John Wiley & Sons, Ltd/Inc., New York. [Cited on p. 11]

Dunn, G. (2004). *Statistical Evaluation of Measurement Errors: Design and Analysis of Reliability Studies*, second edn, Hodder Arnold, London. [Cited on p. 29]

Early Breast Cancer Trialists' Collaborative Group (1998). Tamoxifen for early breast cancer: an overview of the randomised trials, *Lancet* **351**: 1451–1467. [Cited on p. 163]

Efron, B. (1979). Bootstrap methods: another look at the jackknife, *Annals of Statistics* **7**: 1–26. [Cited on p. 19]

Efron, B., Hastie, T., Jonstone, I. and Tibshirani, R. (2004). Least angle regression, *Annals of Statistics* **32**: 407–499. [Cited on p. 44]

Efron, B. and Tibshirani, R. J. (1993). *An introduction to the bootstrap*, Chapman & Hall/CRC, New York. [Cited on p. 185]

Eilers, P. H. C. and Marx, B. D. (1996). Flexible smoothing with B-splines and penalties, *Statistical Science* **11**: 89–121. [Cited on p. 220]

Faes, C., Aerts, M., Geys, H. and Molenberghs, G. (2007). Model averaging using fractional polynomials to estimate a safe level of exposure, *Risk Analysis* **27**: 111–123. [Cited on pp. 185, 250, 252]

Faes, C., Geys, H., Aerts, M. and Molenberghs, G. (2003). Use of fractional polynomials for dose–response modelling and quantitative risk assessment in developmental toxicity studies, *Statistical Modelling* **3**: 109–125. [Cited on pp. 250, 251, 252]

Faes, C., Geys, H., Aerts, M. and Molenberghs, G. (2006a). A hierarchical modeling approach for risk assessment in developmental toxicity studies, *Computational Statistics and Data Analysis* **51**: 1848–1861. [Cited on p. 254]

Faes, C., Hens, N., Aerts, M., Shkedy, Z., Geys, H., Mintiens, K., Boelaert, F. and Laevens, H. (2006b). Effect of clustering on the force of infection: application of bovine herpes virus-1 in Belgian cattle, *Journal of the Royal Statistical Society, Series C* **55**: 593–613. [Cited on p. 254]

Fan, J. and Gijbels, I. (1996). *Local polynomial modelling and its applications*, Chapman & Hall, London. [Cited on p. 15]

Farewell, V. T., Tom, B. D. M. and Royston, P. (2004). The impact of dichotomization on the efficiency of testing for an interaction effect in exponential family models, *Journal of the American Statistical Association* **99**: 822–831. [Cited on p. 153]

Fleiss, J. L. and Cohen, J. (1973). The equivalence of weighted kappa and the intraclass correlation coefficient as measures of reliability, *Educational and Psychological Measurement* **33**: 613–619. [Cited on p. 29]

Fleming, T. R. and Harrington, D. P. (1991). *Counting Processes and Survival Analysis*, John Wiley & Sons, Ltd/Inc., New York. [Cited on p. 267]

Foekens, J., Peters, H., Look, M., Portengen, H., Schmitt, M., Kramer, M., Brunner, N., Jänicke, F., Meijer-van Gelder, M., Henzen-Logmans, S., van Putten, W. and Klijn, J. (2000). The urokinase system of plasminogen activation and prognosis in 2780 breast cancer patients, *Cancer Research* **60**: 636–643. [Cited on p. 28]

Freedman, D. A. (1983). A note on screening regression equations, *The American Statistician* **37**(2): 152–155. [Cited on p. 31]

Furnival, G. M. and Wilson, R. W. (1974). Regression by leaps and bounds, *Technometrics* **16**: 499–511. [Cited on p. 34]

Galea, M. H., Blamey, R. W., Elston, C. E. and Ellis, I. O. (1992). The Nottingham Prognostic Index in primary breast cancer, *Breast Cancer Research and Treatment* **22**: 207–219. [Cited on p. 136]

Ganzach, Y. (1998). Nonlinearity, multicollinearity and the probability of type II error in detecting interaction, *Journal of Management* **24**(5): 615–622. [Cited on p. 172]

Geys, H., Molenberghs, G. and Ryan, L. M. (1999). Pseudolikelihood modelling of multivariate outcomes in developmental toxicology, *Journal of the American Statistical Association* **94**(447): 734–745. [Cited on p. 250]

Gifi, J. (1990). *Nonlinear multivariate analysis*, John Wiley & Sons, Chichester. [Cited on p. 184]

Gong, G. (1982). Some ideas on using the bootstrap in assessing model variability, *in* K. W. Heiner, R. S. Sacher and J. W. Wilkinson (eds), *Computer Science and Statistics: Proceedings of the 14th Symposium on the Interface*, Springer, New York, pp. 169–173. [Cited on p. 184]

Graf, E., Schmoor, C., Sauerbrei, W. and Schumacher, M. (1999). Assessment and comparison of prognostic classification schemes for survival data, *Statistics in Medicine* **18**: 2529–2545. [Cited on p. 135]

Grambsch, P. M. and Therneau, T. M. (1994). Proportional hazards tests and diagnostics based on weighted residuals, *Biometrika* **81**: 515–526. [Cited on pp. 13, 241]

Gray, R. J. (1992). Flexible methods for analyzing survival data using splines, *Journal of the American Statistical Association* **87**: 942–951. [Cited on p. 241]

Green, P. J. and Silverman, B. W. (1994). *Nonparametric regression and generalized linear models: a roughness penalty approach*, Chapman & Hall, London. [Cited on pp. 9, 14, 66, 67, 206, 220]

Greenland, S. (1983). Tests for interaction in epidemiologic studies: a review and a study of power, *Statistics in Medicine* **2**: 243–251. [Cited on p. 172]

Greenland, S. (1989). Modeling and variable selection in epidemiologic analysis, *American Journal of Public Health* **79**: 340–348. [Cited on pp. 29, 133, 172]

Greenland, S. (1993). Basic problems in interaction assessment, *Environmental Health Perspectives Supplements* **101**: 59–66. [Cited on p. 153]

Hand, D. J. (2006). Classifier technology and the illusion of progress (with discussion), *Statistical Science* **21**: 1–14. [Cited on pp. 19, 20, 197, 198]

Hand, D. J. and Vinciotti, V. (2003). Local versus global models for classification problems: fitting models where it matters, *American Statistician* **57**: 124–131. [Cited on p. 9]

Hardin, J. W. and Hilbe, J. M. (2007). *Generalized linear models and extensions*, second edn, Stata Press, College Station, Texes. [Cited on pp. 29, 135]

Harrell, F. E. (2001). *Regression modeling strategies, with applications to linear models, logistic regression, and survival analysis*, Springer, New York. [Cited on pp. 10, 16, 24, 27, 29, 30, 33, 55, 127, 135, 162, 255]

Harrell, F. E., Lee, K. L., Califf, R. M., Pryor, D. B. and Rosati, R. A. (1984). Regression modeling strategies for improved prognostic prediction, *Statistics in Medicine* **3**: 143–152. [Cited on pp. 184, 197]

Harrell, F. E., Lee, K. L. and Mark, D. B. (1996). Multivariable prognostic models: issues in developing models, evaluating assumptions and accuracy, and measuring and reducing errors, *Statistics in Medicine* **15**: 361–387. [Cited on pp. 25, 29, 47]

Harrison, D. and Rubinfeld, D. L. (1978). Hedonic house prices and the demand for clear air, *Journal of Environmental Economics and Management* **5**: 81–102. [Cited on p. 207]

Hastie, T. J., Sleeper, L. and Tibshirani, R. J. (1992). Flexible covariate effects in the proportional hazards model, *Breast Cancer Research and Treatment* **22**: 241–250. [Cited on p. 202]

Hastie, T. J. and Tibshirani R. J. (1990). *Generalized Additive Models*, Chapman & Hall, New York. [Cited on pp. 14, 66, 68, 133, 139, 149, 202, 206]

Hastie, T. J. and Tibshirani, R. J. (1993). Varying-coefficient models (with discussion), *Journal of the Royal Statistical Society, Series B* **55**: 757–796. [Cited on p. 161, 241]

Hauck, W. W. and Miike, R. (1991). A proposal for examining and reporting stepwise regressions, *Statistics in Medicine* **10**: 711–715. [Cited on p. 34]

Haybittle, J. L., Blamey, R. W., Elston, C. W., Johnson, J., Doyle, P. J., Campbell, F. C., Nicholson, R. I. and Griffiths, K. (1982). A prognostic index in primary breast cancer, *British Journal of Cancer* **45**: 361–366. [Cited on p. 136]

He, X. (1992). Quantile curves without crossing, *American Statistician* **51**: 186–192. [Cited on p. 247]

Heinzl, H. and Kaider, A. (1997). Proportional hazards regression models with cubic spline functions, *Computational Methods and Programs in Biomedicine* **54**: 201–218. [Cited on p. 241]

Helland, I. (1987). On the interpretation and use of R^2 in regression analysis, *Biometrics* **43**: 61–69. [Cited on p. 95]

Hess, K. R. (1994). Assessing time-by-covariate interactions in proportional hazards regression models using cubic spline functions, *Statistics in Medicine* **13**: 1045–1062. [Cited on p. 241]

Hess, K. R. (1995). Graphical methods for assessing violations for proportional hazards assumption in Cox regression, *Statistics in Medicine* **14**: 1707–1723. [Cited on p. 241]

Hilsenbeck, S. G. and Clark, G. M. (1996). Practical *p*-value adjustment for optimally selected cutpoints, *Statistics in Medicine* **15**: 103–112. [Cited on p. 59]

Hjorth, U. (1989). On model selection in the computer age, *Journal of Statistical Planning and Inference* **23**: 101–115. [Cited on pp. 47, 49]

Hoerl, A. E. and Kennard, R. W. (1970). Ridge regression: biased estimation of non-orthogonal components, *Technometrics* **12**: 55–67. [Cited on p. 43]

Hoeting, J. A., Madigan, D., Raftery, A. E. and Volinsky, C. T. (1999). Bayesian model averaging: A tutorial, *Statistical Science* **14**: 382–417. [Cited on pp. 28, 251]

Holländer, N. (2002). *Estimating the functional form of the effect of a continuous covariate on survival time*, PhD thesis, University of Dortmund. [Cited on p. 84]

Holländer, N., Augustin, N. H. and Sauerbrei, W. (2006). Investigation on the improvement of prediction by bootstrap model averaging, *Methods of Information in Medicine* **45**: 44–50. [Cited on p. 51]

Holländer, N., Sauerbrei, W. and Schumacher, M. (2004). Confidence intervals for the effect of a prognostic factor after selection of an 'optimal' cutpoint, *Statistics in Medicine* **23**: 1701–1713. [Cited on pp. 43, 48, 58]

Holländer, N. and Schumacher, M. (2006). Estimating the functional form of a continuous covariate's effect on survival time, *Computational Statistics and Data Analysis* **50**: 1131–1151. [Cited on p. 220]

Holm, S. (1979). A simple sequentially rejective multiple test procedure, *Scandinavian Journal of Statistics* **6**: 65–70. [Cited on p. 154]

Hosmer, D. W. and Lemeshow, S. (1999). *Applied Survival Analysis*, John Wiley & Sons, New York. [Cited on pp. 13, 65]

Hosmer, D. W. and Lemeshow, S. (2000). *Applied Logistic Regression*, second edn, John Wiley & Sons, New York. [Cited on p. 12]

Irwin, J. R. and McClelland, G. H. (2003). Negative consequences of dichotomizing continuous predictor variables, *Journal of Marketing Research* **40**: 366–371. [Cited on p. 58]

Johnson, R. W. (1996). Fitting percentage of body fat to simple body measurements, *Journal of Statistics Education* **4**(1). [Cited on pp. 265, 266]

Kattan, M. W. (2003). Judging new markers by their ability to improve predictive accuracy, *Journal of the National Cancer Institute* **4953**: 634–635. [Cited on p. 136]

Koenker, R. W. and Bassett, G. W. (1978). Regression quantiles, *Econometrica* **46**: 33–50. [Cited on p. 247]

Krall, J. M., Uthoff, V. A. and Harley, J. B. (1975). A step-up procedure for selecting variables associated with survival, *Biometrics* **31**: 49–57. [Cited on p. 35]

Kuha, J. (2004). AIC and BIC: Comparison of assumptions and performance, *Sociological Methods & Research* **33**: 188–229. [Cited on pp. 30, 33]

Kuk, A. Y. C. (1984). All subsets regression in a proportional hazards model, *Biometrika* **71**: 587–592. [Cited on pp. 30, 34, 36]

Kullback, S. and Leibler, R. A. (1951). On information and sufficiency, *Annals of Mathematical Statistics* **22**: 79–86. [Cited on p. 33]

Lagakos, S. W. (1988). Effects of mismodelling and mismeasuring explanatory variables on tests of their association with a response variable, *Statistics in Medicine* **7**: 257–274. [Cited on pp. 2, 153]

Lambert, P. C., Smith, L. K., Jones, D. R. and Botha, J. L. (2005). Additive and multiplicative covariate regression models for relative survival incorporating fractional polynomials for time-dependent effects, *Statistics in Medicine* **24**: 3871–3885. [Cited on p. 252]

Lausen, B. and Schumacher, M. (1992). Maximally selected rank statistics, *Biometrics* **48**: 73–85. [Cited on p. 59]

Lausen, B. and Schumacher, M. (1996). Evaluating the effect of optimized cutoff values in the assessment of prognostic factors, *Computational Statistics and Data Analysis* **21**: 307–326. [Cited on p. 58]

Lawless, J. F. and Singhal, K. (1978). Efficient screening of non-normal regression models, *Biometrics* **34**: 318–327. [Cited on p. 34]

Lee, K. L., Harrell, F. E., Tolley, H. D. and Rosati, R. A. (1983). A comparison of test statistics for assessing the effects of concomitant variables in survival analysis, *Biometrics* **93**: 341–350. [Cited on p. 33]

LePage, R. and Billard, L. (1992). *Exploring the Limits of Bootstrap*, John Wiley & Sons, Ltd/Inc., New York. [Cited on p. 50]

Lubinski, D. and Humphreys, L. G. (1990). Assessing spurious 'moderator effects': Illustrated substantively with the hypothesized ('synergistic') relation between spatial and mathematical ability, *Psychological Bulletin* **107**: 385–393. [Cited on p. 173]

Luke, A., Durazo-Arvizu, R. and others (1997). Relation between body mass index and body fat in black population samples from Nigeria, Jamaica, and the United States, *American Journal of Epidemiology* **145**: 620–628. [Cited on pp. 2, 264]

MacCallum, R. C. and Mar, C. M. (1995). Distinguishing between moderator and quadratic effects in multiple regression, *Psychological Bulletin* **118**(3): 405–421. [Cited on pp. 172, 173]

MacCallum, R., Zhang, S., Preacher, K. J. and Rucker, D. D. (2002). On the practice of dichotomization of quantitative variables, *Psychological Methods* **7**: 19–40. [Cited on p. 58]

Mallows, C. L. (1998). The zeroth problem, *American Statistician* **52**: 1–9. [Cited on pp. 25, 26]

Mann, J. I., Lewis, B., Shepherd, J., Winder, A. F., Fenster, S., Rose, L. and Morgan, B. (1988). Blood lipid concentrations and other cardiovascular risk factors: distribution, prevalence and detection in Britain, *British Medical Journal* **296**: 1702–1706. [Cited on p. 249]

Mantel, N. (1970). Why stepdown procedures in variable selection?, *Technometrics* **12**: 621–625. [Cited on pp. 19, 32, 57]

Marcus, R., Peritz, E. and Gabriel, K. R. (1976). On closed test procedures with special reference to ordered analysis of variance, *Biometrika* **76**: 655–660. [Cited on pp. 82, 205]

Marmot, M. G., Shipley, M. J. and Rose, G. (1984). Inequalities in death – specific explanations of a general pattern?, *Lancet* **323**: 1003–1006. [Cited on p. 142]

Martinussen, T., Scheike, T. H. and Skovgaard, I. M. (2002). Efficient estimation of fixed and time-varying covariate effects in multiplicative intensity models, *Scandinavian Journal of Statistics* **29**: 57–74. [Cited on p. 241]

Marubini, E. and Valsecchi, M. G. (1995). *Analysing Survival Data from Clinical Trials and Observational Studies*, John Wiley & Sons, Ltd/Inc., New York, Chichester. [Cited on p. 31]

Mazumdar, M. and Glassman, J. R. (2000). Categorizing a prognostic variable: review of methods, code for easy implementation and applications to decision-making about cancer treatments, *Statistics in Medicine* **19**: 113–132. [Cited on p. 70]

Mazumdar, M., Smith, A. and Bacik, J. (2003). Methods for categorizing a prognostic variable in a multivariable setting, *Statistics in Medicine* **22**: 559–571. [Cited on p. 70]

McCullagh, P. and Nelder, J. A. (1989). *Generalized linear models*, second edn, Chapman & Hall, London. [Cited on p. 14]

McShane, L. M., Altman, D. G., Sauerbrei, W., Taube, S. E., Gion, M. and for the Statistics Subcommittee of the NCI-EORTC Working Group on Cancer Diagnostics, G. M. C. (2005). Reporting recommendations for tumor marker prognostic studies (REMARK), *Journal of the National Cancer Institute* **97**: 1180–1184. [Cited on p. 221]

Medical Research Council Renal Cancer Collaborators (1999). Interferon-α and survival in metastatic renal carcinoma: early results of a randomised controlled trial, *Lancet* **353**: 14–17. [Cited on p. 269]

Meinshausen, N. and Bühlmann, P. (2006). High-dimensional graphs and variable selection with the lasso, *The Annals of Statistics* **147**: 389–425. [Cited on pp. 44, 50]

Miller, A. J. (1984). Selection of subsets of regression variables (with discussion), *Journal of the Royal Statistical Society, Series A* **147**: 389–425. [Cited on p. 40]

Miller, A. J. (2002). *Subset selection in regression*, second edn, Chapman & Hall/CRC, Boca Raton. [Cited on pp. 1, 40, 41, 199]

Miller, R. and Siegmund, A. (1982). Maximally selected chi-square statistics, *Biometrics* **38**: 1011–1016. [Cited on p. 59]

Molenberghs, G. and Ryan, L. M. (1999). An exponential family model for clustered multivariate binary data, *Environmetrics* **10**: 279–300. [Cited on p. 251]

Morgan, T. M. and Elashoff, R. M. (1986). Effect of categorizing a continuous covariate on the comparison of survival time, *Journal of the American Statistical Association* **81**: 917–921. [Cited on p. 2]

Mosteller, F. and Tukey, J. W. (1977). *Data analysis and regression*, Addison-Wesley, New York. [Cited on pp. 3, 73, 100]

Ng'Andu, N. H. (1997). An empirical comparison of statistical tests for assessing the proportional hazards assumption of Cox's model, *Statistics in Medicine* **16**: 611–626. [Cited on p. 241]

Pearce, N. and Greenland, S. (2005). Confounding and interaction, *in* W. Ahrens and I. Pigeot (eds), *Handbook of Epidemiology*, Springer, Berlin, part I.9, pp. 371–397. [Cited on p. 173]

Peduzzi, P., Concato, J., Feinstein, A. R. and Holford, T. R. (1995). Importance of events per independent variable in proportional hazards analysis. II. Accuracy and precision of regression estimates, *Journal of Clinical Epidemiology* **48**: 1503–1510. [Cited on p. 47]

Peduzzi, P., Concato, J., Kemper, E., Holford, T. R. and Feinstein, A. R. (1996). A simulation study of the number of events per variable in logistic regression analysis, *Journal of Clinical Epidemiology* **49**: 1503–1510. [Cited on p. 47]

Pena, D. (2005). A new statistic for influence in linear regression, *Technometrics* **47**: 1–12. [Cited on p. 207]

Pencina, M. J., D'Agostino Sr, R. B., DAgostino Jr, R. B. and Vasan, R. S. (2008). Evaluating the added predictive ability of a new marker: From area under the ROC curve to reclassification and beyond, *Statistics in Medicine*. **27**: 157–172. [Cited on p. 136]

Pfisterer, J., Kommoss, F., Sauerbrei, W., Menzel, D., Kiechle, M., Giese, E., Hilgarth, M. and Pfleiderer, A. (1995). DNA flow cytometry in node positive breast cancer: Prognostic value and correlation to morphological and clinical factors, *Analytical and Quantitative Cytology and Histology* **17**: 406–412. [Cited on p. 58]

Pinsker, I. S., Kipnis, V. and Grechanovsky, E. (1987). The use of conditional cuttoffs in a forward selection procedure, *Communications in Statistics-Theory and Methods* **16**: 2227–2241. [Cited on p. 34]

Qiu, X., Xiao, Y., Gordon, A. and Yakovlev, A. (2006). Assessing stability of gene selection in microarray data analysis, *BMC Bioinformatics* **7**: 50. http://www.biomedcentral.com/ 1471-2105/7/50. [Cited on p. 184]

Raftery, A. E., Madigan, D. and Hoeting, J. (1997). Bayesian model averaging for linear regression models, *Journal of American Statistical Association* **92**: 179–191. [Cited on p. 51]

Richards, F. J. (1959). A flexible growth function for empirical use, *Journal of Experimental Botany* **10**: 290–300. [Cited on p. 112]

Rigby, R. A. and Stasinopoulos, D. M. (2004). Smooth centile curves for skew and kurtotic data modelled using the Box–Cox power exponential distribution, *Statistics in Medicine* **23**: 3053–3076. [Cited on p. 249]

Robertson, C., Boyle, P., Hsieh, C.-C., Macfarlane, G. J. and Maisonneuve, P. (1994). Some statistical considerations in the analysis of case-control studies when the exposure variables are continuous measurements, *Epidemiology* **5**: 164–170. [Cited on p. 92]

Rosenberg, P. S., Katki, H., Swanson, C. A., Brown, L. M., Wacholder, S. and Hoover, R. N. (2003). Quantifying epidemiologic risk factors using nonparametric regression: model selection remains the greatest challenge, *Statistics in Medicine* **22**: 3369–3381. [Cited on pp. 133, 138, 139, 202, 204, 220]

Rossiter, J. E. (1991). Calculating centile curves using kernel density estimation methods with application to infant kidney lengths, *Statistics in Medicine* **10**: 1693–1701. [Cited on p. 247]

Rothman, K. J. and Greenland, S. (1998). *Modern epidemiology*, second edn, Lippincott-Raven, Philadelphia, PA. [Cited on pp. 152, 154]

Royston, P. (1995). Calculation of unconditional and conditional reference intervals for fetal size and growth from longitudinal measurements, *Statistics in Medicine* **14**: 1417–1436. [Cited on p. 259]

Royston, P. (2000). A strategy for modelling the effect of a continuous covariate in medicine and epidemiology, *Statistics in Medicine* **19**: 1831–1847. [Cited on pp. 67, 125]

Royston, P. (2005). Multiple imputation of missing values: update of ICE, *Stata Journal* **5**: 527–536. [Cited on pp. 215, 243, 269]

Royston, P. and Altman D. G. (1994). Regression using fractional polynomials of continuous covariates: parsimonious parametric modelling (with discussion), *Applied Statistics* **43**(3): 429–467. [Cited on pp. 2, 3, 7, 9, 68, 73, 74, 79, 83, 104, 108]

Royston, P. and Altman, D. G. (1997). Approximating statistical functions by using fractional polynomial regression, *The Statistician* **46**: 411–422. [Cited on pp. 73, 109, 253]

Royston, P., Altman, D. G. and Sauerbrei, W. (2006). Dichotomizing continuous predictors in multiple regression: a bad idea, *Statistics in Medicine* **25**: 127–141. [Cited on pp. 8, 58, 65, 70, 137]

Royston, P. and Ambler, G. (1998). Generalized additive models, *Stata Technical Bulletin* **42**: 38–43. [Cited on p. 207]

Royston, P. and Ambler, G. (1999). Nonlinear regression models involving power or exponential functions of covariates, *Stata Technical Bulletin* **49**: 25–30. [Cited on p. 85]

Royston, P., Ambler, G. and Sauerbrei, W. (1999). The use of fractional polynomials to model continuous risk variables in epidemiology, *International Journal of Epidemiology* **28**: 964–974. [Cited on pp. 60, 85, 87, 133, 142, 267]

Royston, P. and Sauerbrei, W. (2003). Stability of multivariable fractional polynomial models with selection of variables and transformations: a bootstrap investigation, *Statistics in Medicine* **22**: 639–659. [Cited on pp. 19, 124, 131, 132, 149, 170, 186, 191, 192, 193, 195, 198, 199, 214, 232, 236, 239, 258]

Royston, P. and Sauerbrei, W. (2004a). A new approach to modelling interactions between treatment and continuous covariates in clinical trials by using fractional polynomials, *Statistics in Medicine* **23**: 2509–2525. [Cited on pp. 20, 153, 154, 155, 157]

Royston, P. and Sauerbrei, W. (2004b). A new measure of prognostic separation in survival data, *Statistics in Medicine* **23**: 723–748. [Cited on p. 135]

Royston, P. and Sauerbrei, W. (2005). Building multivariable regression models with continuous covariates in clinical epidemiology, with an emphasis on fractional polynomials, *Methods of Information in Medicine* **44**: 561–571. [Cited on pp. 19, 148]

Royston, P. and Sauerbrei, W. (2007a). Improving the robustness of fractional polynomial models by preliminary covariate transformation, *Computational Statistics and Data Analysis* **51**: 4240–4253. [Cited on pp. 105, 108, 113, 148, 258]

Royston, P. and Sauerbrei, W. (2007b). Multivariable modeling with cubic regression splines: A principled approach, *Stata Journal* **7**: 45–70. [Cited on pp. 149, 202, 203, 205, 220]

Royston, P., Sauerbrei, W. and Ritchie, A. W. S. (2004). Is treatment with interferon-α effective in all patients with metastatic renal carcinoma? A new approach to the investigation of interactions, *British Journal of Cancer* **23**: 794–799. [Cited on p. 157]

Royston, P. and Thompson, S. G. (1995). Comparing non-nested regression models, *Biometrics* **51**: 114–127. [Cited on p. 113]

Royston, P. and Wright, E. M. (1997). Age-specific reference intervals for normally distributed data, *Stata Technical Bulletin* **38**: 4–9. [Cited on p. 248]

Royston, P. and Wright, E. M. (1998). A method for estimating age-specific reference intervals ('normal ranges') based on fractional polynomials and exponential transformation, *Journal of the Royal Statistical Society, Series A* **161**: 79–101. [Cited on p. 249]

Rubin, D. B. (1987). *Multiple Imputation for Nonresponse in Surveys*, John Wiley & Sons, Ltd/Inc., New York. [Cited on p. 260]

Runge, C. (1901). Über empirische Funktionen und die Interpolation zwischen äquidistanten Ordinaten ([on empirical functions and the interpolation between equidistant ordinates], *Zeitschrift Mathematische Physik* **46**: 224–243. [Cited on p. 68]

Ruppert, D., Wand, M. P. and Carroll, R. J. (2003). *Semiparametric Regression*, Cambridge University Press, Cambridge. [Cited on p. 182]

Ryan, T. P. (1997). *Modern regression methods*, John Wiley & Sons, New York. [Cited on p. 123]

Samarov, A. M. (1993). Exploring regression structure using nonparametric functional estimation, *Journal of the American Statistical Association* **88**: 836–847. [Cited on p. 207]

Sasieni, P. and Royston, P. (1998). Pointwise confidence intervals for `running`, *Stata Technical Bulletin* **41**: 17–23. [Cited on pp. 3, 11]

Sasieni, P., Royston, P. and Cox, N. J. (2005). Symmetric nearest neighbour linear smoothers, *Stata Journal* **5**: 285. [Cited on p. 15]

Sauerbrei, W. (1992). Variablenselektion in Regressionsmodellen unter besonderer Berücksichtigung medizinischer Fragestellungen [Variable selection in regression models with special reference to application in medical research]. PhD Dissertation, University of Dortmund. [Cited on pp. 19, 31, 41, 45, 47, 48]

Sauerbrei, W. (1993). Comparison of variable selection procedures in regression models – a simulation study and practical examples, *in* J. Michaelis, G. Hommel and S. Wellek (eds), *Europäische Perspectiven der Medizinischen Informatik, Biometrie und Epidemiologie*, Munich: MMV Medizin, pp. 108–113. [Cited on pp. 19, 31, 41]

Sauerbrei, W. (1999). The use of resampling methods to simplify regression models in medical statistics, *Applied Statistics* **48**: 313–329. [Cited on pp. 19, 39, 40, 43, 45, 46, 50, 57, 146, 184, 197, 198, 258]

Sauerbrei, W., Holländer, N., Buchholz, A. (2008). Investigation about a screening step in model selection, *Statistics and Computing.* **18**: 195–208. [Cited on p. 51]

Sauerbrei, W., Holländer, N., Riley, R. D. and Altman, D. G. (2006a). Evidence based assessment and application of prognostic markers: the Long way from single studies to meta-analysis, *Communications in Statistics – Theory and Methods* **35**: 1333–1342. [Cited on p. 221]

Sauerbrei, W., Hübner, K., Schmoor, C., Schumacher, M. and the German Breast Cancer Study Group (1997). Validation of existing and development of new prognostic classification schemes in node negative breast cancer, *Breast Cancer Research and Treatment* **42**: 149–163. [Cited on p. 6]

Sauerbrei, W., Madjar, H. and Prömpeler, H. J. (1998). Differentiation of benign and malignant breast tumors by logistic regression and a classification tree using Doppler flow signals, *Methods of Information in Medicine* **37**: 226–234. [Cited on p. 191]

Sauerbrei, W., Meier-Hirmer, C., Benner, A. and Royston, P. (2006b). Multivariable regression model building by using fractional polynomials: description of SAS, STATA and R programs, *Computational Statistics and Data Analysis* **50**: 3464–3485. [Cited on p. 269]

Sauerbrei, W. and Royston, P. (1999). Building multivariable prognostic and diagnostic models: transformation of the predictors using fractional polynomials, *Journal of the Royal Statistical Society, Series A* **162**: 71–94. [Cited on pp. 7, 8, 9, 16, 27, 44, 46, 103, 108, 113, 116, 124, 129, 148, 190, 195, 257, 258, 265]

Sauerbrei, W. and Royston, P. (2002). Corrigendum: Building multivariable prognostic and diagnostic models: transformation of the predictors using fractional polynomials, *Journal of the Royal Statistical Society, Series A* **165**: 399–400. [Cited on p. 82]

Sauerbrei, W. and Royston, P. (2007). Modelling to extract more information from clinical trials data – on some roles for the bootstrap, *Statistics in Medicine* **26**: 4989–5001. [Cited on p. 155]

Sauerbrei, W., Royston, P. and Binder, H. (2007a). Selection of important variables and determination of functional form for continuous predictors in multivariable model-building, *Statistics in Medicine* **26**: 5512–5528. [Cited on pp. 19, 26, 30, 50, 127, 148, 220, 239, 255]

Sauerbrei, W., Royston, P. and Binder, H. (2007b). Variable selection for multivariable model building, with an emphasis on functional form for continuous covariates (FDM-Preprint 98). [Cited on p. 47]

Sauerbrei, W., Royston, P., Bojar, H., Schmoor, C., Schumacher, M. and the German Breast Cancer Study Group (1999). Modelling the effects of standard prognostic factors in node positive breast cancer, *British Journal of Cancer* **79**: 1752–1760. [Cited on pp. 6, 8]

Sauerbrei, W., Royston, P. and Look, M. (2007c). A new proposal for multivariable modelling of time-varying effects in survival data based on fractional polynomial time-transformation, *Biometrical Journal* **49**: 453–473. [Cited on pp. 20, 242, 243, 245]

Sauerbrei, W., Royston, P. and Schumacher, M. (2005). Bootstrap methods for developing predictive models [letter], *American Statistician* **59**: 116–118. [Cited on p. 184]

Sauerbrei, W., Royston, P. and Zapien, K. (2007d). Detecting an interaction between treatment and a continuous covariate: a comparison of two approaches, *Computational Statistics and Data Analysis* **51**: 4054–4063. [Cited on pp. 157, 171]

Sauerbrei, W. and Schumacher, M. (1992). A bootstrap resampling procedure for model building: application to the Cox regression model, *Statistics in Medicine* **11**: 2093–2109. [Cited on pp. 19, 50, 51, 184, 185, 186, 187, 197, 198, 236, 266]

Schafer, J. L. (1997). *Analysis of Incomplete Multivariate Data*, Chapman & Hall, London. [Cited on p. 259]

Schemper, M. (1993). The relative importance of prognostic factors in studies of survival, *Statistics in Medicine* **12**: 2377–2382. [Cited on p. 134]

Schmoor, C., Sauerbrei, W. and Schumacher, M. (2000). Sample size considerations for the evaluation of prognostic factors in survival analysis, *Statistics in Medicine* **19**: 441–452. [Cited on p. 25]

Schumacher, M., Holländer, N. and Sauerbrei, W. (1997). Resampling and cross-validation techniques: a tool to reduce bias caused by model building?, *Statistics in Medicine* **16**: 2813–2827. [Cited on pp. 40, 43, 45, 48]

Schumacher, M., Holländer, N., Schwarzer, G. and Sauerbrei, W. (2006). Prognostic factor studies, *in* J. Crowley and D. Ankerst (eds), *Handbook of Statistics in Clinical Oncology*, second edn, Chapman & Hall/CRC, Boca Raton, FL, pp. 289–333. [Cited on pp. 47, 59, 220]

Schwarz, G. (1978). Estimating the dimension of a model, *Annals of Statistics* **6**(2): 461–464. [Cited on p. 30]

Shkedy, Z., Aerts, M., Molenberghs, G., Beutels, P. and van Damme, P. (2006). Modelling age-dependent force of infection from prevalence data using fractional polynomials, *Statistics in Medicine* **25**: 1577–1591. [Cited on pp. 79, 254]

Simon, R. M., Korn, E. L., McShane, L. M., Wright, M. D. R. G. W. and Zhao, Y. (2003). *Design and Analysis of DNA Microarray Investigations*, Springer, New York. [Cited on p. 28]

Sockett, E. B., Daneman, D., Clarson, C. and Ehrich, R. M. (1987). Factors affecting and patterns of residual insulin secretion during first year of Type I (insulin-dependent) diabetes mellitus in children, *Diabetologia* **30**: 453–459. [Cited on p. 139]

Stamey, T. A., Kabalin, J. N., McNeal, J. E., Johnstone, I. M., Freiha, F., Redwine, E. A. and Yang, N. (1989). Prostate specific antigen in the diagnosis and treatment of adenocarcinoma of the prostate. ii. radical prostatectomy treated patients, *Journal of Urology* **141**: 1076–1083. [Cited on p. 266]

StataCorp (2007). *Stata Statistical Software: Release 10*, StataCorp, College Station, Texas. [Cited on pp. 32, 269]

Steyerberg, E., Eijkemans, M., Harrell, F. and Habbema, J. (2000). Prognostic modelling with logistic regression analysis: a comparison of selection and estimation methods in small data sets, *Statistics in Medicine* **19**: 1059–1079. [Cited on pp. 25, 47]

Steyerberg, E., Eijkemans, M., Harrell, F. and Habbema, J. (2001). Prognostic modelling with logistic regression analysis: in search of a sensible strategy in small data sets, *Medical Decision Making* **21**: 45–56. [Cited on p. 47]

Teräsvirta, T. and Mellin, I. (1986). Model selection criteria and model selection tests in regression models, *Scandinavian Journal of Statistics* **13**: 159–171. [Cited on pp. 19, 30, 33, 49]

Therneau, T. M. and Grambsch, P. M. (2000). *Modeling survival data: extending the Cox model*, Springer, Heidelberg. [Cited on pp. 13, 65, 241]

Therneau, T. M., Grambsch, P. M. and Fleming, T. R. (1990). Martingale-based residuals for survival models, *Biometrika* **77**: 147–160. [Cited on p. 13]

Thomas, D. C., Siemiatycki, J., Dewar, R., Robins, J., Goldberg, M. and Armstrong, B. G. (1985). The problem of multiple inference in studies designed to generate hypotheses, *American Journal of Epidemiology* **122**: 1080–1095. [Cited on p. 29]

Tian, L., Zucker, D. and Wei, L. J. (2005). On the Cox model with time-varying regression coefficients, *Journal of the American Statistical Association* **100**: 172–183. [Cited on p. 241]

Tibshirani, R. (1996). Regression shrinkage and selection via the lasso, *Journal of the Royal Statistical Society, Series B* **58**: 267–288. [Cited on p. 43]

Tibshirani, R. (1997). The lasso method for variable selection in the Cox model, *Statistics in Medicine* **16**: 385–395. [Cited on p. 44]

Tukey, J. W. (1957). On the comparative anatomy of transformations, *Annals of Mathematical Statistics* **28**: 602–632. [Cited on pp. 3, 73]

Ulm, K., Schmoor, C., Sauerbrei, W., Kemmler, G., Aydemir, U., Müller, B. and Schumacher, M. (1989). Strategien zur Auswertung einer Therapiestudie mit der Überlebenszeit als Zielkriterium, *Biometrie und Informatik in Medizin und Biologie* **20**: 171–205. [Cited on p. 266]

Vach, K., Sauerbrei, W. and Schumacher, M. (2001). Variable selection and shrinkage: comparison of some approaches, *Statistica Neerlandica* **55**: 53–75. [Cited on p. 43]

van Buuren, S., Boshuizen, H. C. and Knook, D. L. (1999). Multiple imputation of missing blood pressure covariates in survival analysis, *Statistics in Medicine* **18**: 681–694. [Cited on p. 243]

van Houwelingen, H. C. (2000). Validation, calibration, revision and combination of prognostic survival models, *Statistics in Medicine* **19**: 3401–3415. [Cited on p. 137] van Houwelingen, J. C. and le Cessie, S. (1990). Predictive value of statistical models, *Statistics in Medicine* **9**: 1303–1325. [Cited on p. 43, 44, 45]

Verweij, P. J. M. and van Houwelingen, J. C. (1993). Cross-validation in survival analysis, *Statistics in Medicine* **12**: 2305–2315. [Cited on p. 44, 45]

Verweij, P. and van Houwelingen, H. (1995). Time-dependent effects of fixed covariates in Cox regression, *Biometrics* **51**: 1550–1556. [Cited on p. 241]

Vidakovic, B. (1999). *Statistical Modeling by Wavelets*, Wiley Interscience, New York. [Cited on p. 67]

Vittinghof, E., Glidden, D. V., Shiboski, S. C. and McCulloch, C. E. (2005). *Regression Methods in Biostatistics. Linear, Logistic, Survival, and Repeated Measures Models*, Springer, New York. [Cited on pp. 10, 152, 180]

Vittinghoff, E. and McCulloch, C. E. (2007). Relaxing the rule of ten events per variable in logistic and Cox regression, *American Journal of Epidemiology* **165**: 710–718. [Cited on pp. 25, 47]

Volinsky, C. T. and Raftery, A. E. (2000). Bayesian information criterion for censored survival models, *Biometrics* **56**: 256–262. [Cited on pp. 33, 163]

von Elm, E., Altman, D. G., Egger, M., Pocock, S. J., Gotzsche, P. C. and Vandenbroucke, J. P. for the STROBE Initiative, (2007). The Strengthening the Reporting of Observational Studies in Epidemiology (STROBE) statement: guidelines for reporting observational studies, *Preventive Medicine* **45**: 247–251. [Cited on p. 221]

Vorschraegen, C., Vinh-Hung, V., Cserni, G., Gordon, R., Royce, M. E., Vlastos, G., Tai, P. and Storme, G. (2005). Modeling the effect on tumor size in early breast cancer, *Annals of Surgery* **241**: 309–318. [Cited on p. 195]

Walter, S. D., Feinstein, A. R. and Wells, C. K. (1987). Coding ordinal independent variables in multiple regression analyses, *American Journal of Epidemiology* **125**(2): 319–323. [Cited on p. 57]

Weinberg, C. R. (1995). How bad is categorization? (Editorial), *Epidemiology* **6**: 345–347. [Cited on p. 60]

Weisberg, S. (2005). *Applied linear regression*, third edn, John Wiley & Sons, Ltd/Inc., Hoboken, NJ. [Cited on pp. 10, 11]

Wood, A., White, I. R. and Royston, P. (2008). How should variable selection be performed with multiply imputed data?, *Statistics in Medicine* **00**: 00–00. [Cited on p. 260]

Wood, S. N. (2006). *Generalized Additive Models. An Introduction with R*, Chapman & Hall/CRC, Boca Raton. [Cited on pp. 202, 203]

Woodward, M. (1999). *Epidemiology: Study Design and Data Analysis*, Chapman & Hall/CRC, Boca Raton. [Cited on p. 152]

Wright, E. M. and Royston, P. (1996). Age-specific reference intervals ('normal ranges'), *Stata Technical Bulletin* **34**: 24–34. [Cited on p. 249]

Wyatt, J. and Altman, D. G. (1995). Prognostic models: clinically useful or quickly forgotten?, *British Medical Journal* **311**: 1539–1541. [Cited on pp. 18, 67]

Yuan, M. and Lin, Y. (2005). Efficient empirical Bayes variable selection and estimation in linear models, *Journal of the American Statistical Association* **100**: 1215–1225. [Cited on p. 207]

Zou, H. and Hastie, T. (2005). Regularization and variable selection via the elastic net, *Journal of the Royal Statistical Society, Series B* **67**: 301–320. [Cited on pp. 44, 50]

Index

Multivariable Model-Building   Patrick Royston, Willi Sauerbrei
© 2008 John Wiley & Sons, Ltd

WILEY SERIES IN PROBABILITY AND STATISTICS

Established by WALTER A. SHEWHART and SAMUEL S. WILKS

Editors: *David J. Balding, Noel A. C. Cressie, Garrett M. Fitzmaurice, Iain M. Johnstone, Geert Molenberghs, David W. Scott, Adrian F. M. Smith, Ruey S. Tsay, Sanford Weisberg*
Editors Emeriti: *Vic Barnett, J. Stuart Hunter, David G. Kendall, Jozef L. Teugels*

The *Wiley Series in Probability and Statistics* is well established and authoritative. It covers many topics of current research interest in both pure and applied statistics and probability theory. Written by leading statisticians and institutions, the titles span both state-of-the-art developments in the field and classical methods.

Reflecting the wide range of current research in statistics, the series encompasses applied, methodological and theoretical statistics, ranging from applications and new techniques made possible by advances in computerized practice to rigorous treatment of theoretical approaches.

This series provides essential and invaluable reading for all statisticians, whether in academia, industry, government, or research.

---

* Now avilable in a lower priced paperback edition in the Wiley Classics Library.
† Now avilable in a lower priced paperback edition in the Wiley–Interscience Paperback Series.

BELSLEY · Conditioning Diagnostics: Collinearity and Weak Data in Regression

† BELSLEY, KUH, and WELSCH · Regression Diagnostics: Identifying Influential Data and Sources of Collinearity

BENDAT and PIERSOL · Random Data: Analysis and Measurement Procedures, *Third Edition*

BERRY, CHALONER, and GEWEKE · Bayesian Analysis in Statistics and Econometrics: Essays in Honor of Arnold Zellner

BERNARDO and SMITH · Bayesian Theory

BHAT and MILLER · Elements of Applied Stochastic Processes, *Third Edition*

BHATTACHARYA and WAYMIRE · Stochastic Processes with Applications

BILLINGSLEY · Convergence of Probability Measures, *Second Edition*

BILLINGSLEY · Probability and Measure, *Third Edition*

BIRKES and DODGE · Alternative Methods of Regression

BISWAS, DATTA, FINE, and SEGAL · Statistical Advances in the Biomedical Sciences: Clinical Trials, Epidemiology, Survival Analysis, and Bioinformatics

BLISCHKE AND MURTHY (editors) · Case Studies in Reliability and Maintenance

BLISCHKE AND MURTHY · Reliability: Modeling, Prediction, and Optimization

BLOOMFIELD · Fourier Analysis of Time Series: An Introduction, *Second Edition*

BOLLEN · Structural Equations with Latent Variables

BOLLEN and CURRAN · Latent Curve Models: A Structural Equation Perspective

BOROVKOV · Ergodicity and Stability of Stochastic Processes

BOULEAU · Numerical Methods for Stochastic Processes

BOX · Bayesian Inference in Statistical Analysis

BOX · R. A. Fisher, the Life of a Scientist

BOX and DRAPER · Response Surfaces, Mixtures, and Ridge Analyses, *Second Edition*

* BOX and DRAPER · Evolutionary Operation: A Statistical Method for Process Improvement

BOX and FRIENDS · Improving Almost Anything, *Revised Edition*

BOX, HUNTER, and HUNTER · Statistics for Experimenters: Design, Innovation, and Discovery, *Second Edition*

BOX and LUCEÑO · Statistical Control by Monitoring and Feedback Adjustment

BRANDIMARTE · Numerical Methods in Finance: A MATLAB-Based Introduction

† BROWN and HOLLANDER · Statistics: A Biomedical Introduction

BRUNNER, DOMHOF, and LANGER · Nonparametric Analysis of Longitudinal Data in Factorial Experiments

BUCKLEW · Large Deviation Techniques in Decision, Simulation, and Estimation

CAIROLI and DALANG · Sequential Stochastic Optimization

CASTILLO, HADI, BALAKRISHNAN, and SARABIA · Extreme Value and Related Models with Applications in Engineering and Science

CHAN · Time Series: Applications to Finance

CHARALAMBIDES · Combinatorial Methods in Discrete Distributions

CHATTERJEE and HADI · Regression Analysis by Example, *Fourth Edition*

CHATTERJEE and HADI · Sensitivity Analysis in Linear Regression

CHERNICK · Bootstrap Methods: A Guide for Practitioners and Researchers, *Second Edition*

CHERNICK and FRIIS · Introductory Biostatistics for the Health Sciences

CHILÈS and DELFINER · Geostatistics: Modeling Spatial Uncertainty

CHOW and LIU · Design and Analysis of Clinical Trials: Concepts and Methodologies, *Second Edition*

CLARKE and DISNEY · Probability and Random Processes: A First Course with Applications, *Second Edition*

---

* Now avilable in a lower priced paperback edition in the Wiley Classics Library.

† Now avilable in a lower priced paperback edition in the Wiley–Interscience Paperback Series.

* COCHRAN and COX · Experimental Designs, *Second Edition*

CONGDON · Applied Bayesian Modelling

CONGDON · Bayesian Models for Categorical Data

CONGDON · Bayesian Statistical Modelling

CONOVER · Practical Nonparametric Statistics, *Third Edition*

COOK · Regression Graphics

COOK and WEISBERG · Applied Regression Including Computing and Graphics

COOK and WEISBERG · An Introduction to Regression Graphics

CORNELL · Experiments with Mixtures, Designs, Models, and the Analysis of Mixture Data, *Third Edition*

COVER and THOMAS · Elements of Information Theory

COX · A Handbook of Introductory Statistical Methods

* COX · Planning of Experiments

CRESSIE · Statistics for Spatial Data, *Revised Edition*

CSÖRGŐ and HORVÁTH · Limit Theorems in Change Point Analysis

DANIEL · Applications of Statistics to Industrial Experimentation

DANIEL · Biostatistics: A Foundation for Analysis in the Health Sciences, *Eighth Edition*

* DANIEL · Fitting Equations to Data: Computer Analysis of Multifactor Data, *Second Edition*

DASU and JOHNSON · Exploratory Data Mining and Data Cleaning

DAVID and NAGARAJA · Order Statistics, *Third Edition*

* DEGROOT, FIENBERG, and KADANE · Statistics and the Law

DEL CASTILLO · Statistical Process Adjustment for Quality Control

DeMARIS · Regression with Social Data: Modeling Continuous and Limited Response Variables

DEMIDENKO · Mixed Models: Theory and Applications

DENISON, HOLMES, MALLICK and SMITH · Bayesian Methods for Nonlinear Classification and Regression

DETTE and STUDDEN · The Theory of Canonical Moments with Applications in Statistics, Probability, and Analysis

DEY and MUKERJEE · Fractional Factorial Plans

DILLON and GOLDSTEIN · Multivariate Analysis: Methods and Applications

DODGE · Alternative Methods of Regression

* DODGE and ROMIG · Sampling Inspection Tables, *Second Edition*

* DOOB · Stochastic Processes

DOWDY, WEARDEN, and CHILKO · Statistics for Research, *Third Edition*

DRAPER and SMITH · Applied Regression Analysis, *Third Edition*

DRYDEN and MARDIA · Statistical Shape Analysis

DUDEWICZ and MISHRA · Modern Mathematical Statistics

DUNN and CLARK · Basic Statistics: A Primer for the Biomedical Sciences, *Third Edition*

DUPUIS and ELLIS · A Weak Convergence Approach to the Theory of Large Deviations

EDLER and KITSOS · Recent Advances in Quantitative Methods in Cancer and Human Health Risk Assessment

* ELANDT-JOHNSON and JOHNSON · Survival Models and Data Analysis

ENDERS · Applied Econometric Time Series

† ETHIER and KURTZ · Markov Processes: Characterization and Convergence

EVANS, HASTINGS, and PEACOCK · Statistical Distributions, *Third Edition*

FELLER · An Introduction to Probability Theory and Its Applications, Volume I, *Third Edition*, Revised; Volume II, *Second Edition*

* Now avilable in a lower priced paperback edition in the Wiley Classics Library.

† Now avilable in a lower priced paperback edition in the Wiley–Interscience Paperback Series.

FISHER and VAN BELLE · Biostatistics: A Methodology for the Health Sciences

FITZMAURICE, LAIRD, and WARE · Applied Longitudinal Analysis

* FLEISS · The Design and Analysis of Clinical Experiments

FLEISS · Statistical Methods for Rates and Proportions, *Third Edition*

† FLEMING and HARRINGTON · Counting Processes and Survival Analysis

FULLER · Introduction to Statistical Time Series, *Second Edition*

† FULLER · Measurement Error Models

GALLANT · Nonlinear Statistical Models

GEISSER · Modes of Parametric Statistical Inference

GELMAN and MENG · Applied Bayesian Modeling and Causal Inference from Incomplete-Data Perspectives

GEWEKE · Contemporary Bayesian Econometrics and Statistics

GHOSH, MUKHOPADHYAY, and SEN · Sequential Estimation

GIESBRECHT and GUMPERTZ · Planning, Construction, and Statistical Analysis of Comparative Experiments

GIFI · Nonlinear Multivariate Analysis

GIVENS and HOETING · Computational Statistics

GLASSERMAN and YAO · Monotone Structure in Discrete-Event Systems

GNANADESIKAN · Methods for Statistical Data Analysis of Multivariate Observations, *Second Edition*

GOLDSTEIN and LEWIS · Assessment: Problems, Development, and Statistical Issues

GREENWOOD and NIKULIN · A Guide to Chi-Squared Testing

GROSS and HARRIS · Fundamentals of Queueing Theory, *Third Edition*

* HAHN and SHAPIRO · Statistical Models in Engineering

HAHN and MEEKER · Statistical Intervals: A Guide for Practitioners

HALD · A History of Probability and Statistics and their Applications Before 1750

HALD · A History of Mathematical Statistics from 1750 to 1930

† HAMPEL · Robust Statistics: The Approach Based on Influence Functions

HANNAN and DEISTLER · The Statistical Theory of Linear Systems

HEIBERGER · Computation for the Analysis of Designed Experiments

HEDAYAT and SINHA · Design and Inference in Finite Population Sampling

HEDEKER and GIBBONS · Longitudinal Data Analysis

HELLER · MACSYMA for Statisticians

HINKELMANN and KEMPTHORNE · Design and Analysis of Experiments, Volume 1: Introduction to Experimental Design, *Second Edition*

HINKELMANN and KEMPTHORNE · Design and Analysis of Experiments, Volume 2: Advanced Experimental Design

HOAGLIN, MOSTELLER, and TUKEY · Exploratory Approach to Analysis of Variance

* HOAGLIN, MOSTELLER, and TUKEY · Exploring Data Tables, Trends and Shapes

* HOAGLIN, MOSTELLER, and TUKEY · Understanding Robust and Exploratory Data Analysis

HOCHBERG and TAMHANE · Multiple Comparison Procedures

HOCKING · Methods and Applications of Linear Models: Regression and the Analysis of Variance, *Second Edition*

HOEL · Introduction to Mathematical Statistics, *Fifth Edition*

HOGG and KLUGMAN · Loss Distributions

HOLLANDER and WOLFE · Nonparametric Statistical Methods, *Second Edition*

HOSMER and LEMESHOW · Applied Logistic Regression, *Second Edition*

HOSMER, LEMESHOW, and MAY · Applied Survival Analysis: Regression Modeling of Time-to-Event Data, *Second Edition*

† HUBER · Robust Statistics

HUBERTY · Applied Discriminant Analysis

HUBERTY and OLEJNIK · Applied MANOVA and Discriminant Analysis, *Second Edition*

HUNT and KENNEDY · Financial Derivatives in Theory and Practice, *Revised Edition*

HURD and MIAMEE · Periodically Correlated Random Sequences: Spectral Theory and Practice

HUSKOVA, BERAN, and DUPAC · Collected Works of Jaroslav Hajek—with Commentary

HUZURBAZAR · Flowgraph Models for Multistate Time-to-Event Data

IMAN and CONOVER · A Modern Approach to Statistics

† JACKSON · A User's Guide to Principle Components

JOHN · Statistical Methods in Engineering and Quality Assurance

JOHNSON · Multivariate Statistical Simulation

JOHNSON and BALAKRISHNAN · Advances in the Theory and Practice of Statistics: A Volume in Honor of Samuel Kotz

JOHNSON and BHATTACHARYYA · Statistics: Principles and Methods, Fifth Edition

JOHNSON and KOTZ · Distributions in Statistics

JOHNSON and KOTZ (editors) · Leading Personalities in Statistical Sciences: From the Seventeenth Century to the Present

JOHNSON, KOTZ, and BALAKRISHNAN · Continuous Univariate Distributions, Volume 1, *Second Edition*

JOHNSON, KOTZ, and BALAKRISHNAN · Continuous Univariate Distributions, Volume 2, *Second Edition*

JOHNSON, KOTZ, and BALAKRISHNAN · Discrete Multivariate Distributions

JOHNSON, KEMP, and KOTZ · Univariate Discrete Distributions, *Third Edition*

JUDGE, GRIFFITHS, HILL, LÜTKEPOHL, and LEE · The Theory and Practice of Econometrics, *Second Edition*

JUREČKOVÁ and SEN · Robust Statistical Procedures: Aymptotics and Interrelations

JUREK and MASON · Operator-Limit Distributions in Probability Theory

KADANE · Bayesian Methods and Ethics in a Clinical Trial Design

KADANE AND SCHUM · A Probabilistic Analysis of the Sacco and Vanzetti Evidence

KALBFLEISCH and PRENTICE · The Statistical Analysis of Failure Time Data, *Second Edition*

KARIYA and KURATA · Generalized Least Squares

KASS and VOS · Geometrical Foundations of Asymptotic Inference

† KAUFMAN and ROUSSEEUW · Finding Groups in Data: An Introduction to Cluster Analysis

KEDEM and FOKIANOS · Regression Models for Time Series Analysis

KENDALL, BARDEN, CARNE, and LE · Shape and Shape Theory

KHURI · Advanced Calculus with Applications in Statistics, *Second Edition*

KHURI, MATHEW, and SINHA · Statistical Tests for Mixed Linear Models

KLEIBER and KOTZ · Statistical Size Distributions in Economics and Actuarial Sciences

KLUGMAN, PANJER, and WILLMOT · Loss Models: From Data to Decisions, *Second Edition*

KLUGMAN, PANJER, and WILLMOT · Solutions Manual to Accompany Loss Models: From Data to Decisions, *Second Edition*

KOTZ, BALAKRISHNAN, and JOHNSON · Continuous Multivariate Distributions, Volume 1, *Second Edition*

KOVALENKO, KUZNETZOV, and PEGG · Mathematical Theory of Reliability of Time-Dependent Systems with Practical Applications

---

* Now avilable in a lower priced paperback edition in the Wiley Classics Library.

† Now avilable in a lower priced paperback edition in the Wiley–Interscience Paperback Series.

KOWALSKI and TU · Modern Applied U-Statistics

KROONENBERG · Applied Multiway Data Analysis

KVAM and VIDAKOVIC · Nonparametric Statistics with Applications to Science and Engineering

LACHIN · Biostatistical Methods: The Assessment of Relative Risks

LAD · Operational Subjective Statistical Methods: A Mathematical, Philosophical, and Historical Introduction

LAMPERTI · Probability: A Survey of the Mathematical Theory, *Second Edition*

LANGE, RYAN, BILLARD, BRILLINGER, CONQUEST, and GREENHOUSE · Case Studies in Biometry

LARSON · Introduction to Probability Theory and Statistical Inference, *Third Edition*

LAWLESS · Statistical Models and Methods for Lifetime Data, *Second Edition*

LAWSON · Statistical Methods in Spatial Epidemiology

LE · Applied Categorical Data Analysis

LE · Applied Survival Analysis

LEE and WANG · Statistical Methods for Survival Data Analysis, *Third Edition*

LePAGE and BILLARD · Exploring the Limits of Bootstrap

LEYLAND and GOLDSTEIN (editors) · Multilevel Modelling of Health Statistics

LIAO · Statistical Group Comparison

LINDVALL · Lectures on the Coupling Method

LIN · Introductory Stochastic Analysis for Finance and Insurance

LINHART and ZUCCHINI · Model Selection

LITTLE and RUBIN · Statistical Analysis with Missing Data, *Second Edition*

LLOYD · The Statistical Analysis of Categorical Data

LOWEN and TEICH · Fractal-Based Point Processes

MAGNUS and NEUDECKER · Matrix Differential Calculus with Applications in Statistics and Econometrics, *Revised Edition*

MALLER and ZHOU · Survival Analysis with Long Term Survivors

MALLOWS · Design, Data, and Analysis by Some Friends of Cuthbert Daniel

MANN, SCHAFER, and SINGPURWALLA · Methods for Statistical Analysis of Reliability and Life Data

MANTON, WOODBURY, and TOLLEY · Statistical Applications Using Fuzzy Sets

MARCHETTE · Random Graphs for Statistical Pattern Recognition

MARDIA and JUPP · Directional Statistics

MASON, GUNST, and HESS · Statistical Design and Analysis of Experiments with Applications to Engineering and Science, *Second Edition*

McCULLOCH and SEARLE · Generalized, Linear, and Mixed Models

McFADDEN · Management of Data in Clinical Trials, *Second Edition*

* McLACHLAN · Discriminant Analysis and Statistical Pattern Recognition

McLACHLAN, DO, and AMBROISE · Analyzing Microarray Gene Expression Data

McLACHLAN and KRISHNAN · The EM Algorithm and Extensions, *Second Edition*

McLACHLAN and PEEL · Finite Mixture Models

McNEIL · Epidemiological Research Methods

MEEKER and ESCOBAR · Statistical Methods for Reliability Data

MEERSCHAERT and SCHEFFLER · Limit Distributions for Sums of Independent Random Vectors: Heavy Tails in Theory and Practice

MICKEY, DUNN, and CLARK · Applied Statistics: Analysis of Variance and Regression, *Third Edition*

---

* Now avilable in a lower priced paperback edition in the Wiley Classics Library.

† Now avilable in a lower priced paperback edition in the Wiley–Interscience Paperback Series.

* MILLER · Survival Analysis, *Second Edition*

MONTGOMERY, PECK, and VINING · Introduction to Linear Regression Analysis, *Fourth Edition*

MORGENTHALER and TUKEY · Configural Polysampling: A Route to Practical Robustness

MUIRHEAD · Aspects of Multivariate Statistical Theory

MULLER and STOYAN · Comparison Methods for Stochastic Models and Risks

MURRAY · X-STAT 2.0 Statistical Experimentation, Design Data Analysis, and Nonlinear Optimization

MURTHY, XIE, and JIANG · Weibull Models

MYERS and MONTGOMERY · Response Surface Methodology: Process and Product Optimization Using Designed Experiments, *Second Edition*

MYERS, MONTGOMERY, and VINING · Generalized Linear Models. With Applications in Engineering and the Sciences

† NELSON · Accelerated Testing, Statistical Models, Test Plans, and Data Analyses

† NELSON · Applied Life Data Analysis

NEWMAN · Biostatistical Methods in Epidemiology

OCHI · Applied Probability and Stochastic Processes in Engineering and Physical Sciences

OKABE, BOOTS, SUGIHARA, and CHIU · Spatial Tesselations: Concepts and Applications of Voronoi Diagrams, *Second Edition*

OLIVER and SMITH · Influence Diagrams, Belief Nets and Decision Analysis

PALTA · Quantitative Methods in Population Health: Extensions of Ordinary Regressions

PANJER · Operational Risk: Modeling and Analytics

PANKRATZ · Forecasting with Dynamic Regression Models

PANKRATZ · Forecasting with Univariate Box-Jenkins Models: Concepts and Cases

* PARZEN · Modern Probability Theory and Its Applications

PEÑA, TIAO, and TSAY · A Course in Time Series Analysis

PIANTADOSI · Clinical Trials: A Methodologic Perspective

PORT · Theoretical Probability for Applications

POURAHMADI · Foundations of Time Series Analysis and Prediction Theory

POWELL · Approximate Dynamic Programming: Solving the Curses of Dimensionality

PRESS · Bayesian Statistics: Principles, Models, and Applications

PRESS · Subjective and Objective Bayesian Statistics, *Second Edition*

PRESS and TANUR · The Subjectivity of Scientists and the Bayesian Approach

PUKELSHEIM · Optimal Experimental Design

PURI, VILAPLANA, and WERTZ · New Perspectives in Theoretical and Applied Statistics

† PUTERMAN · Markov Decision Processes: Discrete Stochastic Dynamic Programming

QIU · Image Processing and Jump Regression Analysis

* RAO · Linear Statistical Inference and Its Applications, *Second Edition*

RAUSAND and HØYLAND · System Reliability Theory: Models, Statistical Methods, and Applications, *Second Edition*

RENCHER · Linear Models in Statistics

RENCHER · Methods of Multivariate Analysis, *Second Edition*

RENCHER · Multivariate Statistical Inference with Applications

* RIPLEY · Spatial Statistics

* RIPLEY · Stochastic Simulation

ROBINSON · Practical Strategies for Experimenting

ROHATGI and SALEH · An Introduction to Probability and Statistics, *Second Edition*

ROLSKI, SCHMIDLI, SCHMIDT, and TEUGELS · Stochastic Processes for Insurance and Finance

ROSENBERGER and LACHIN · Randomization in Clinical Trials: Theory and Practice

* Now avilable in a lower priced paperback edition in the Wiley Classics Library.

† Now avilable in a lower priced paperback edition in the Wiley–Interscience Paperback Series.

302

ROSS · Introduction to Probability and Statistics for Engineers and Scientists
ROSSI, ALLENBY, and McCULLOCH · Bayesian Statistics and Marketing
† ROUSSEEUW and LEROY · Robust Regression and Outlier Detection
* RUBIN · Multiple Imputation for Nonresponse in Surveys
RUBINSTEIN and KROESE · Simulation and the Monte Carlo Method, *Second Edition*
RUBINSTEIN and MELAMED · Modern Simulation and Modeling
RYAN · Modern Engineering Statistics
RYAN · Modern Experimental Design
RYAN · Modern Regression Methods
RYAN · Statistical Methods for Quality Improvement, *Second Edition*
SALEH · Theory of Preliminary Test and Stein-Type Estimation with Applications
* SCHEFFE · The Analysis of Variance
SCHIMEK · Smoothing and Regression: Approaches, Computation, and Application
SCHOTT · Matrix Analysis for Statistics, *Second Edition*
SCHOUTENS · Levy Processes in Finance: Pricing Financial Derivatives
SCHUSS · Theory and Applications of Stochastic Differential Equations
SCOTT · Multivariate Density Estimation: Theory, Practice, and Visualization
† SEARLE · Linear Models for Unbalanced Data
† SEARLE · Matrix Algebra Useful for Statistics
† SEARLE, CASELLA, and McCULLOCH · Variance Components
SEARLE and WILLETT · Matrix Algebra for Applied Economics
SEBER · A Matrix Handbook For Statisticians
† SEBER · Multivariate Observations
SEBER and LEE · Linear Regression Analysis, *Second Edition*
† SEBER and WILD · Nonlinear Regression
SENNOTT · Stochastic Dynamic Programming and the Control of Queueing Systems
* SERFLING · Approximation Theorems of Mathematical Statistics
SHAFER and VOVK · Probability and Finance: It's Only a Game!
SILVAPULLE and SEN · Constrained Statistical Inference: Inequality, Order, and Shape
    Restrictions
SMALL and McLEISH · Hilbert Space Methods in Probability and Statistical Inference
SRIVASTAVA · Methods of Multivariate Statistics
STAPLETON · Linear Statistical Models
STAPLETON · Models for Probability and Statistical Inference: Theory and Applications
STAUDTE and SHEATHER · Robust Estimation and Testing
STOYAN, KENDALL, and MECKE · Stochastic Geometry and Its Applications, *Second Edition*
STOYAN and STOYAN · Fractals, Random Shapes and Point Fields: Methods of Geometrical Statistics
STREET and BURGESS · The Construction of Optimal Stated Choice Experiments: Theory and Methods
STYAN · The Collected Papers of T. W. Anderson: 1943–1985
SUTTON, ABRAMS, JONES, SHELDON, and SONG · Methods for Meta-Analysis in Medical
    Research
TAKEZAWA · Introduction to Nonparametric Regression
TANAKA · Time Series Analysis: Nonstationary and Noninvertible Distribution Theory
THOMPSON · Empirical Model Building
THOMPSON · Sampling, *Second Edition*
THOMPSON · Simulation: A Modeler's Approach
THOMPSON and SEBER · Adaptive Sampling

---

* Now available in a lower priced paperback edition in the Wiley Classics Library.
† Now available in a lower priced paperback edition in the Wiley–Interscience Paperback Series.

THOMPSON, WILLIAMS, and FINDLAY · Models for Investors in Real World Markets

TIAO, BISGAARD, HILL, PEÑA, and STIGLER (editors) · Box on Quality and Discovery: with Design, Control, and Robustness

TIERNEY · LISP-STAT: An Object-Oriented Environment for Statistical Computing and Dynamic Graphics

TSAY · Analysis of Financial Time Series, *Second Edition*

UPTON and FINGLETON · Spatial Data Analysis by Example, Volume II: Categorical and Directional Data

VAN BELLE · Statistical Rules of Thumb

VAN BELLE, FISHER, HEAGERTY, and LUMLEY · Biostatistics: A Methodology for the Health Sciences, *Second Edition*

VESTRUP · The Theory of Measures and Integration

VIDAKOVIC · Statistical Modeling by Wavelets

VINOD and REAGLE · Preparing for the Worst: Incorporating Downside Risk in Stock Market Investments

WALLER and GOTWAY · Applied Spatial Statistics for Public Health Data

WEERAHANDI · Generalized Inference in Repeated Measures: Exact Methods in MANOVA and Mixed Models

WEISBERG · Applied Linear Regression, *Third Edition*

WELSH · Aspects of Statistical Inference

WESTFALL and YOUNG · Resampling-Based Multiple Testing: Examples and Methods for *p*-Value Adjustment

WHITTAKER · Graphical Models in Applied Multivariate Statistics

WINKER · Optimization Heuristics in Economics: Applications of Threshold Accepting

WONNACOTT and WONNACOTT · Econometrics, *Second Edition*

WOODING · Planning Pharmaceutical Clinical Trials: Basic Statistical Principles

WOODWORTH · Biostatistics: A Bayesian Introduction

WOOLSON and CLARKE · Statistical Methods for the Analysis of Biomedical Data, *Second Edition*

WU and HAMADA · Experiments: Planning, Analysis, and Parameter Design Optimization

WU and ZHANG · Nonparametric Regression Methods for Longitudinal Data Analysis

YANG · The Construction Theory of Denumerable Markov Processes

YOUNG, VALERO-MORA, and FRIENDLY · Visual Statistics: Seeing Data with Dynamic Interactive Graphics   ZELTERMAN · Discrete Distributions—Applications in the Health Sciences

* ZELLNER · An Introduction to Bayesian Inference in Econometrics

ZHOU, OBUCHOWSKI, and McCLISH · Statistical Methods in Diagnostic Medicine

---

* Now avilable in a lower priced paperback edition in the Wiley Classics Library.

† Now avilable in a lower priced paperback edition in the wiley–Interscience Paperback Series.